Digital Signal Compression

With clear and easy-to-understand explanations, this book covers the fundamental concepts and coding methods of signal compression, while still retaining technical depth and rigor. It contains a wealth of illustrations, step-by-step descriptions of algorithms, examples, and practice problems, which make it an ideal textbook for senior undergraduate and graduate students, as well as a useful self-study tool for researchers and professionals.

Principles of lossless compression are covered, as are various entropy coding techniques, including Huffman coding, arithmetic coding, run-length coding, and Lempel–Ziv coding. Scalar and vector quantization, and their use in various lossy compression systems, are thoroughly explained, and a full chapter is devoted to mathematical transformations, including the Karhunen–Loeve transform, discrete cosine transform (DCT), and wavelet transforms. Rate control is covered in detail, with several supporting algorithms to show how to achieve it. State-of-the-art transform and subband/wavelet image and video coding systems are explained, including JPEG2000, SPIHT, SBHP, EZBC, and H.264/AVC. Also, unique to this book is a chapter on set partition coding that sheds new light on SPIHT, SPECK, EZW, and related methods.

William A. Pearlman is a Professor Emeritus in the Electrical, Computer, and Systems Engineering Department at the Rensselear Polytechnic Institute (RPI), where he has been a faculty member since 1979. He has more than 35 years of experience in teaching and researching in the fields of information theory, data compression, digital signal processing, and digital communications theory, and he has written about 250 published works in these fields. He is a Fellow of the IEEE and the SPIE, and he is the co-inventor of two celebrated image compression algorithms: SPIHT and SPECK.

Amir Said is currently a Master Researcher at Hewlett-Packard Laboratories, where he has worked since 1998. His research interests include multimedia communications, coding and information theory, image and video compression, signal processing, and optimization, and he has more than 100 publications and 20 patents in these fields. He is co-inventor with Dr. Pearlman of the SPIHT image compression algorithm and co-recipient, also with Dr. Pearlman, of two Best Paper Awards, one from the IEEE Circuits and Systems Society and the other from the IEEE Signal Processing Society.

Digital Signal Compression

Principles and Practice

WILLIAM A. PEARLMAN
Rensselear Polytechnic Institute, New York

AMIR SAID
Hewlett-Packard Laboratories, Palo Alto, California

CAMBRIDGE
UNIVERSITY PRESS

Shaftesbury Road, Cambridge CB2 8EA, United Kingdom

One Liberty Plaza, 20th Floor, New York, NY 10006, USA

477 Williamstown Road, Port Melbourne, VIC 3207, Australia

314–321, 3rd Floor, Plot 3, Splendor Forum, Jasola District Centre, New Delhi – 110025, India

103 Penang Road, #05–06/07, Visioncrest Commercial, Singapore 238467

Cambridge University Press is part of Cambridge University Press & Assessment,
a department of the University of Cambridge.

We share the University's mission to contribute to society through the pursuit of
education, learning and research at the highest international levels of excellence.

www.cambridge.org
Information on this title: www.cambridge.org/9780521899826

© Cambridge University Press & Assessment 2011

First published 2011

A catalogue record for this publication is available from the British Library

Library of Congress Cataloging-in-Publication data
Pearlman, William A. (William Abraham)
Digital signal compression : principles and practice / William A. Pearlman, Amir Said.
    p. cm.
Includes bibliographical references and index.
ISBN 978-0-521-89982-6 (hardback)
1. Data compression (Telecommunication)   2. Signal processing – Digital techniques.
I. Said, Amir. II. Title.
TK5102.92.P43   2011
005.74´6–dc23

                                                                                    2011012972

ISBN   978-0-521-89982-6   Hardback

Additional resources for this publication at www.cambridge.org/9780521899826

To Eleanor
To Celli and Ricardo

Contents

Preface

This book is an outgrowth of a graduate level course taught for several years at Rensselaer Polytechnic Institute (RPI). When the course started in the early 1990s, there were only two textbooks available that taught signal compression, Jayant and Noll[1] and Gersho and Gray.[2] Certainly these are excellent textbooks and valuable references, but they did not teach some material considered to be necessary at that time, so the textbooks were supplemented with handwritten notes where needed. Eventually, these notes grew to many pages, as the reliance on published textbooks diminished. The lecture notes remained the primary source even after the publication of the excellent book by Sayood,[3] which served as a supplement and a source of some problems. While the Sayood book was up to date, well written, and authoritative, it was written to be accessible to undergraduate students, so lacked the depth suitable for graduate students wanting to do research or practice in the field. The book at hand teaches the fundamental ideas of signal compression at a level that both graduate students and advanced undergraduate students can approach with confidence and understanding. The book is also intended to be a useful resource to the practicing engineer or computer scientist in the field. For that purpose and also to aid understanding, the 40 algorithms listed under *Algorithms* in the Index are not only fully explained in the text, but also are set out step-by-step in special algorithm format environments.

This book contains far more material than can be taught in a course of one semester. As it was being written, certain subjects came to light that begged for embellishment and others arose that were needed to keep pace with developments in the field. One example of this additional material, which does not appear in any other textbook, is Chapter 14 on "Distributed source coding," a subject which has received considerable attention lately. The intent was to present the fundamental ideas there, so that the student can understand and put into practice the various methods being proposed that use this technique.

The two longest chapters in the book are Chapters 10 and 11, entitled "Set partition coding" and "Subband/wavelet coding systems," respectively. They were actually the first chapters written and were published previously as a monograph in two parts.[4] The versions appearing here are updated with some minor errors corrected. Being the inventors of SPIHT and proponents and pioneers of set partition coding, we felt that its fundamental principles were not expounded in the technical literature. Considering the great interest in SPIHT, as evidenced by the thousands of inquiries received by us over the years since its origin in 1995 (at this writing 94,200 hits on Google), we were eager to publish a true tutorial on the fundamental concepts of this algorithm. We believe that Chapter 10 fulfills this intent. Other books usually present only the SPIHT algorithm, almost always by working an example without revealing the underlying principles. Chapter 11 describes more image wavelet coding systems than any other book, including the JPEG2000 standard, fully scalable SPIHT, SBHP, and EZBC. The last three are set partition coders, while JPEG2000 contains auxiliary algorithms that use set partitioning. Furthermore, this chapter explains how to embed features of scalability and random access in coded bitstreams.

Besides distributed source coding, some preliminary material are also firsts in this book. They are: analysis of null-zone quantization, rate allocation algorithms, and the link between filters and wavelets. The aforementioned link is explained in Chapter 7 on "Mathematical transformations," in a way that requires only some knowledge of discrete-time Fourier transforms and linear system analysis. The treatment avoids the concepts of functional analysis and the use of polyphase transforms with little compromise of rigor. The intent was to make the book accessible to advanced undergraduates, who would likely not have exposure to these subjects. Also to serve this purpose, prior exposure to information theory is not a prerequisite, as the book teaches the relevant aspects needed to grasp the essential ideas.

One criticism that might be levied at this book is its emphasis on compression of images. Certainly, that reflects the main field of research and practice of the authors. However, image compression is possible only by understanding and applying the principles of compression that pertain to all source data. In fact, the material of the first eight chapters is generic and dimension independent. The notation is one-dimensional for the most part, and the generalization to higher dimensions is mostly obvious and hence unnecessary. Although the applications are mostly to images in the remainder of the book, except for the generic Chapter 14, the corresponding one-dimensional signal methods are mentioned when applicable and even included in some problems. The standards and state-of-the-art systems of image compression are treated in some detail, as they represent the most straightforward application of basic principles. The standard speech and audio coding systems require additional complications of perceptual and excitation models, and echo cancellation. Their inclusion would make the book too long and cumbersome and not add much to its primary objective. Nevertheless, the material on image compression systems in Chapters 9, 11, and 12 is comprehensive enough to meet the secondary objective of serving as a good tutorial and reference for image compression researchers and practitioners.

Chapter 12 treats the subject of lossless image compression. Lossless image compression is used only for archiving of image data. There seems to be no call for lossless compression of any sources for the purpose of multimedia communication, as the data transfer would be too slow or require too much bandwidth or both. For example, there is no compression of audio or speech in the WAV storage format for compact discs. MP3 is a compressed format for audio and speech transmission and recording; the compressed format of JPEG is standard in every digital camera for consumer use; all digital video is compressed by MPEG or ITU standard formats. Images seem to be the only sources that are subject to lossless compression. The standard methods described in Chapter 12 serve as good examples of how basic principles are put into practice.

The book did not seem complete without a chapter on how compression is applied to three-dimensional sources, such as color images, volume medical images, and video. At the time of writing, there are no other textbooks that teach three-dimensional wavelet coding methods. Therefore, we wrote Chapter 13 with the intent to show the reader how the methods of the earlier chapters are extended to these higher dimensional sources. We purposely omitted detailed descriptions of video standards. We just explained the general framework in these systems upon which compression is applied and a little about the compression methods, which are mostly covered in detail in earlier chapters.

We urge potential buyers or browsers to read Chapters 1 and 2, which discuss the motivation to learn signal compression and take a brief tour through the book. This book turned out to be different in many respects from what was taught in the RPI course. Roughly, the coverage of that course was all of Chapters 3, 4, 5, and 6, which was deemed essential material. One can be selective in Chapter 7, for example, skipping the lapped orthogonal transform and some parts of Section 7.7, "Subband transforms," especially the detailed development of the connection between wavelet theory and FIR filters. Likewise, in Chaper 8, some of the rate allocation algorithms may be skipped, as well as the detailed derivations of optimal rate allocation and coding gain. One can skim through Section 9.3 in the next chapter, and skip Section 9.4 on H.264/AVC intra coding, which did not exist when the course was last taught. Time may not allow anything in Chapter 10, other than set partitioning for SPIHT and the accompanying coding example, and in Chapter 11 only a sketch of the JPEG2000 coding system. Lossless image compression in Chapter 12 could be covered earlier in the course, perhaps after Chapter 4, "Entropy coding techniques." Certainly, there is enough material to accommodate different choices of emphasis and objective.

For students with a background in information theory and signal processing or those more interested in computer science or actual implementation, an instructor may skip some of the preliminary material in the early chapters and teach all of the rate allocation algorithms in Chapter 8 and cover Chapters 10 and 11 in their entirety. Chapter 11 contains much practical material on implementation of coding systems. In fact, we think that approach would be appropriate for a short course.

The book contains many figures and problems. The problems in many cases have to be worked using a calculation and plotting program, such as MATLAB, and sometimes by making a computer program. Some datasets and basic software in C or C++ and MATLAB m-files, will be made freely available on the course website: http://www.cambridge.org/pearlman. Also freely available on the website are Powerpoint animations of the SPIHT and SPECK algorithms. Figures and problem solutions will be made available to instructors.

Notes

1. N. S. Jayant and P. Noll, *Digital Coding of Waveforms*, Prentice-Hall 1984.
2. A. Gersho and R. M. Gray, *Vector Quantization and Signal Compression*, Kluwer Academic Publishers 1992.
3. K. Sayood, *Introduction to Data Compression*, Morgan Kaufmann Publishers 1996, 2000, 2006.
4. Look under **Books** in http://www.cipr.rpi.edu/~pearlman/.

Acknowledgments

I wish to acknowledge my debt and gratitude over the years to all my students from whom I have learned so much. Without the experience of interacting with them, I could not have written this book. There is not enough space to list all the contributors and their contributions, as I would like. Instead, I shall mention only those who were directly involved in the content of this book. I wish to thank Sungdae Cho, who created most of the video and three-dimensional SPIHT software and who encouraged me to write this book. He also created the SPIHT animation that is available on the book's website. I am grateful to Ying Liu, who developed the higher dimensional SBHP algorithms and graciously helped me with my programming difficulties. My gratitude goes out to Asad Islam, who developed the SPECK algorithm, and Xiaoli Tang, who extended it to higher dimensions. I also wish to thank Matthias Narroschke, who created the SPECK animation while I was visiting the University of Hannover, and Emmanuel Christophe, who developed resolution-scalable SPIHT, while he was a visiting student to RPI from the University of Toulouse and TeSA (Telecom Spatiales Aeronautiques). I also wish to thank Professor James Fowler of Mississippi State University for checking my explanation of bisection in two-dimensional block partitioning. I am grateful to Alessandro Dutra, who helped me write the solutions manual for the book's problems.

A special acknowledgment goes to Amir Said, my co-author and long-time collaborator, who, while a doctoral student at RPI working under Professor John Anderson, came to me to do a special project on image compression. From this project, by virtue of Amir's intellectual brilliance, creativity, and prodigious programming skills, the SPIHT algorithm was born. There are no words that adequately express my thanks to him.

Lastly, I wish to thank my wife, Eleanor, who suffered from having an often distracted and inattentive husband during more than three years while I was writing this book. This book is dedicated to her, as I could not manage without her love and support.

William A. Pearlman

1 Motivation

1.1 The importance of compression

It is easy to recognize the importance of data compression technology by observing the way it already pervades our daily lives. For instance, we currently have more than a billion users [1] of digital cameras that employ JPEG image compression, and a comparable number of users of portable audio players that use compression formats such as MP3, AAC, and WMA. Users of video cameras, DVD players, digital cable or satellite TV, hear about MPEG-2, MPEG-4, and H.264/AVC. In each case, the acronym is used to identify the type of compression. While many people do not know what exactly compression means or how it works, they have to learn some basic facts about it in order to properly use their devices, or to make purchase decisions.

Compression's usefulness is not limited to multimedia. An increasingly important fraction of the world's economy is in the transmission, storage, and processing of all types of digital information. As Negroponte [2] succinctly put it, economic value is indeed moving "from atoms to bits." While it is true that many constraints from the physical world do not affect this "digital economy," we cannot forget that, due to the huge volumes of data, there has to be a large physical infrastructure for data transmission, processing, and storage. Thus, just as in the traditional economy it is very important to consider the efficiency of transportation, space, and material usage, the efficiency in the representation of digital information also has great economic importance.

This efficiency is the subject of this book. Data compression encompasses the theory and practical techniques that are used for representing digital information in its most efficient format, as measured (mostly) by the number of bits used for storage or telecommunication (bandwidth). Our objective is to present, in an introductory text, all the important ideas required to understand how current compression methods work, and how to design new ones.

A common misconception regarding compression is that, if the costs of storage and bandwidth fall exponentially, compression should eventually cease to be useful. To see why this is not true, one should first note that some assumptions about costs are not universal. For example, while costs of digital storage can indeed fall exponentially, wireless telecommunications costs are constrained by the fact that shared radio spectrum is a resource that is definitely limited, land line communications may need large new investments, etc. Second, it misses the fact that the value of compression is unevenly distributed according to applications, type of user, and in time. For instance, compression is commonly essential to enable the early adoption of new technologies – like it was for digital photography – which would initially be too expensive without it. After a while, it becomes less important, but since the infrastructure to use it is already in place, there is little incentive to stop using it. Furthermore, there is the aspect of relative cost. While even cheap digital cameras may already have more memory than will ever be needed by their owners, photo storage is still a very important part of the operational costs for companies that store the photos and videos of millions of users. Finally, it also ignores the fact that the costs for generating new data, in large amounts, can also decrease exponentially, and we are just beginning to observe an explosion not only in the volume of data, but also in the number and capabilities of the devices that create new data (cf. Section 1.4 and reference [1]).

In conclusion, we expect the importance of compression to keep increasing, especially as its use moves from current types of data to new applications involving much larger amounts of information.

1.2   Data types

Before presenting the basics of the compression process, it is interesting to consider that the data to be compressed can be divided into two main classes, with different properties.

1.2.1   Symbolic information

We can use the word *text* broadly to describe data that is typically arranged in a sequence of arbitrary symbols (or *characters*), from a predefined *alphabet* or *script* (writing system) of a given size. For example, the most common system for English text is the set of 8-bit ASCII characters, which include all letters from the Latin script, plus some extra symbols. It is being replaced by 16-bit UNICODE characters, which include all the important scripts currently used, and a larger set of special symbols.

Normally we cannot exploit the numerical value of the digital representation of symbolic information, since the identification of the symbols, and ultimately their meaning, depend on the convention being used, and the character's context. This means that the compression of symbolic data normally has to be *lossless*, i.e., the information that is recovered after decompression has to be identical to the original. This is usually what is referred to as *text compression* or *data compression*.

Even when all information must be preserved, savings can be obtained by removing a form of *redundancy* from the representation. Normally, we do not refer to compression as the simplest choices of more economical representations, such as converting text that is known to be English from 16-bit UNICODE to 8-bit ASCII. Instead, we refer to compression as the techniques that exploit the fact that some symbols, or sequences of symbols, are much more commonly used than others. As we explain later, it is more efficient to use a smaller number of bits to represent the most common characters (which necessarily requires using a larger number for the less common). So, in its simplest definition, lossless compression is equivalent simply to reducing "wasted space." Of course, for creating effective compression systems, we need a rigorous mathematical framework to define what exactly "wasted" means. This is the subject of *information theory* that is covered from Chapter 2.

1.2.2 Numerical information

The second important type of data corresponds to information obtained by measuring some physical quantity. For instance, audio data is created by measuring sound intensity in time; images are formed by measuring light intensity in a rectangular area; etc. For convenience, physical quantities are commonly considered to be real numbers, i.e., with infinite precision. However, since any actual measurement has limited precision and accuracy, it is natural to consider how much of the measurement data needs to be preserved. In this case, compression savings are obtained not only by eliminating redundancy, but also by removing data that we know are *irrelevant* to our application. For instance, if we want to record a person's body temperature, it is clear that we only need to save data in a quite limited range, and up to a certain precision. This type of compression is called *lossy* because some information—the part deemed irrelevant—is discarded, and afterwards the redundancy of the remaining data is reduced.

Even though just the process of discarding information is not by itself compression, it is such a fundamental component that we traditionally call this combination *lossy compression*. Methods that integrate the two steps are much more efficient in keeping the essential information than those that do not. For instance, the reader may already have observed that popular lossy media compression methods such as the JPEG image compression, or MP3 audio compression, can achieve one or two orders of magnitude in size reduction, with very little loss in perceptual quality.

It is also interesting to note that measurement data commonly produce relatively much larger amounts of data than text. For instance, hundreds of text words can be represented with the number of bits required to record speech with the single word "hello." This happens because text is a very economical representation of spoken words, but it excludes many other types of information. From recorded speech, on the other hand, we can identify not only the spoken words, but possibly a great deal more, such as the person's sex, age, mood, accents, and even very reliably identify the person speaking (e.g., someone can say "this is definitely my husband's voice!").

Similarly, a single X-ray image can use more data than that required to store the name, address, medical history, and other textual information of hundreds of patients. Thus, even though lossless compression of text is also important, in this book the emphasis is on compression of numerical information, since it commonly needs to be represented with a much larger number of bits.

1.3   Basic compression process

In practical applications we frequently have a mixture of data types that have to be compressed together. For example, in video compression we have to process a sequence of images together with their corresponding (synchronized) multichannel audio components. However, when starting the study of compression, it is much better to consider separately each component, in order to properly understand and exploit its particular properties. This approach is also used in practice (video standards do compress image and audio independently), and it is commonly easy to extend the basic approach and models to more complex situations.

A convenient model for beginning the study of compression is shown in Figure 1.1: we have a *data source* that generates a sequence of data elements $\{x_1, x_2, \ldots\}$, where all x_i are of the same type and each belongs to a set $\mathcal{A}$. For example, for compressing ASCII text, we can have $\mathcal{A} = \{a, b, c, \ldots, A, B, C, \ldots, 0, 1, 2, \ldots\}$. However, this representation is not always convenient, and even though this is a symbolic data type (instead of numerical), it is frequently better to use for x_i the numerical value of the byte representing the character, and have $\mathcal{A} = \{0, 1, 2, \ldots, 255\}$. For numerical data we can use as $\mathcal{A}$ intervals in the set of integer or real numbers.

The compression or *encoding* process corresponds to mapping the sequence of source symbols into the sequence $\{c_1, c_2, \ldots\}$, where each c_i belongs to a set $\mathcal{C}$ of compressed data symbols. The most common data symbols are bits ($\mathcal{C} = \{0, 1\}$) or, less frequently, bytes ($\mathcal{C} = \{0, 1, \ldots, 255\}$). The decompression or *decoding* process maps the compressed data sequence back to a sequence $\{\tilde{x}_1, \tilde{x}_2, \ldots\}$ with elements from $\mathcal{A}$. With lossless compression we always have $x_i = \tilde{x}_i$ for all i, but not necessarily with lossy compression.

There are many ways data can be organized before being encoded. Figure 1.2 shows some examples. In the case of Figure 1.2(a), groups with a fixed number of source symbols are mapped to groups with a fixed number of compressed data symbols. This approach is employed in some mathematical proofs, but is not very common. Better compression is achieved by allowing groups of different sizes. For instance, we can create a simple text compression method by using 16 bits as indexes to all text characters plus about 64,000 frequently used words (a predefined set). This corresponds to the variable-to-fixed scheme of Figure 1.2(b), where a variable number of source symbols are parsed into characters or words, and each is coded with the same number of bits. Methods for coding numerical data commonly use the method of Figure 1.2(c) where groups with a fixed number of data symbols are coded with a variable number of bits. While these fixed schemes are useful for introducing coding concepts, in practical applications it is interesting to have the maximum degree of freedom in organizing both the

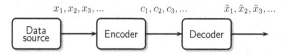

Figure 1.1   Basic data encoding and decoding model.

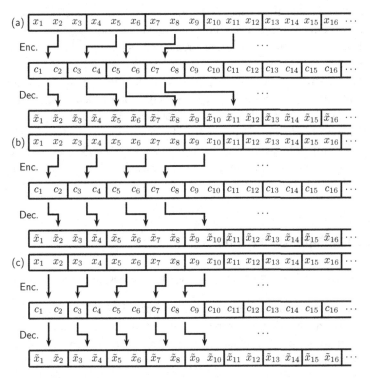

Figure 1.2 Examples of some forms in which the source and compressed data symbols can be organized for compression.

source and the compressed data symbols, so a variable-to-variable approach is more common.

1.4 Compression applications

The first popular compression applications appeared with the transition from analog to digital media formats. A great deal of research activity, and the development of the first compression standards, happened because there was a lot of analog content, such as movies, that could be converted. This is still happening, as we observe that only now is television broadcast in the USA transitioning to digital.

Another wave of applications is being created by the improvement, cost reduction, and proliferation of data-generating devices. For instance, much more digital voice traffic, photos, and even video, are being created by cell phones than with previous devices, simply because many more people carry them everywhere they go, and thus have many more opportunities to use them. Other ubiquitous personal communications and collaboration systems, such as videoconference, also have compression as a fundamental component.

The development of new sensor technologies, with the increase in resolution, precision, diversity, etc., also enables the creation of vast amounts of new digital information.

For example, in medical imaging, three-dimensional scans, which produce much more data than two-dimensional images, are becoming much more common. Similarly, better performance can be obtained by using large numbers of sensors, such as microphone or antenna arrays.

Another important trend is the explosive growth in information exchanged between devices only, instead of between devices and people. For instance, much of surveillance data is commonly stored without ever being observed by a person. In the future an even larger volume of data will be gathered to be automatically analyzed, mostly without human intervention. In fact, the deployment of sensor networks can produce previously unseen amounts of data, which may use communications resources for local processing, without necessarily being stored.

1.5   Design of compression methods

When first learning about compression, one may ask questions like:

- Why are there so many different types of compression?
- What makes a method superior in terms of performance and features?
- What compromises (if any) should be taken into account when designing a new compression method?

From the practical point of view, the subject of data compression is similar to many engineering areas, i.e., most of the basic theory had been established for many years, but research in the field is quite active because the practice is a combination of both art and science. This happens because while information theory results clearly specify optimal coding, and what are the coding performance limits, it commonly assumes the availability of reasonably accurate statistical models for the data, including model parameter values. Unfortunately, the most interesting data sources, such as text, audio, video, etc., are quite difficult to model precisely.[1]

In addition, some information theory results are asymptotic, i.e., they are obtained only when some parameter, which may be related to computational complexity or size of the data to be compressed, goes to infinity. In conclusion, while information theory is essential in providing the guidance for creating better compression algorithms, it frequently needs to be complemented with practical schemes to control computational complexity, and to create good statistical models.

It has been found that there can be great practical advantages in devising schemes that implicitly exploit our knowledge about the data being compressed. Thus, we have a large number of compression methods that were created specifically for text, executable code, speech, music, photos, tomography scans, video, etc. In all cases, assumptions were made about some typical properties of the data, i.e., the way to compress the data is in itself an implicit form of modeling the data.

Data statistical modeling is a very important stage in the design of new compression methods, since it defines what is obviously a prime design objective: reducing the number of bits required to represent the data. However, it is certainly not the only objective. Typical additional design objectives include

- low computational complexity;
- fast search or random access to compressed data;
- reproductions scalable by quality, resolution, or bandwidth;
- error resiliency.

The need to control computational complexity is a main factor in making practical compression differ from more abstract and general methods derived from information theory. It pervades essential aspects of all the compression standards widely used, even though it is not explicitly mentioned. This also occurs throughout this book: several techniques will be presented that were motivated by the need to optimize the use of computational resources.

Depending on the type of data and compression method, it can be quite difficult to identify in compressed data some types of information that are very easily obtained in the uncompressed format. For instance, there are very efficient algorithms for searching for a given word in an uncompressed text file. These algorithms can be easily adapted if the compression method always uses the same sequence of bits to represent the letters of that word. However, advanced compression methods are not based on the simple replacement of each symbol by a fixed set of bits. We have many other techniques, such as

- data is completely rearranged for better compression;
- numerical data goes through some mathematical transformation before encoding;
- the bits assigned for each character may change depending on context and on how the encoder automatically learns about the data characteristics;
- data symbols can be represented by a fractional number of bits;
- data may be represented by pointers to other occurrences of the same (or similar) data in the data sequence itself, or in some dynamic data structures (lists, trees, stacks, etc.) created from it.

The consequence is that, in general, fast access to compressed information cannot be done in an ad hoc manner, so these capabilities can only be supported if they are planned when designing the compression method. Depending on the case, the inclusion of these data access features may degrade the compression performance or complexity, and this presents another reason for increasing the number of compression methods (or modes in a given standard).

Resiliency to errors in the compressed data can be quite important because they can have dramatic consequences, producing what is called *catastrophic error propagation*. For example, modifying a single bit in a compressed data file may cause all the subsequent bits to be decoded incorrectly. Compression methods can be designed to include techniques that facilitate error detection, and that constrains error propagation to well-defined boundaries.

1.6   Multi-disciplinary aspect

As explained above, developing and implementing compression methods involves taking many practical factors, some belonging to different disciplines, into account. In fact, one interesting aspect of practical compression is that it tends to be truly multidisciplinary. It includes concepts from coding and information theory, signal processing, computer science, and, depending on the material, specific knowledge about the data being compressed. For example, for coding image and video it is advantageous to have at least some basic knowledge about some features of the human vision, color perception, etc. The best audio coding methods exploit psychophysical properties of human hearing.

Learning about compression also covers a variety of topics, but as we show in this book, it is possible to start from basic material that is easy to understand, and progressively advance to more advanced topics, until reaching the state-of-the-art. As we show, while some topics can lead to a quite complex and advanced theory, commonly only some basic facts are needed to be used in compression.

In the next chapter we present a quick overview of the topics included in this book.

Note

1. Note that we are referring to statistical properties only, and excluding the much more complex semantic analysis.

References

1. J. F. Gantz, C. Chite, A. Manfrediz, S. Minton, D. Reinsel, W. Schliditing, and A. Torcheva, "The diverse and exploding digital universe," International Data Corporation (IDC), Framingham, MA, White paper, Mar. 2008, (http://www.emc.com/digital_universe).
2. N. Negroponte, *Being Digital*. New York, NY: Alfred A. Knopf, Inc., 1995.

2 Book overview

2.1 Entropy and lossless coding

Compression of a digital signal source is just its representation with fewer information bits than its original representation. We are excluding from compression cases when the source is trivially over-represented, such as an image with gray levels 0 to 255 written with 16 bits each when 8 bits are sufficient. The mathematical foundation of the discipline of *signal compression,* or what is more formally called *source coding,* began with the seminal paper of Claude Shannon [1, 2], entitled "A mathematical theory of communication," that established what is now called *Information Theory.* This theory sets the ultimate limits on achievable compression performance. Compression is theoretically and practically realizable even when the reconstruction of the source from the compressed representation is identical to the original. We call this kind of compression *lossless coding.* When the reconstruction is not identical to the source, we call it *lossy coding.* Shannon also introduced the discipline of *Rate-distortion Theory* [1–3], where he derived the fundamental limits in performance of lossy coding and proved that they were achievable. Lossy coding results in loss of information and hence distortion, but this distortion can be made tolerable for the given application and the loss is often necessary and unavoidable in order to satisfy transmission bandwidth and storage constraints. The payoff is that the degree of compression is often far greater than that achievable by lossless coding.

In this book, we attempt to present the principles of compression, the methods motivated by these principles, and various compression (source coding) systems that utilize these methods. We start with the theoretical foundations as laid out by Information Theory. This theory sets the framework and the language, motivates the methods of coding, provides the means to analyze these methods, and establishes ultimate bounds in their performance. Many of the theorems are presented without proof, since this book is not a primer on Information Theory, but in all cases, the consequences for compression are explained thoroughly. As befits any useful theory, the source models are simplifications of practical ones, but they point the way toward the treatment of compression of more realistic sources.

The main focus of the theoretical presentation is the definition of information entropy and its role as the smallest size in bits achievable in lossless coding of a source. The data source emits a sequence of random variables, $X_1, X_2, \ldots$, with respective probability mass functions, $q_{X_1}(x_1), q_{X_2}(x_2), \ldots$ These variables are called *letters* or *symbols* and their range of values is called the *alphabet*. This range is assumed to be discrete and finite. Such a source is said to be *discrete*. We start with the simplest kind of discrete data source, one that emits its letters independently, each with the same probability distribution, a so-called *i.i.d.* (independently and identically distributed) source. Therefore, we consider just a single random variable or letter X that takes values $a_1, a_2, \ldots, a_K$ with probabilities $P(a_1), P(a_2), \ldots, P(a_K)$, respectively. The entropy of X is defined to be

$$H(X) = \sum_{k=1}^{K} P(a_k) \log \frac{1}{P(a_k)}, \tag{2.1}$$

where the base of the logarithm is 2 unless otherwise specified. Note that $H(X)$ depends on the probability distribution, not on the letters themselves. An important property of $H(X)$ is that

$$0 \le H(X) \le \log K. \tag{2.2}$$

The entropy is non-negative and is upper bounded by the base-2 logarithm of the alphabet size K. This upper bound is achieved if and only if the letters are equally probable and is the least number of bits required in a natural binary representation of the indices $1, 2, \ldots, K$. Especially important is that the entropy $H(X)$ is the fewest bits per source letter or lowest rate to which an i.i.d. source can be encoded without loss of information and can be decoded perfectly.

We then present several methods of lossless coding, most of which are provably capable of reaching the entropy limit in their compression. Therefore, such methods are referred to as *entropy coding* methods. Examples of these methods are *Huffman coding* and *arithmetic coding*. However, these methods can only achieve the lower rate limit of entropy as the length N of the data sequence tends to infinity. Their average codeword length per source symbol can reach within a tolerance band of $1/N$ or $2/N$ bits anchored at $H(X)$, so sometimes a small source length N is good enough for the application. Lossless or entropy coding is extremely important in all coding systems, as it is a component of almost every lossy coding system, as we shall see.

2.2 Quantization

Almost all physical sources are adequately modeled with a range of continuous values having a continuous probability distribution or density function. Such sources are said to be *continuous*. Discrete sources are often *quantized* samples of a continuous source. *Quantization* is the approximation of values of a source with a given smaller set of values, known at the source and destination. The values in this smaller set are indexed, so that only the indices have to be encoded. For example, rounding values to the nearest integer is quantization. Analog to digital (A/D) conversion is quantization followed by binary representation or encoding of the quantized value, as depicted in Figure 2.1. Quantization is absolutely necessary, because the number of bits needed to represent

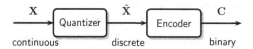

Figure 2.1   Source letter quantization followed by binary coding.

a continuous variable is infinite. That is, infinite precision requires an infinite number of bits. Clearly, infinite precision is beyond the capability of any physical device, but we wish to capture this finite precision, whatever it is, using fewer bits than its normal representation. We can think of the source as emitting either scalar (single) or vector (blocks of) random variables at each time. Whether scalar or vector, we seek to minimize the number of bits needed to encode the quantized values for a given distortion measure between the original and quantized values. The number of bits is determined either by its raw representation or its entropy when entropy coding is used. The idea is to choose the set of quantization values and the ranges of the random variables that are mapped to these values, so that the required minimum is attained. Quantizing vectors rather than scalars leads to smaller numbers of bits per source letter for a given distortion, but at the cost of larger memory to hold the quantization values and more computation to find the quantization value closest to the source value.

2.3   Source transformations

2.3.1   Prediction

As mentioned above, encoding blocks of source letters is more efficient than encoding the letters individually. This is true whether or not the letters are statistically independent. However, when the source letters are dependent, i.e., they have memory, one tries to remove the memory or transform the sequence mathematically to produce an independent sequence for encoding. Clearly such memory removal or transformation must be completely reversible. The motivation is that one can prove that memory removal achieves a reduction in distortion for a given bit rate for lossy coding and a reduction in bit rate in lossless coding. The ratio of distortions in lossy coding and ratio of rates in lossless coding between similar coding with and without memory removal is called *coding gain*. If the mathematical operation of memory removal or transformation is linear, then the output sequence is statistically independent only if the source is Gaussian and stationary. If non-Gaussian, the sequence is uncorrelated, which is a weaker property than independence. Nonetheless, these same linear operations are used, because coding gains are obtained even when the outputs of the memory removal are uncorrelated or approximately uncorrelated.

Various kinds of memory removal operation are used. The simplest is prediction of the current source value from past values to produce the error value, called the residual. A linear minimum mean squared error (MMSE) prediction of each member of the source sequence produces an uncorrelated residual sequence. Each member of the residual sequence is then encoded independently. Such coding is called *predictive*

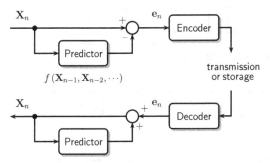

Figure 2.2 Lossless predictive coding system: top is encoder; bottom is decoder.

coding. When the encoding of the residuals is lossless, the source sequence can be reconstructed perfectly. A block diagram of a lossless predictive coding system appears in Figure 2.2. The decoder receives the current residual and is able to predict the current source value from perfect reconstructions of past source values. The current residual added to the prediction gives the exact current source value. Lossy predictive coding systems are slightly more complicated, having a quantizer directly following the residual in the encoder, and will be treated in detail in a later chapter. There, we shall study various means of prediction, and quantization and encoding of the residual sequence.

2.3.2 Transforms

2.3.2.1 Principal components and approximations

Mathematical transformations decorrelate the source sequence directly, whereas the prediction operation in predictive coding systems produces an uncorrelated residual sequence. The output of the transformation is a sequence of coefficients that represents the source sequence in a new basis. One such transformation is the Karhunen–Loève or *principal components* transformation. This transform depends on the statistical characteristics of the source, which is often unknown beforehand, so most often is used as an approximately decorrelating transform, such as the discrete cosine transform (DCT), that is not source dependent. A block diagram of such a transform coding system is shown in Figure 2.3. After transformation of the source sequence, the transform coefficients are individually quantized and encoded. In the decoder side, the coefficient codes are decoded to produce the quantized coefficients, which are then inverted by the inverse transform to yield a lossy reconstruction of the source sequence. Another property of these transforms besides uncorrelatedness is that the transform coefficients with the largest variances (energy) always appear at the smallest indices. Therefore, retaining only a few of these low index coefficients normally yields a fairly accurate reconstruction upon inverse transformation. When encoding such a transform with a target bit budget, bits are allocated unevenly, so that the highest energy coefficients get the most bits and the lowest very few or zero bits. Various bit allocation procedures have been formulated to allocate the bits optimally, so that the distortion is smallest for the given bit budget. When the bit allocation is done properly, it leads to coding gains, as shall be proved by analysis and practice.

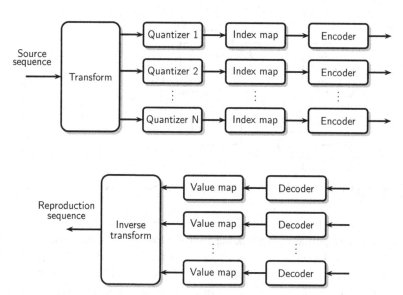

Figure 2.3   Transform coding system: encoder on top; decoder at bottom.

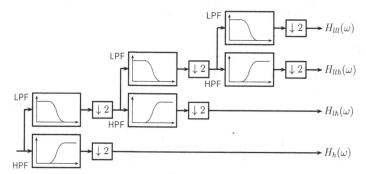

Figure 2.4   Example of how lowpass (LPF) and highpass (HPF) filters can be used for computing the subband transform.

2.3.2.2   Subband transforms

Certain mathematical transformations, called *subband transforms,* can be implemented by a series of lowpass and highpass filtering plus decimation. For example, for time sequences we can implement the subband transform with temporal filtering using a dynamic system as shown in Figure 2.4. The different outputs correspond to different responses for different frequency intervals or *subbands,* as shown in the example of Figure 2.5 (which corresponds to the filters of Figure 2.4). In images the subband transforms are done in a similar manner, but with two-dimensional spatial filtering. Commonly the filters are defined so that filtering can be done for each dimension separately, and they produce two-dimensional subbands, which are logically classified as shown in the top-left diagram of Figure 2.6, but that are actually obtained from two-dimensional filter responses as shown in the other graphs of Figure 2.6. Note that in all

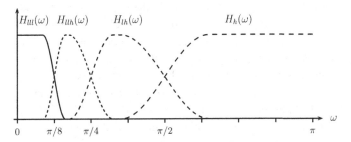

Figure 2.5   Example of a type of frequency response produced by filters used in the subband transform.

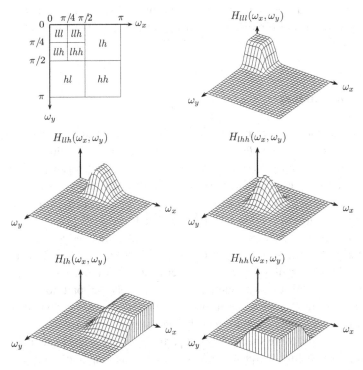

Figure 2.6   Example logical division of subbands (top left) and types of frequency response produced by some of the filters used in the two-dimensional subband transform.

cases the filter responses do not correspond to ideal filters, but nevertheless it is possible to use special sets of easy-to-implement filters that make the subband transform perfectly reversible.

When the filters are derived from a wavelet kernel, they are called *wavelet filters* and their subbands are called *wavelet subbands*. There are various kinds of wavelet filter and ways to implement them very efficiently, among them being a method called lifting, which will be explained in this book. There are filters that are reversible only if we assume floating-point computation, but there are forms of computation that are also completely reversible while maintaining a fixed, finite precision for all operations, so that there are no roundoff errors of any kind. Most of the high-magnitude or high-energy coefficients appear in the lowest frequency subbands and the coefficients in all subbands, except the lowest one, tend to be nearly uncorrelated. The subbands are often encoded independently, since they are independent for a linearly filtered, stationary Gaussian source. In all cases, the subbands have different variances and are treated as having the same probability distribution function, except perhaps the lowest frequency subband. Therefore, as with the non-subband transforms above,[1] the target bit budget has to be unevenly allocated among the subbands to minimize the distortion in lossy coding. This kind of bit allocation results in coding gain. Lossless coding of the subbands requires no bit allocation and gives perfect reconstruction of the source when reversible, fixed-precision filters are used for the subband transform. There is even a theoretical justification for coding gains in lossless coding, as it has been proved that a subband transform reduces the entropy of the source and therefore the rate for lossless coding [4].

2.4   Set partition coding

One of the unique aspects of this book is a coherent exposition of the principles and methods of *set partition coding* in Chapter 10. This kind of coding is fundamentally lossless, but is not entropy coding in the same category as other general purpose coding methods that can be used for any type of data. It generally works best upon transforms of a source because it can more efficiently exploit some properties that are difficult to model. As explained in Section 1.3, coding methods divide the data to be coded in various ways to obtain better compression. For example, the fixed-to-variable scheme of Figure 1.2(c) is commonly used for coding quantized transform coefficients. Grouping source symbols for coding can lead to better results, but it tends to produce exponential growth in complexity with the number of symbols. Set partition coding employs a variable-to-variable approach, but one in which it adaptively subdivides the source symbols, in order to obtain simultaneously better compression and maintain low computational complexity. More specifically, set partition coding refers to methods that divide the source (or transform) sequence into sets of different amplitude ranges.

Figure 2.7 illustrates sequential partitioning of a two-dimensional block of data samples into sets of decreasing maximum value. The "dark" samples exceed in magnitude

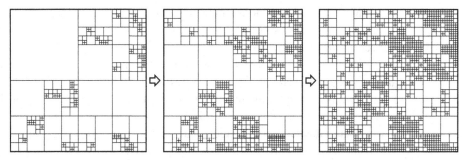

Figure 2.7   Example of how a set of image pixels is sequentially subdivided for more efficient compression using set partition coding.

a given threshold while samples in the "white" areas do not. When the threshold is lowered, every white area containing one or more above-threshold samples is recursively quadri-sected until all these above-threshold samples are located and hence darkened. The threshold is again lowered and the process repeats. Once a source or transform sample is located within a set, the most bits needed to represent it are the base-2 logarithm of the associated range. In transforms, the sets with a small range tend to be large, so their coefficients can be encoded with very few bits. All-zero sets require zero bits once they are located. The combination of the set location bits and the range bits often provides excellent compression.

Methods of set partitioning are fundamentally both simple and powerful. They require only simple mathematical and computer operations, and produce efficient compression, even without subsequent entropy coding of the range and location bits. Furthermore, these methods are natural companions for *progressive coding*. One usually defines the sets as non-overlapping with given upper and lower limits. When one searches them in order of decreasing upper limit, then the elements in ranges of larger magnitudes are encoded before those in any range of smaller magnitude. Therefore, in the decoder, the fidelity of the reconstruction increases progressively and as quickly as possible as more bits are decoded.

2.5   Coding systems

2.5.1   Performance criteria

At first glance, the performance criterion for a coding system seems obvious. One would like the system to produce the fewest bits possible to reconstruct the source either perfectly or within a given distortion. If perfect reproduction is unnecessary or unattainable, then one needs to define the measure of distortion. The most common measure is mean squared error (MSE) between the reproduction and the source. The reason is that it is easy to measure and is analytically tractable in that it allows derivation of formulas for compression performance, optimal bit allocation, and other aspects of a coding system. There are various arguments against this distortion criterion, but the chief one is that it is inappropriate for particular kinds of exploitation of the data. For example, if the source is audio, MSE is only indirectly related to the aural perception. Similarly for images, in that visual reproduction artifacts may not be adequately measured by MSE, especially if the artifacts are only locally contained. Also, one may only be interested in detection of objects or classification of vegetation in images, so the accuracy of these tasks cannot be related directly to MSE. Nonetheless, when a source is compressed for general use, MSE seems to be the best criterion. Practically all coding systems are created and proven for their performance in MSE or the related peak signal-to-noise ratio (PSNR), which is a translation of the logarithm of MSE.

Beyond ultimate performance, there are other attributes sought for a coding system. We have already mentioned progressive coding. There, decoding can stop when the user is satisfied with the fidelity of the reconstruction. A related attribute is *embedded coding,* when the system's compressed file of any lower rate resides within its larger compressed file of larger rate. Therefore, one compressed bitstream can serve the needs of various users with different terminal capabilities. Another attribute is resolution scalability, when source reconstructions of reduced resolution can be decoded directly from the compressed bitstream of the full resolution source.

An attribute often more important even than minimum possible rate (bits per source letter) is low complexity. Low complexity manifests itself in low numbers of mathematical or computational operations, high speed in encoding and decoding, and small, compact hardware with low power requirements. One most often must sacrifice compression performance to meet the demands of lower complexity or high speed. For example, according to industry professionals, physicians will not wait more than one or two seconds to view a decompressed X-ray image. So as long as the compression is adequate, the method with the fastest decoding speed is the winner for physicians. Anyway, in creating a coding system for a particular application, the various attributes and requirements of the compression must be carefully considered.

2.5.2   Transform coding systems

2.5.2.1   JPEG image compression

The most ubiquitous coding system, the JPEG image compression standard, is a transform coding system. The source image is divided into contiguous 8×8 blocks, which are then independently transformed with the DCT. The DC or $(0,0)$ indexed coefficients are encoded by a lossless predictive scheme and the non-DC (AC) are quantized with uniform range intervals depending on their coordinates within their block. A linear sequence of indices of the quantization intervals is formed through a zigzag scan of the block. This sequence is then encoded losslessly by a combination of run-length symbols and Huffman coding. The details may be found in Chapter 11. The JPEG baseline mode, which seems to be the only one in use, has no attributes of progressiveness, embeddedness, or scalability in resolution, but the complexity is quite low.

2.5.2.2   H.264/AVC intraframe standard

The standard video compression systems all have an intraframe or single image frame mode for the beginning frame of a group. The H.264/AVC Video Standard's intraframe mode is a transform coding method. The frame is divided into either 4×4, 4×8, or 8×8 blocks, depending on a measure of activity. For example, 8×8 blocks have the lowest activity and 4×4 the highest. The blocks are individually transformed by a simple integer approximation to the DCT. The resulting DCT blocks are encoded differently than JPEG, as intra-block prediction and context-based, adaptive arithmetic bitplane coding are employed. More details will be revealed in Chapter 9. The method has no resolution scalability, but it does have an optional limited fidelity progressive mode attained by encoding two residual layers.

2.5.3   Subband coding systems

Subband coding of images has received much attention since 1986, when Woods and O'Neil [5] published the first journal paper on the subject. However, the seminal paper in subband coding of speech by Crochiere *et al.* [6] appeared 10 years earlier. Really surprising was that the image coding results were far better than what standard analysis predicted. Specifically, standard analysis revealed that optimal predictive coding in subbands was no better than optimal predictive coding within the original image. In other words, there was supposedly no coding gain when predictive coding was used in subbands versus the same being used on the original source. A later article resolved the contradiction between theory and practice [7, 8]. It proved that coding in subbands using non-optimal finite order prediction, as what must be done in practice, does indeed show gains over coding using the same order prediction within the source. This fact is true regardless of the dimensionality of the source.

Systems employing coding of subbands seem to fall into two categories: block-based and tree-based. We shall be exploring several of them in both categories. Block-based systems include the classical systems that encode the subbands independently. The bit rates for the subbands are assigned with a bit allocation procedure based on variances. Currently popular systems assign bit rates based on the actual amplitude of coefficients in the subbands. Such block-based systems are the JPEG2000 still image standard with its embedded block coding with optimized truncation (EBCOT) coding engine, set partitioning embedded block (SPECK) and its variants subband block hierarchical partitioning (SBHP) and embedded zero block coder (EZBC), and amplitude and group partitioning (AGP). Tree-based systems encode spatial orientation trees (SOTs), or what are called erroneously *zerotrees,* in the wavelet transform.[2] These trees are rooted in the lowest frequency subband and branch successively to coefficients having the same spatial orientation in the higher subbands. A SOT of an image (two-dimensional) wavelet transform is shown in Figure 2.8. Embedded zerotree wavelet (EZW) and set partitioning in hierarchical trees (SPIHT) are the two best-known tree-based coders. Of the coders mentioned, all except JPEG2000 (and EBCOT) and EZW are set partitioning

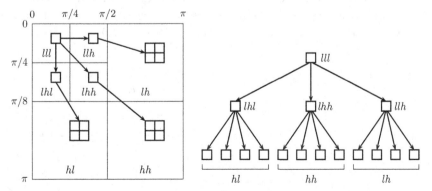

Figure 2.8   Transform coefficients in spatial frequency wavelet subbands (left) are organized to form spatial orientation trees (SOTs, right).

coders. Other set partitioning coders are group testing of wavelets (GTW), which will be
be described later, and significance linked connected component analysis (SLCCA) [9].
The interested reader may consult the given reference for information about the latter
method.

The coding methods just mentioned were originally developed for compressing
images. Many of these methods can be extended in an obvious way to higher dimensional data sources, such as hyperspectral and tomographic data, video, and three-
dimensional motion images, such as fMRI. They also can be utilized for compression
of one-dimensional signals, such as audio, speech, and biological signals. Audio and
speech require integration and synthesis of a perceptual model to obtain efficient
compression without annoying artifacts.

2.6   Distributed source coding

Distributed source coding is a subject that has come into prominence recently. At the
time of writing, there are no textbooks that treat this subject. A system employing
distributed source coding encodes correlated sources independently and decodes them
jointly and attains performance nearly or exactly as good as if they were encoded jointly
at the source. Such a system is illustrated in Figure 2.9 for two sources.

Handheld devices, such as mobile phones, that capture and transmit video, require
coding that does not consume much energy and does not cost too much to purchase.
Most of the cost and complexity of video encoding are linked to motion estimation, so
through distributed coding, the decoding in the base station absorbs the major brunt of
the complexity, with little compromise in performance. A network of sensors, each of
which transmits images or video of the same scene, cannot cooperate easily in source
coding, so a processing node or base station can reconstruct the transmitted data with
nearly the same accuracy as if the sensors cooperated in encoding.

For the example of a two-source system, the sum of the rate for the two sources need
only be the joint entropy $H(X, Y)$ to achieve lossless recovery, as long as certain lower
limits on the bit rates of the two sources are obeyed. The way the system might work
is roughly as follows. One of the sources X is encoded losslessly with rate $H(X)$. The
other source is encoded with rate $H(Y/X)$ $(H(X) + H(Y/X) = H(X, Y))$. The input

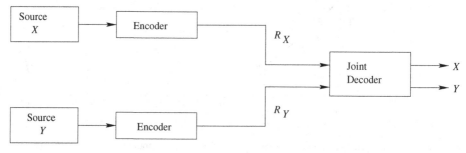

Figure 2.9   Distributed source coding: independent encoding and joint decoding of correlated sources.

space is partitioned into $H(Y/X)$ subsets (called *cosets* of a channel code) of dispersed elements. Instead of encoding Y itself, the source conveys the coset into which Y falls using $H(Y/X)$ bits. Since X is correlated with Y, it occurs with high probability close to Y. So the decoder declares Y to be the member of that coset that is closest to X. The details and justification are explained in Chapter 14.

The intention of the treatment of distributed source coding in this textbook is to explain the principles both for lossless and lossy source coding. In a practical system, one cannot avoid a small probability of recovery failure, so these probabilities are derived for some source correlation models. There is no attempt to describe the numerous application scenarios, as they lie outside the purview and objective of this book, which is to concentrate on the teaching of fundamental principles.

Notes

1. One can show that these mathematical transforms are special cases of subband transforms, but we make this distinction here, because they are usually not treated like subband transforms for encoding purposes.
2. The term *zerotree* means that all nodes in the SOT are tagged with zero to signify that they are insignificant. The term *SOT* just designates the structure of the tree without assigning values to the nodes.

References

1. C. E. Shannon, "A mathematical theory of communication," *Bell Syst. Technol. J.*, vol. 27, pp. 379–423 and 632–656, July and Oct. 1948.
2. C. E. Shannon and W. Weaver, *The Mathematical Theory of Communication.* Urbana, IL: University of Illinois Press, 1949.
3. C. E. Shannon, "Coding theorems of a discrete source with a fidelity criterion," in *Information and Decision Processes*, ed. R. E. Machol New York, NY: McGraw-Hill Publishing Co., 1960, pp. 93–126.
4. R. P. Rao and W. A. Pearlman, "On entropy of pyramid structures," *IEEE Trans. Inf. Theory*, vol. 37, no. 2, pp. 407–413, Mar. 1991.
5. J. W. Woods and S. D. O'Neil, "Subband coding of images," *IEEE Trans. Acoust. Speech Signal Process.*, vol. 34, no. 5, pp. 1278–1288, Oct. 1986.
6. R. E. Crochiere, S. A. Webber, and J. L. Flanagan, "Digital coding of speech in sub-bands," in *Proceedings of the IEEE International Conference on Acoustics Speech Signal Processing*, Philadelphia, PA, Apr. 1976, pp. 233–236.
7. W. A. Pearlman, "Performance bounds for subband coding," in *Subband Image Coding*, ed. J. W. Woods Norwell, MA: Kluwer Academic Publishers, 1991, ch. 1.
8. S. Rao and W. A. Pearlman, "Analysis of linear prediction, coding, and spectral estimation from subbands," *IEEE Trans. Inf. Theory*, vol. 42, no. 4, pp. 1160–1178, July 1996.
9. B.-B. Chai, J. Vass, and X. Zhuang, "Significance-linked connected component analysis for wavelet image coding," *IEEE Trans. Image Process.*, vol. 8, no. 6, pp. 774–784, June 1999.

Further reading

B. Girod, A. M. Aaron, S. Rane, and D. Rebollo-Monedero, "Distributed video coding," *Proc. IEEE*, vol. 91, no. 1, pp. 71–83, Jan. 2005.

E. S. Hong and R. E. Ladner, "Group testing for image compression," *IEEE Trans. Image Process.*, vol. 11, no. 8, pp. 901–911, Aug. 2002.

W. A. Pearlman, A. Islam, N. Nagaraj, and A. Said, "Efficient, low-complexity image coding with a set-partitioning embedded block coder," *IEEE Trans. Circuits Syst. Video Technol.*, vol. 14, no. 11, pp. 1219–1235, Nov. 2004.

S. S. Pradhan, J. Kusuma, and K. Ramchandran, "Distributed compression in a dense microsensor network," *IEEE Signal Process. Mag.*, pp. 51–60, Mar. 2002.

A. Said and W. A. Pearlman, "A new, fast, and efficient image codec based on set partitioning in hierarchical trees," *IEEE Trans. Circuits Syst. Video Technol.*, vol. 6, no. 3, pp. 243–250, June 1996.

J. M. Shapiro, "Embedded image coding using zerotrees of wavelet coefficients," *IEEE Trans. Signal Process.*, vol. 41, no. 12, pp. 3445–3462, Dec. 1993.

D. S. Taubman, "High performance scalable image compression with EBCOT," *IEEE Trans. Image Process.*, vol. 9, no. 7, pp. 1158–1170, July 2000.

3 Principles of lossless compression

3.1 Introduction

Source coding began with the initial development of information theory by Shannon in 1948 [1] and continues to this day to be influenced and stimulated by advances in this theory. Information theory sets the framework and the language, motivates the methods of coding, provides the means to analyze the methods, and establishes the ultimate bounds in performance for all methods. No study of image coding is complete without a basic knowledge and understanding of the underlying concepts in information theory.

In this chapter, we shall present several methods of lossless coding of data sources, beginning with the motivating principles and bounds on performance based on information theory. This chapter is not meant to be a primer on information theory, so theorems and propositions will be presented without proof. The reader is referred to one of the many excellent textbooks on information theory, such as Gallager [2] and Cover and Thomas [3], for a deeper treatment with proofs. The purpose here is to set the foundation and present lossless coding methods and assess their performance with respect to the theoretical optimum when possible. Hopefully, the reader will derive from this chapter both a knowledge of coding methods and an appreciation and understanding of the underlying information heory.

The notation in this chapter will indicate a scalar source on a one-dimensional field, i.e., the source values are scalars and their locations are on a one-dimensional grid, such as a regular time or space sequence. Extensions to multi-dimensional fields, such as images or video, and even to vector values, such as measurements of weather data (temperature, pressure, wind speed) at points in the atmosphere, are often obvious once the scalar, one-dimensional field case is mastered.

3.2 Lossless source coding and entropy

Values of data are not perfectly predictable and are only known once they are emitted by a source. However, we have to know something about the source statistics to characterize the compressibility of the source. Therefore, the values are modeled as random variables with a given probability distribution. In that vein, consider that an information source emits a sequence of N random variables $(X_1, X_2, \ldots, X_N)$. We group these random variables to form the random vector $\mathbf{X} = (X_1, X_2, \ldots, X_N)$. Each random variable or vector element X_n, $n = 1, 2, \ldots, N$, is called a source letter or source symbol. In addition, we assume that the values taken by these random variables or source letters are on a discrete and finite set $\mathcal{A} = \{a_1, a_2, \ldots, a_K\}$ called the alphabet of the source and that the random vector has a known probability distribution (mass function) denoted by $q_{\mathbf{X}}(x_1, x_2, \ldots, x_N)$. A random vector from the source together with its probability distribution (density or mass function) for any arbitrary finite length N is called a *random process* or *ensemble*.

Assume that the source is a stationary random process, so that the probability distribution is the same for any contiguous subgroup of the same size from these N letters and and for every length N vector emitted at any time. We say that a stationary source is *memoryless* when the letters are statistically independent. In this case, the probability distribution takes the form:

$$q_{\mathbf{X}}(x_1, x_2, \ldots, x_N) = q_X(x_1)q_X(x_2)\cdots q_X(x_N), \tag{3.1}$$

where each letter has the same probability distribution $q_X(x)$ for $x \in \mathcal{A}$. Such a source sequence of N letters is also said to be independent and identically distributed (i.i.d.). Stationary sources with memory have letters that are statistically dependent. The most common such source is the first-order Markov source whose probability distribution takes the general form

$$q_{\mathbf{X}}(x_1, x_2, \ldots, x_N) = q_{X_1}(x_1)q_{X_2|X_1}(x_2|x_1)q_{X_3|X_2}(x_3|x_2)\cdots \tag{3.2}$$
$$\cdots q_{X_N|X_{N-1}}(x_N|x_{N-1}),$$

where $q_{X_i|X_{i-1}}(x_i|x_{i-1})$, $i = 2, 3, \ldots, N$ is the conditional probability distribution of X_i, given that X_{i-1}. When the source is stationary, these conditional distributions are the same for each i. Note that the Markov source is characterized by statistical dependence only on the preceding letter.

An important attribute of a (discrete value) source is its entropy, defined as:

$$H(\mathbf{X}) = -\sum_{x_N \in \mathcal{A}} \cdots \sum_{x_1 \in \mathcal{A}} q_{\mathbf{X}}(x_1, x_2, \ldots, x_N) \log_2 q_{\mathbf{X}}(x_1, x_2, \ldots, x_N). \tag{3.3}$$

The logarithmic base of 2 provides an information measure in bits and will be understood as the base in all logarithms that follow, unless otherwise specified. It is often more meaningful to refer to entropy per source letter, defined as:

$$H_N(\mathbf{X}) \equiv \frac{1}{N} H(\mathbf{X})$$

When the source is stationary and memoryless (i.i.d.), substitution of Equation (3.1) into the per-letter entropy $H_N(\mathbf{X})$ yields

$$H_N(\mathbf{X}) = -\frac{1}{N} \sum_{i=1}^{N} \sum_{x_i \in \mathcal{A}} q_X(x_i) \log q_X(x_i) \tag{3.4}$$

$$= -\sum_{x \in \mathcal{A}} q_X(x) \log q_X(x) \tag{3.5}$$

The last equality follows from the identical distributions of the component letters X_i, giving identical sums for each i.

The per-letter entropy of the source, denoted as

$$H(X) = -\sum_{x \in \mathcal{A}} q_X(x) \log q_X(x),$$

is the entropy of the single random variable X with probability distribution $q_X(x)$, consistent with the sequence or vector entropy of Equation (3.3). A more explicit expression for this entropy in terms of the letter values $a_1, a_2, \ldots, a_K$ is

$$H(X) = -\sum_{k=1}^{K} P(a_k) \log P(a_k), \tag{3.6}$$

where $P(a_k)$ denotes the probability of a_k. An intuitive interpretation of entropy of the random variable X is average uncertainty of its letter values. It is also information in the sense that it requires $H(X)$ bits of information to remove the uncertainty. The quantity $-\log P(a_k)$ measures the uncertainty of occurrence (or surprise) of a_k, because the smaller the probability $P(a_k)$, the larger the surprise or uncertainty of its occurrence and vice versa. $H(X)$ is the average or expectation of the random variable $-\log P(a_k)$. The same interpretation holds true for the vector entropy $H(\mathbf{X})$ in Equation (3.3). It is worth noting and remembering that the entropies $H(X)$ and $H(\mathbf{X})$ are functions only of probabilities and not of values of the random variables.

Through the non-negativity of $-\log P(a_k)$ for all k and the inequality $\log_e x \leq x - 1$, it is easy to prove that

$$H(X) \geq 0$$

and

$$H(X) \leq \log K, \tag{3.7}$$

with equality in the latter if and only if the probabilities $P(a_k)$ are equal.

In the case when the source alphabet is binary, we can have only two symbol probabilities, p and $1 - p$. The entropy of this source is defined by the binary entropy function

$$H_b(p) = -p \log p - (1 - p) \log(1 - p), \tag{3.8}$$

which is plotted in Figure 3.1.

Note that its maximum value is 1 bit, and that its value decreases very quickly near $p = 1$. Figure 3.2 shows how the entropy of a source with a ternary alphabet varies with their probabilities. We can observe the same property, which deserves to be considered more carefully, since it can help us to understand where to focus our efforts when designing compression methods.

For any data source, if $p_{\max}$ represents the maximum symbol probability

$$p_{\max} = \max_k \{P(a_k)\}, \tag{3.9}$$

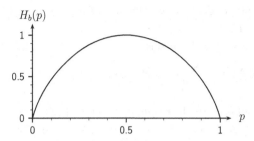

Figure 3.1   Binary entropy function.

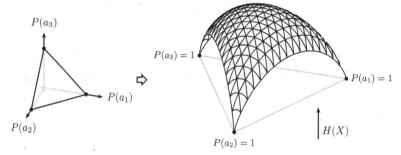

Figure 3.2   Entropy function for source alphabets with three symbols.

then we have the following upper and lower bounds on the entropy function

$$H(X) \leq H_b(p_{max}) + (1 - p_{max}) \log(K - 1) \qquad (3.10)$$

$$H(X) \geq \log(1/p_{max}), \; p_{max} \neq 0. \qquad (3.11)$$

Figure 3.3 shows a graph of these bounds in the range of p_{max} from 0.2 to 1.0.

The bounds make no sense for a very small p_{max}, because then the rest of the probabilities, being smaller, would have to be spread over a very large alphabet. Whatever the alphabet size, the bounds converge toward each other only near $p_{max} = 1$. What these bounds on the entropy function show is that, independent of the alphabet size and the probabilities of the other symbols, if p_{max} is near one, then the entropy has to be small. Conversely, low entropy values can only be achieved when coding is applied to a source that has one symbol with probability near one.

When the source is not memoryless, it is fairly obvious that the entropy $H(X)$ of the vector **X** follows Equation (3.7) when K is interpreted as the number of values of **X** with non-zero probability. It can also be shown that

$$H_N(X) \equiv \frac{1}{N}H(X) \leq \frac{1}{N}\sum_{i=1}^{N} H(X_i) = H(X), \qquad (3.12)$$

which means that the uncertainty per source letter is reduced when there is memory or dependence between the individual letters. Furthermore, as N tends towards infinity,

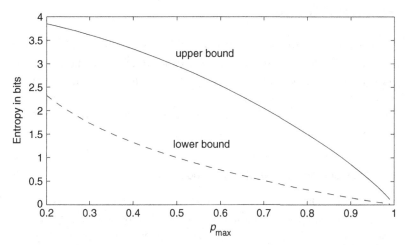

Figure 3.3 Upper and lower bounds on the entropy function. Solid line corresponds to Equation (3.10) for $K = 16$ and dashed line to Equation (3.11).

$H_N(\mathbf{X})$ goes monotonically down to a limit $H_\infty(\mathbf{X})$, known as the entropy rate of the source. The following source coding theorem can now be stated:

THEOREM 3.1 *For any $\epsilon > 0, \delta > 0$, there exists N sufficiently large that a vector of N source letters can be put into one-to-one correspondence with binary sequences of length $L = N[H_\infty(\mathbf{X}) + \epsilon]$ except for a set of source sequences occurring with probability less than δ. Conversely, if $\frac{L}{N} < H_\infty(\mathbf{X})$, the set of source sequences having no binary codewords, approaches one as N grows sufficiently large.*

Note that when the source is memoryless $H_\infty(\mathbf{X}) = H(X)$. The ramification of this theorem is that we can select the $K = 2^{N[H_\infty(\mathbf{X})+\epsilon]}$ vectors from the source which occur with probability greater than $1 - \delta$ (i.e., arbitrarily close to one), and index each of them with a unique binary codeword of length $L = N[H_\infty(\mathbf{X}) + \epsilon]$. If we transmit the binary index of one of these vectors to some destination where the same correspondences between the K indices and vectors are stored, then the original source vector is perfectly reconstructed. When the source emits a vector which is not among the K indexed ones, an erasure sequence is transmitted with no recovery possible at the destination. The probability of this error event is less than δ. The set of K binary sequences is called a code with rate in bits per source letter of $R = \frac{1}{N} \log_2 K = H_\infty(\mathbf{X}) + \epsilon$. The converse of the theorem means that $H_\infty(\mathbf{X})$ is the smallest possible rate in bits per source letter for a code that enables perfect reconstruction of the source vectors at the destination. For a stationary memoryless (i.i.d.) source, the theorem means that one can construct a code having no more than $H(X)$ bits per source letter and reconstruct it perfectly at the destination with probability close to one. However, whether or not the source has memory, the probability of the erasure or error event is guaranteed to be small only for large codeword lengths N, which may be impractical.

3.3   Variable length codes

Consider now the case of an i.i.d. source. If one is willing to transmit binary codeword sequences of variable length, one can theoretically eliminate the error event associated with fixed length codeword sequences. In practice, however, one needs to utilize a fixed length buffer which may overflow or become empty with a finite probability when operating for a finite length of time. For the moment, we shall ignore the buffering problems by assuming an infinite length buffer.

3.3.1   Unique decodability and prefix-free codes

Fixed length codewords are joined one after another in the codestream without separating markers, because the decoder knows the common codeword length and can separate them easily. When variable length codewords are strung together and transmitted, there is no assurance that they can be uniquely separated and decoded at the destination unless an extra mark symbol is inserted between each codeword. As insertion of an extra symbol for each codeword is uneconomical, one seeks codes (sets of codewords) which are uniquely decodable when strung together and of minimum average length. A code is uniquely decodable when any string of codewords in a codestream corresponds to one distinct sequence of source symbols. An important class of uniquely decodable codes is the prefix-free (or prefix) codes. A code is said to be prefix-free when no codeword is a prefix of any longer codeword in the code. This means that once a string of code symbols is found in the codestream that matches a codeword, it can be decoded immediately, since the following code symbols belong to the next codeword. This means that decoding a prefix-free code is instantaneous in that it requires no searching backward or forward in the string to decode a codeword.

An example of a prefix-free code is the binary code shown in Table 3.1: the code string 0110111100110 can be uniquely parsed into 0-110-111-10-0-110 and decoded as ACDBAC.

3.3.2   Construction of prefix-free codes

We shall now show how to construct a prefix-free code. The code alphabet in the above example is binary with symbols (called *digits*) 0 and 1. Because this is the most common and simplest case to illustrate and promote understanding, we shall continue to assume a binary {0,1} code alphabet. The idea is to associate nodes of a binary tree with binary strings and terminal nodes with codewords. Consider the binary branching tree in Figure 3.4, drawn to four levels of branchings for space reasons. Starting from the root node at the top, we associate a "0" to the next node in branching to the left and a "1" to the next node in branching to the right. We continue this same association of code symbols when branching from any node. Therefore, the binary string associated with the node marked "A" in Figure 3.4 is 010. Note that all nodes in the subtree branching from node "A" have associated binary strings with the prefix 010, because the paths to

Table 3.1 Binary prefix-free code

Letter	Codeword
A	0
B	10
C	110
D	111

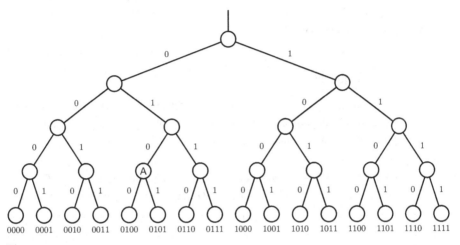

Figure 3.4 Binary tree.

them pass through node "A". Note also that no nodes other than those in the subtree rooted at node "A" have 010 as a prefix to its associated binary string. Therefore, if we remove the subtree descending from node "A", leaving it as a terminal node, we are left with the binary string 010 that is a prefix of no other node string in the tree. Therefore, a set of terminal nodes in a binary tree comprises a set of associated binary strings in which none is a prefix of another. We have thus a method to construct a prefix-free code.

Consider the code in Table 3.1. The tree associated with this code is shown in Figure 3.5.

We can visualize a receiver decoding a prefix-free code by observing a sequence of code digits and tracing a path down the code tree, going to the left for a "0" and to the right for a "1". At each node, the next code digit tells which branch, left or right, to take. When it reaches a terminal node, it decodes the code string.

For a general alphabet of D possible symbols (still called *digits*) or a D-ary alphabet, trees with D branches at each node are constructed. For example, the tree branches of an octal code will be labeled with digits $0, 1, 2, \ldots, 7$. The terminal nodes of such a tree comprise a prefix-free octal code.

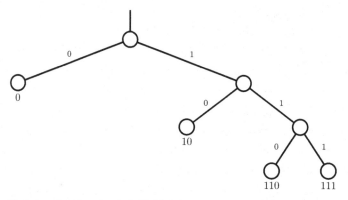

Figure 3.5   Code tree for binary code in Table 3.1.

3.3.3   Kraft inequality

There are uniquely decodable codes which are not prefix-free, but prefix-free codes are uniquely decodable and also instantaneously decodable. An important property of prefix-free codes is stated in the following Kraft Inequality (KI).

THEOREM 3.2 *Kraft Inequality For any prefix-free code over a D-ary alphabet, the codeword lengths $\ell_1, \ell_2, \ldots, \ell_K$ (in D-ary symbols) must satisfy the Kraft Inequality*

$$\sum_{k=1}^{K} D^{-\ell_k} \leq 1. \tag{3.13}$$

Conversely, given a set of codeword lengths $\ell_1, \ell_2, \ldots, \ell_K$ that satisfy the above equality, there exists a prefix-free code with these codeword lengths.

This theorem means that the lengths of codewords in prefix-free codes must satisfy the KI in Theorem 3.2 and that prefix-free codes can be constructed with lengths satisfying this inequality. In particular, the set of lengths $\ell_k = -\log_D P(a_k)$ satisfy this inequality, if we ignore, for now, the fact that these lengths may not be integers.

We shall offer two methods of proof for Theorem 3.2, because their constuctions are the bases of two well-known codes to be presented later.

Proof 1. Given a prefix-free code with lengths $\ell_1, \ell_2, \ldots, \ell_K$, construct a full tree of $\ell_{\max} = \max_k \ell_k$ branch levels. It has $D^{\ell_{\max}}$ terminal nodes, at least one of which corresponds to a codeword. Emanating from an intermediate node at level $\ell_k < \ell_{\max}$, corresponding to one of the codewords, is a subtree with $D^{\ell_{\max} - \ell_k}$ terminal nodes. Therefore, $D^{\ell_{\max} - \ell_k}$ terminal nodes were eliminated from the full tree to produce the prefix-free codeword at the corresponding intermediate node. So, too, for all the other codewords associated with nodes at levels less than $\ell_{\max}$. The sum of all eliminated terminal nodes must not exceed the number in the full tree. Therefore,

$$\sum_{k=1}^{K} D^{\ell_{\max}-\ell_k} \leq D^{\ell_{\max}}. \tag{3.14}$$

Dividing by $D^{\ell_{\max}}$ yields the stated result of

$$\sum_{k=1}^{K} D^{-\ell_k} \leq 1. \tag{3.15}$$

The proof of the converse statement has almost been revealed already. First, re-index the lengths so that $\ell_1 \leq \ell_2 \leq \ell_3 \cdots \leq \ell_K = \ell_{\max}$. Grow a tree to level ℓ_1 and designate one of the terminal nodes as a codeword node. Grow the tree from all the other nodes to level ℓ_2 and designate one of the resulting terminal nodes as a codeword node. Continue in this way to grow the tree from the non-designated codeword nodes until all lengths ℓ_k are associated with a terminal node at that tree level. The resulting set of codewords is prefix-free.

This procedure produced the tree in Figure 3.5 for the code in Table 3.1. The ordered lengths of the codewords are $\ell_1 = 1$, $\ell_2 = 2$, $\ell_3 = 3$, and $\ell_4 = 3$. Starting from the root of a binary tree, grow $\ell_1 = 1$ level and mark the left (0) node as a terminal node. Grow the right (1) node to level $\ell_2 = 2$ and mark the left (10) node as terminal. Grow the right (11) node to level $\ell_3 = \ell_4 = 3$ and mark both nodes (110 and 111) as terminal to complete the tree.

Proof 2. A codeword with length ℓ_k can be represented as $y_1 y_2 \cdots y_{\ell_k}$, where y_j, $j = 1, 2, \ldots, \ell_k$ are the D-ary digits $0, 1, 2, \ldots, D-1$. Associate with each codeword the real D-ary fraction

$$0.y_1 y_2 \cdots y_{\ell_k} = \sum_{j=1}^{\ell_k} y_j D^{-j}, \quad k = 1, 2, \ldots, K. \tag{3.16}$$

The set of all D-ary fractions beginning with $y_1 y_2 \cdots y_{\ell_1}$ lies in the interval $[0.y_1 y_2 \cdots y_{\ell_1}, 0.y_1 y_2 \cdots y_{\ell_1} + D^{-\ell_1})$, because a sequence of digits beyond ℓ_1, no matter how long, adds less than $D^{-\ell_1}$. Therefore, the codeword associates to this interval and the next codeword of length $\ell_2 \geq \ell_1$ belongs to a disjoint interval, since it cannot have the codeword $y_1 y_2 \cdots y_{\ell_1}$ as a prefix. Similarly, all the other codewords associate to distinct, disjoint intervals. Since all these intervals are non-overlapping intervals within the unit interval $[0,1)$, the union of these intervals lies in $[0,1]$ and the sum of their lengths cannot exceed one. So we have, as above,

$$\sum_{k=1}^{K} D^{-\ell_k} \leq 1.$$

One can construct a prefix-free code with lengths satisfying the KI (3.13) by associating intervals to codewords, as in the forward part of this proof. Start with 0, the left end of $[0,1)$ with the length ℓ_1 all-0 codeword and mark off the open-ended interval $[0, D^{-\ell_1})$. Then take the D-ary fraction $D^{-\ell_1}$ of length ℓ_2 and use its digits as the next

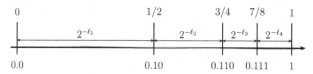

Figure 3.6   Binary code construction for lengths $\ell_1 = 1, \ell_2 = 2, \ell_3 = 3, \ell_4 = 3$ by association to binary fractions.

codeword. Mark off the open-ended interval of length $D^{-\ell_2}$ from $D^{-\ell_1}$ and choose the fraction $D^{-\ell_1} + D^{-\ell_2}$ with ℓ_3 digits and use these digits for the next codeword. The process continues until codewords are associated with all the lengths. Note that the codewords are the ℓ_k digits of the fractions 0.0 and $\sum_{j=1}^{k-1} D^{-\ell_j}$. The code is prefix-free, because the codewords are associated with contiguous disjoint intervals in $[0,1)$, the sum of whose lengths, $\sum_{j=1}^{m} D^{-\ell_j}$, is no more than one, as given.

In Figure 3.6, we illustrate the construction of the binary code in Table 3.1 through association with binary fractions. We are given the codeword lengths $\ell_1 = 1, \ell_2 = 2, \ell_3 = 3, \ell_4 = 3$ that satisfy KI: $2^{-1} + 2^{-2} + 2^{-3} + 2^{-3} = 1$. The partial sums $\sum_{j=1}^{k-1} 2^{-\ell_j}$ give the binary fractions 0.0, 0.10, 0.110, 0.111, respectively. The codewords are therefore 0, 10, 110, 111.

3.4   Optimality of prefix-free codes

We now see, as was stated earlier, that use of variable length codes removes the complication of an improbable event of a sequence of digits that cannot be decoded. Now we wish to settle the issue of optimality of such codes. First, we need the following theorem.

THEOREM 3.3 *Let a D-ary code have codeword lengths $\ell_1, \ell_2, \ldots, \ell_K$. If the code is uniquely decodable, the Kraft Inequality, (3.13), must be satisfied.*

We shall not prove this theorem here and we leave it to the curious reader to consult the textbook by Gallager [2] or Cover and Thomas [3]. We remark that the theorem also says that if the codeword lengths of a code do not satisfy the KI, the code is not uniquely decodable.

Let $\ell_1, \ell_2, \ldots, \ell_K$ be lengths of codewords in a D-ary prefix-free code for the i.i.d. source X. The letters of X corresponding to these lengths are $a_1, a_2, \ldots, a_K$ with respective probabilities $P(a_1), P(a_2), \ldots, P(a_K)$. The average codeword length

$$L = \sum_{k=1}^{K} P(a_k)\ell_k$$

satisfies

$$L < H_D(X) + 1. \tag{3.17}$$

Furthermore, for any uniquely decodable set of codewords,

$$L \geq H_D(X), \tag{3.18}$$

where $H_D(X)$ is the base-D entropy of X defined by

$$H_D(X) = - \sum_{k=1}^{K} P(a_k) \log_D P(a_k).$$

For the binary case, the base of the logarithm is $D = 2$, so that

$$H_D(X) = H(X) = - \sum_{k=1}^{K} P(a_k) \log_2 P(a_k)$$

in units of bits.

Proof. Consider the lower bound to L in (3.18) first.

$$H_D(X) - L = - \sum_{k=1}^{K} P(a_k) \log_D P(a_k) - \sum_{k=1}^{K} P(a_k) \ell_k$$

$$= \sum_{k=1}^{K} P(a_k) \log_D \frac{D^{-\ell_k}}{P(a_k)}$$

$$\leq \log_D e \sum_{k=1}^{K} P(a_k) \left(\frac{D^{-\ell_k}}{P(a_k)} - 1 \right)$$

$$= \log_D e \left(\sum_{k=1}^{K} D^{-\ell_k} - \sum_{k=1}^{K} P(a_k) \right)$$

$$\leq \log_D e(1 - 1) = 0$$

Therefore,

$$H_D(X) \leq L,$$

which proves the lower bound. The first upper bound in the proof in 3.19 follows from the well-known inequality

$$\log_D x \leq (\log_D e)(x - 1) \tag{3.19}$$

with equality if and only if $x = 1$. The last upper bound results from substitution of the upper bound of 1 in the KI, which pertains here since the given code is uniquely decodable.

The lower bound on L of $H_D(X)$ is an absolute limit. A code with $L < H_D(X)$ cannot be uniquely decodable as a consequence of Theorem 3.3, because the codeword lengths would violate the KI. To prove this fact, just assume that $L < H_D(X)$ for the given set of codeword lengths. Then, the following must be true:

$$\sum_k P(a_k)\ell_k < -\sum_k P(a_k)\log P(a_k)$$

$$\sum_k P(a_k)\log \frac{P(a_k)}{D^{-\ell_k}} < 0$$

$$\left(\sum_k P(a_k)\right)\log \frac{\sum_k P(a_k)}{\sum_k D^{-\ell_k}} \leq \sum_k P(a_k)\log \frac{P(a_k)}{D^{-\ell_k}} < 0 \qquad (3.20)$$

$$\log \frac{1}{\sum_k D^{-\ell_k}} < 0$$

$$\sum_k D^{-\ell_k} > 1 \qquad (3.21)$$

The base of the logarithm is understood to be D and the sums on k from 1 to K. The left-hand inequality in (3.20) expresses the log-sum inequality. Substituting $\sum_k P(a_k)=1$ leads to the last inequality in (3.21), which states that the codeword lengths violate the KI. So by Theorem 3.3, the code with the given lengths cannot be uniquely decodable and therefore cannot be prefix-free.

To prove the upper bound and provide a set of lengths which satisfy the lower bound, for each $k = 1, 2, \ldots, K$, choose an integer codeword length ℓ_k for the letter a_k such that

$$-\log_D P(a_k) \leq \ell_k < -\log_D P(a_k) + 1. \qquad (3.22)$$

Certainly such an integer exists for every k, but one must show that these choices of lengths are compatible with a prefix-free code. The lower limit says the following

$$-\log_D P(a_k) \leq \ell_k \qquad (3.23)$$

$$D^{-\ell_k} \leq P(a_k) \qquad (3.24)$$

$$\sum_{k=1}^K D^{-\ell_k} \leq \sum_{k=1}^K P(a_k) = 1 \qquad (3.25)$$

The last statement above says that the KI is satisfied by the choices of these codeword lengths. Therefore, a prefix-free code exists with these lengths. Now multiplying the inequality series in (3.22) by $P(a_k)$ and summing over k yields

$$-\sum_{k=1}^K P(a_k)\log_D P(a_k) \leq \sum_{k=1}^K P(a_k)\ell_k$$

$$< -\sum_{k=1}^K P(a_k)(\log_D P(a_k) + 1),$$

which is re-stated as

$$H_D(X) \leq L < H_D(X) + 1. \qquad (3.26)$$

For the case of binary code digits,

$$H(X) \le L < H(X) + 1 \quad \text{bits.} \tag{3.27}$$

The minimum average codeword length of entropy $H(X)$ $(H_D(X))$ can be achieved if $-\log P(a_k)$ is an integer for all k. The extra 1 bit (Dit or D-ary digit) is the maximum overhead due to some non-integer $-\log P(a_k)$'s. When $H(X) \gg 1$, this overhead is usually tolerable. Otherwise, this overhead, although at most 1 bit per source symbol, might be too expensive for the application. The remedy is to develop a code for blocks of source symbols.

The average length limits and codeword lengths above depend only on probabilities, so it does not matter whether the source to be encoded is a single random variable X or a vector $\mathbf{X} = (X_1, X_2, \ldots, X_N)$ of N random variables. We just have to restate the entropies above in terms of the vector $\mathbf{X}$ and its probability distribution $q_{\mathbf{X}}(\mathbf{x}) = q_{\mathbf{X}}(x_1, x_2, \ldots, x_N)$, in which the components x_n take values in the alphabet $\mathcal{A} = \{a_1, a_2, \ldots, a_K\}$. The number of possible values of the N-vector is K^N. Therefore, for the vector $\mathbf{X}$ of N component random variables, the bounds on average length take the analogous form

$$H(\mathbf{X}) \le L < H(\mathbf{X}) + 1$$

where

$$H(\mathbf{X}) = - \sum_{x_1 \in \mathcal{A}} \sum_{x_2 \in \mathcal{A}} \cdots \sum_{x_N \in \mathcal{A}} q_{\mathbf{X}}(x_1, x_2, \ldots, x_N) \log q_{\mathbf{X}}(x_1, x_2, \ldots, x_N).$$

For the i.i.d. source, substitution of

$$q_{\mathbf{X}}(x_1, x_2, \ldots, x_N) = q_X(x_1) q_X(x_2) \cdots q_X(x_N)$$

reduces the entropy expression for $H(\mathbf{X})$ to simply

$$H(\mathbf{X}) = N H(X).$$

The average length limits become in this case

$$N H(X) \le L < N H(X) + 1.$$

The average length L is that of the code for N source symbols. The average length per source symbol is L/N. Dividing by N in the above expression gives the conclusion

$$H(X) \le \frac{L}{N} < H(X) + \frac{1}{N}. \tag{3.28}$$

The maximum overhead is now $1/N$ bits per source symbol, meaning that the 1 bit overhead has been spread over the N source symbols. By making N sufficiently large, we can reduce the overhead to be as small as we wish. However, the penalty is the exponential growth of the source alphabet with N.

3.4.1   Sources with memory

For sources with memory, we can still choose a binary codeword of integer length $\ell(\mathbf{x})$ for each source vector realization $\mathbf{x}$ satisfying

$$-\log q_{\mathbf{X}}(\mathbf{x}) \leq \ell(\mathbf{x}) < -\log q_{\mathbf{X}}(\mathbf{x}) + 1.$$

As proved before, the lower limit leads to satisfaction of the KI. When we average over the source probability distribution and divide by N, we obtain the vector analogue of (3.28) in the form

$$H_N(\mathbf{X}) \leq \frac{L}{N} < H_N(\mathbf{X}) + \frac{1}{N},$$

where we have defined $H_N(\mathbf{X}) \equiv \frac{1}{N} H(\mathbf{X})$.

The advantage is that when the source has memory,

$$H_N(\mathbf{X}) < H(X)$$

so that the minimum average length of the code becomes smaller. As N becomes larger, the advantage in coding N-tuples becomes progressively greater. However, since every vector value must have a unique codeword, the number of codewords grows exponentially in the dimension N. In the limit of large N, $H_N(\mathbf{X})$ decreases monotonically to the absolute lower limit of the entropy rate $H_\infty(\mathbf{X})$.

When the source is stationary Markov, the N-tuple has the probability distribution in Equation (3.2). In this case, the entropy rate becomes

$$H_\infty(\mathbf{X}) = \lim_{N \to \infty} H_N(\mathbf{X}) = H(X_2|X_1), \tag{3.29}$$

$$H(X_2|X_1) = -\sum_{x_1 \in \mathcal{A}} \sum_{x_2 \in \mathcal{A}} q_{X_2,X_1}(x_2, x_1) \log q_{X_2|X_1}(x_2|x_1). \tag{3.30}$$

$H(X_2|X_1)$ is called the conditional entropy of X_2 given X_1. It is the average uncertainty associated with the conditional probability distribution of X_2, given X_1. The average codeword length bounds for the stationary Markov source become

$$H(X_2|X_1) \leq \frac{L}{N} < H(X_2|X_1) + \frac{1}{N}.$$

So the lower limit of average codeword length per source symbol is the conditional entropy $H(X_2|X_1)$. This limit is generally lower than for the i.i.d. source by virtue of the inequality

$$H(X_2|X_1) \leq H(X_2) = H(X), \tag{3.31}$$

with equality if and only if X_1 and X_2 are statistically independent, i.e., not strictly Markov. Therefore, we gain the possibility of using fewer bits if we take into account the dependence between successive source symbols. To do so and operate optimally, one would have to assign codewords using conditional probabilities for every source pair (x_2, x_1) in the product alphabet $\mathcal{A} \times \mathcal{A}$ having lengths $\ell(x_2|x_1)$ such that

$$-\log q_{X_2|X_1}(x_2|x_1) \leq \ell(x_2|x_1) < -\log q_{X_2|X_1}(x_2|x_1) + 1. \tag{3.32}$$

There is a total of K^2 codewords, because there must be a unique codeword for each of the K^2 pairs (x_2, x_1). To enact this scheme for a finite N, we must generate codes for $X_1 = X$ using the probability distribution $q_X(x)$ and for X_2 given X_1 using the conditional probability distribution $q_{X_2|X_1}(x_2|x_1)$. We then code X_1 with the code for X, X_2 with the code for X_2 given X_1, X_3 with the code for X_3 given X_2, which is the same as that for X_2 given X_1, and so forth until encoding X_N given X_{N-1} with the code for X_2 given X_1. So the total number of codewords for finite $N \geq 2$ is $K + K^2$.

The conditional entropy inequality in Equation (3.31) holds in general for conditioning between any two random variables, X and Y, i.e.,

$$H(X|Y) \leq H(X), \tag{3.33}$$

with equality if and only if X and Y are statistically independent.

Let $X = X_n$ be a symbol in the source sequence and Y be a function of some other symbols in the sequence, i.e., $Y = f(\text{some } X_i's, i \neq n)$. In this case, Y is often called the *context* of X_n. The context Y can also be a vector with components that are functions of other symbols in the sequence. The tighter the dependence of X_n upon the context Y, the smaller is $H(X_n|Y)$ and the higher the potential for bit-rate savings.

Thus, when sources have memory, the use of conditioning has the potential to produce codes of lower average length at the expense of a larger code size and hence larger computational and storage complexity. The objective of an efficient lossless source coding scheme is to operate as close as possible, within complexity constraints, to the entropy rate of the source.

3.5   Concluding remarks

The material in this chapter reflects the underlying philosophy of the authors that a grasp of fundamental concepts in information theory is essential toward successful study and application of signal compression. Toward this end, we have presented some theory and proofs that will prove to be relevant to principles and methods explained in later chapters. The authors gave proofs only when they would lend the insight needed to understand the succeeding material. The reader may read the missing proofs in an information theory textbook, such as Gallager [2] or Cover and Thomas [3]. This chapter explained the fundamental concept of entropy and its connection to minimum possible code rates. The methods of certain proofs will be used later in some code constructions, so their derivations in this chapter serve a dual purpose. The next chapter's presentation of practical codes having bit rates approaching the entropy limit serves as a validation of our approach.

Problems

3.1   The letters of the discrete random variable (ensemble) X are $a_1, a_2, \ldots, a_K$ with respective probabilities $p_1, p_2, \ldots, p_K$. Using the inequality $\log_e z \leq z - 1$, prove the upper bound on entropy that

$$H(X) \leq \log K,$$

with equality if and only if all the probabilities equal $1/K$.

3.2 Given that random variables X and Y are statistically independent, i.e., $P_{X,Y}(x, y) = P(x)P(y)$ for all x and y, prove that the joint entropy $H(X, Y)$ equals $H(X) + H(Y)$, the sum of the individual entropies of X and Y.

3.3 The letters of the discrete random variable (ensemble) X are $a_1, a_2, \ldots, a_K$ with respective probabilities $p_1, p_2, \ldots, p_K$. Let $p_{max} = \max_k p_k$. Define the entropy of the ensemble with one maximum probability letter removed as $H(X^*)$. (The probabilities in the expression for $H(X^*)$ must sum to 1.)
(a) Prove that

$$H(X) = H_b(p_{max}) + (1 - p_{max})H(X^*).$$

(b) Prove the upper bound on entropy in Equation (3.10), which states that

$$H(X) \leq H_b(p_{max}) + (1 - p_{max}) \log(K - 1).$$

3.4 Consider a binary sequence generated according to the formula

$$x_{n+1} = x_n \oplus w_n,$$

where $x_0 = 0$ and w_n is an independent binary noise sequence such that for every $n \geq 1$,

$$Pr\{w_n = 1\} = 3/4, \ Pr\{w_n = 0\} = 1/4.$$

By computer, generate at least 100 000 samples of the sequence x_n and
(a) calculate empirically the first-order entropy per symbol;
(b) calculate empirically the second-order (letter pairs) entropy per symbol.
You are required to estimate probabilities from relative frequencies in the sequence x_n. Comment on the difference between the entropies you calculate above.

3.5 In the text are the following statements.

> The conditional entropy $H(X/Y)$ is always less than or equal to the unconditional entropy $H(X)$.
>
> The two entropies are equal only when the two random variables X and Y are statistically independent.

Colloquially, these statements mean that, on average, statistical knowledge reduces uncertainty, except when it is irrelevant. (In real life, knowledge sometimes brings even more uncertainty.) Prove that $H(X/Y) \leq H(X)$ with equality if and only if X and Y are statistically independent ($P_{X,Y}(x, y) = P(x)P(y)$ for all x and y).
Hint: Prove that $H(X/Y) - H(X) \leq 0$ using the inequality $\log_e z \leq z - 1$, which is satisfied with equality only when $z = 1$. By the way, for information theorists, the quantity $H(X) - H(X/Y)$ is defined as $I(X; Y)$ and is called average mutual information (between X and Y). $I(X; Y)$ is the amount by

which knowledge of Y reduces the uncertainty in X (on average) and is always non-negative via this proof.

3.6 Calculate the entropy $H(X)$ for the following probability distributions on a source X with the 5 letters a_1, a_2, a_3, a_4, a_5.

(a) $P(a_1) = P(a_2) = P(a_3) = P(a_4) = P(a_5) = 1/5$.

(b) $P(a_1) = P(a_2) = 1/4, P(a_3) = P(a_4) = P(a_5) = 1/6$.

(c) $P(a_1) = 1/2, P(a_2) = 1/4, P(a_3) = 1/8, P(a_4) = 1/16, P(a_5) = 1/16$.

(d) $P(a_1) = 3/8, P(a_2) = 3/8, P(a_3) = 1/8, P(a_4) = 1/16, P(a_5) = 1/16$.

Notice that the first two probabilities of the last set are the average of the first two probabilities of the next-to-last set. Equalizing probabilities in this way always reduces $H(X)$.

3.7 The lengths of the binary codewords of the source X are known to be

$$\ell_1 = 1, \ell_2 = 2, \ell_3 = 3, \ell_4 = 4, \ell_5 = 4.$$

(a) Prove that a prefix-free code with these codeword lengths exists for this source.

(b) Construct a prefix-free code by association to terminal nodes of a tree.

(c) Construct a prefix-free code by association to binary fractions.

3.8 Consider the two-letter source X with letters $a_1 = 1$ and $a_2 = -1$ occurring with equal probability. In transmission, this source is corrupted by additive, independent integer noise W, so that $Y = X + W$ is observed in the receiver. The noise W has the probability distribution:

$$P_W(w) = \begin{cases} 0.4, & w = 0 \\ 0.2, & w = \pm 1 \\ 0.1, & w = \pm 2 \end{cases}$$

Calculate the entropy $H(Y)$ and conditional entropy $H(Y/X)$.

3.9 Derive the lower bound on entropy in Equation (3.11).

References

1. C. E. Shannon, "A mathematical theory of communication," *Bell Syst. Technol. J.*, vol. 27, pp. 379–423 and 632–656, July and Oct. 1948.
2. R. G. Gallager, *Information Theory and Reliable Communication*. New York, NY: John Wiley & Sons, 1968.
3. T. M. Cover and J. A. Thomas, *Elements of Information Theory*. New York, NY: John Wiley & Sons, 1991, 2006.

4 Entropy coding techniques

4.1 Introduction

In the previous chapter, Chapter 3, we presented the theory of lossless coding and derived properties for optimality of uniquely decodable, prefix-free source codes. In particular, we showed that entropy is the absolute lower rate limit of a prefix-free code and presented tree and arithmetic structures that support prefix-free codes. In this chapter, we shall present coding methods that utilize these structures and whose rates approach the entropy limit. These methods are given the generic name of *entropy coding*. Huffman coding is one common form of entropy coding. Another is *arithmetic coding* and several adaptive, context-based enhancements are parts of several standard methods of data compression. Nowadays, lossless codes, whether close to optimal or not, are often called entropy codes. In addition to Huffman and arithmetic coding, we shall develop other important lossless coding methods, including run-length coding, Golomb coding, and Lempel–Ziv coding.

4.2 Huffman codes

The construction invented by Huffman [1] in 1952 yields the minimum length, prefix-free code for a source with a given set of probabilities. First, we shall motivate the construction discovered by Huffman. We consider only binary ($D = 2$) codes in this chapter, since extensions to non-binary are usually obvious from the binary case and binary codes are predominant in practice and the literature. We have learned that a prefix-free code can be constructed, so that its average length is no more than 1 bit from the source entropy, which is the absolute lower limit. In fact, such average length can be achieved when the codeword length ℓ_k of every source letter a_k satisfies

$$-\log P(a_k) \geq \ell_k < -\log P(a_k) + 1,$$

where $P(a_k)$ is the probability of a_k. Suppose that the probabilities are exactly integer powers of 1/2,

$$P(a_k) = 2^{-n_k}, n_k \text{ an integer},$$

Table 4.1 Binary prefix-free code

Letter	Codeword
a_1	0
a_2	10
a_3	110
a_4	111

then, if $\ell_k = n_k = -\log P(a_k)$, the average codeword length is

$$\bar{\ell} \equiv \sum_{k=1}^{K} P(a_k)\ell_k$$

$$= -\sum_{k=1}^{K} P(a_k) \log P(a_k) = H(X). \qquad (4.1)$$

The average length is exactly equal to the source entropy $H(X)$ in this case.

For example, consider the four-letter source with probabilities 1/2, 1/4, 1/8, 1/8. Calculating its entropy,

$$H(X) = \frac{1}{2} \log 2 + \frac{1}{4} \log 4 + \frac{1}{8} \log 8 + \frac{1}{8} \log 8 = 1.75 \text{ bits.}$$

Using the lengths $\ell_1 = \log 2 = 1$, $\ell_2 = \log 4 = 2$, $\ell_3 = \ell_4 = \log 8 = 3$, we see exactly the same terms in the expression for the average length $\bar{\ell}$. Therefore,

$$\bar{\ell} = 1.75 = H(X),$$

and the average length and entropy are identical. The code for this source appears in Table 4.1 and the tree that constructed it in Figure 4.1. This code table and tree are repeated from the previous chapter with just a change in the letter notation. The tree was constructed from prior knowledge of the codeword lengths. Now we start with the source probabilities that dictate these lengths and demonstrate another way to construct this code.

The code tree emulates the actions of the decoder. The decoder reads a bit from the bitstream and moves in the "0" or "1" direction from any node starting at the root, according to the bit it reads. When it reaches a terminal node, it declares the corresponding source letter. In order to minimize the average length of a variable length code, the codewords of the least probable source letters should have the longest lengths. Because we want a prefix-free code, we associate the least probable codewords to the terminal nodes of the longest paths through the tree. This suggests to start constructing the code tree from the end or leaves and continue toward the root. For our four-letter source, the least probable letters, a_3 and a_4 have probabilities of 1/8, so they are placed on two terminal nodes of the tree, one of them on the "0" branch and the other on the "1" branch. These branches emanate from a node at the next lower level and are called *siblings*. The

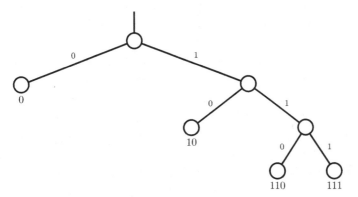

Figure 4.1   Code tree for binary code in Table 4.1.

node from which these branches emanate, the *parent node*, may be viewed as belonging to a superletter a_3a_4 of probability 1/4, because either a_3 or a_4 will be chosen by the decoder after it reaches this node. Counting superletters, the two least probable letters, a_2 and a_3a_4, have probability 1/4. Therefore we make them sibling nodes by placing a_2 on the node of the branch that emanates from the same lower level node as the superletter branch. The two probability of 1/4 nodes, corresponding to a_2 and a_3a_4, combine to form a superletter $a_2a_3a_4$ of probability 1/2. Therefore, we put a_1 with probability 1/2 on the sibling node of $a_2a_3a_4$ with probability 1/2. That exhausts the source letters, so these last two nodes emanate from the root of the tree, which corresponds to $a_1a_2a_3a_4$, the full alphabet. Notice that "1" and "0" can be interchanged everywhere, so the code is not unique. We end up with the same tree as in Figure 4.1 or an equivalent one with the same codeword lengths and average length. Either code is optimal, because $\bar{\ell} = H(X)$.

Notice that when the letter probabilities are integer powers of 1/2, the smallest probability must occur in pairs. Otherwise the probabilities will not sum to one, as they must. In the procedure above, even when forming superletters, we found a matching probability at each level. Suppose that the probabilities are not powers of 1/2. Take, for example, the set of probabilities, $\{3/20, 4/20, 8/20, 2/20, 3/20\}$. Let us round them down to the nearest power of 1/2 to get the set $\{1/8, 1/8, 1/4, 1/16, 1/8\}$ and construct a tree for this set of rounded probabilities. These rounded probabilities do not sum to one, but if we associate them to terminal nodes of a tree, we do get a prefix-free code. We can use the same procedure as above, but now the "probabilities" do not occur in pairs. The remedy is to invent a dummy probability to complete a pair. Therefore, we place 1/16 and a dummy 1/16 at the ends of the tree to start. Combine to form a superletter of probability 1/8 at the parent node and take its sibling as any one of the probability 1/8 letters. Two probability 1/8 letters remain as the least probable, so they form a separate pair at the same level of the tree. Each of these two pairs combine to form superletters of probability 1/4 at the next lower level. There are now three letters or superletters remaining with probability 1/4. Combine any two to form a probability 1/2 superletter on the next lower level node. The smallest probability of 1/4 remaining belongs to a single letter, so

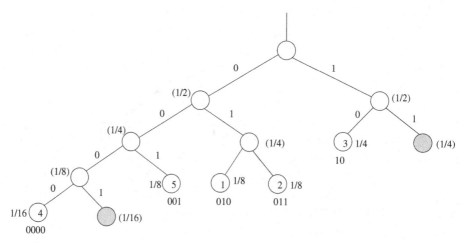

Figure 4.2   Non-optimal code tree using the rounded probabilities $\{1/8, 1/8, 1/4, 1/16, 1/8\}$. The shaded nodes are dummy nodes and probabilities in parentheses belong to dummy letters or superletters.

we pair it with a dummy 1/4. We combine them to obtain a probability 1/2 superletter at the next lower level node. Now we combine the two 1/2 probability letters to form the root of the tree. This code tree is shown in Figure 4.2, where the real and dummy nodes are distinguished by shading. The resulting code of $\{010, 011, 10, 0000, 001\}$ is prefix-free, but clearly not optimal, because extra nodes had to be added. The average length $\bar{\ell} = 2.70$ compared to the source entropy of $H(X) = 2.146$. Notice that we can collapse the two terminal sibling pairs that contain a dummy node to obtain a shorter prefix-free code. The above procedure produces the codeword lengths

$$\ell_k = \lceil -\log_2 P(a_k) \rceil, \ k = 1, 2, \ldots, K \tag{4.2}$$

A length ℓ_k is the same as the level of depth in the code tree. Although the average length is always within 1 bit of the entropy, the lengths of some codewords are unnecessarily long for probabilities that are not (integer) powers of two. An egregious example is the binary source with probabilities 0.99 and 0.01. The resulting codeword lengths are 1 and 7, when it is obvious that the lengths 1 and 1 and codewords 0 and 1 are the solutions for the optimal code. One can never do better than 1 bit per source letter for a scalar binary source.

There is an optimal procedure, called the Huffman code, that produces a prefix-free code whose average codeword length is as close to entropy as possible for any given set of probabilities. The Huffman code uses the same basic method just described, but uses the two smallest probabilities at each stage, whether or not they match. The general steps of the method are stated in Algorithm 4.1.

Let us return to the same five-letter source as above and build its code tree using the Huffman procedure. The steps are depicted in detail in Figure 4.3. The resulting code is $\{010, 000, 1, 011, 001\}$, whose average length $\bar{\ell} = 2.20$ is quite close to the source

ALGORITHM 4.1

The Huffman coding procedure

1. Put the letters (counting superletters) with the two smallest probabilities as a sibling pair on terminal nodes at the highest level of the tree (farthest from the root).
2. Combine letters with the two smallest probabilities to form a superletter having probability equal to the sum of the constituent probabilities.
3. Put superletter on the parent node of the sibling pair in Step 1.
4. If there are more than one letter or superletter remaining, return to Step 1. Otherwise, continue. (You have formed the root and completed the code tree.)
5. Starting at the root, read sequences of code digits 0 or 1 on paths to terminal nodes as the codewords of the code.

entropy. Only if the probabilities are integer powers of 1/2 can the average length equal the entropy.

The fact that the Huffman code is optimal stems from three basic propositions.

1. In an optimal code, the codewords in order of increasing length must correspond to letters in order of decreasing probability.
2. For any given source with $K \geq 2$ letters, an optimum binary code exists in which the two least probable code words have the same length and differ only in the last digit.
3. Suppose that we create a reduced source by combining the two least probable letters to form a superletter of the probability of their sum. Then, if the code of the original source is optimal, the code of the reduced source is optimal.

These propositions mean that an optimal code for the reduced source must also have its two least likely letters terminating the tree with the same length, one with the last digit 0 and the other with last digit 1. Therefore, we can repeat the reduction of the source in this way until all letters are exhausted. At each stage, the two least likely letters on the two nodes of longest path constitute part of the final optimal code.

The proofs of these propositions are straightforward. The first proposition states that if $P(a_i) < P(a_j)$, then $\ell_i > \ell_j$ for all indices $i \neq j$. If not, one can show easily that the average length $\bar{\ell}$ increases. The second proposition must be true, because if there were only one word of longest length corresponding to the least probable letter, then there would be an intermediate node of lower level with only a single branch stemming from it. The least probable word resides at the end of this branch. This branch can be trimmed until reaching the level of the sibling node of the next higher probability letter without violating the prefix condition. Therefore, the codewords of the two least probable letters have the same length. To prove the last proposition, consider the original and reduced letter ensembles, reordered according to decreasing probability, as follows:

$$\{a_1, a_2, \ldots, a_{K-1}, a_K\} \text{ and } \{a_1, a_2, \ldots, a'_{K-1}\},$$

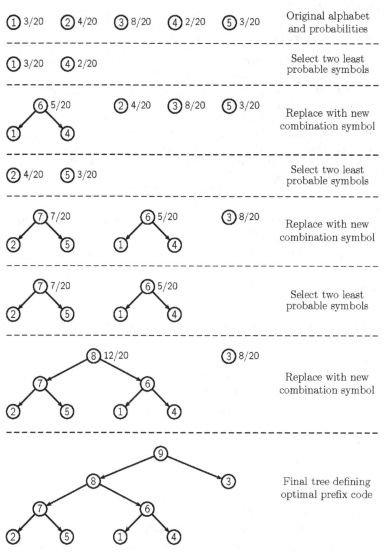

Figure 4.3   Huffman coding procedure.

where a_{K-1} and a_K are combined to produce the superletter a'_{K-1}, so that

$$P(a'_{K-1}) = P(a_{K-1}) + P(a_K) \text{ and } \ell_{K-1} = \ell_K.$$

When we calculate the average lengths $\bar{\ell}$ and $\bar{\ell}'$ of the original and reduced source codes, respectively, we obtain

$$\bar{\ell} = \bar{\ell}' + (P(a_{K-1}) + P(a_K)) \tag{4.3}$$

The difference of their average lengths depends only on the sum of the two smallest probabilities and not on the structure of the reduced code tree. Therefore, if the original code tree is optimal, so is the reduced code tree.

4.3 Shannon–Fano–Elias codes

In the previous chapter, we proved that disjoint intervals of the unit interval $[0,1)$ can be associated with a prefix-free D-ary code. The Shannon–Fano–Elias (SFE) code, which is the foundation of the arithmetic code, is built on this fact. In this chapter, we shall consider only $D = 2$ binary codes. This SFE code procedure is attributed to its three names only by word of mouth, because the work itself is unpublished. It is analyzed in the book by Jelinek [2]. We shall explain this code and the nearness of its rate to the entropy limit.

Consider that the source alphabet $\mathcal{A}$ consists of the K letters $a_1, a_2, \ldots, a_K$. We identify the letters now only by their indices $1, 2, \ldots, K$. These correspondences to indices impose a numerical order on the letters, but otherwise there is no loss of generality. So, let us assume that the source alphabet is $\mathcal{X} = \{1, 2, \ldots, K\}$ with probabilities $P(x)$, $x = 1, 2, \ldots, K$. The cumulative distribution function (CDF), by definition, is

$$F(x) = \sum_{n \leq x} P(n).$$

This ascending staircase function is illustrated in Figure 4.4, where the height of the rise of the step at every horizontal coordinate x is $P(x)$. These rises span disjoint intervals whose union is the unit interval. Now let us define a new function, $\overline{F}(x)$, called a modified CDF, whose values are the midpoints of the rises. Accordingly,

$$\overline{F}(x) = \sum_{n < x} P(n) + \frac{1}{2}P(x). \tag{4.4}$$

It is easy to see that if two letters x and y are different, then $F(x) \neq F(y)$ and $\overline{F}(x) \neq \overline{F}(y)$. In fact, given $\overline{F}(x)$, x is uniquely determined. One can say that $\overline{F}(x)$ is a codeword for x, but being a real number specified by possibly an infinite number of bits, it is certainly not a good codeword. If we make an approximation to $\overline{F}(x)$ that stays within the including rise, this approximation will also specify x uniquely. Therefore, we truncate the binary expansion of $\overline{F}(x)$ to $\ell(x)$ binay digits and denote it by $\lfloor \overline{F}(x) \rfloor_{\ell(x)}$. The error in this approximation is

$$\overline{F}(x) - \lfloor \overline{F}(x) \rfloor_{\ell(x)} < 2^{-\ell(x)}. \tag{4.5}$$

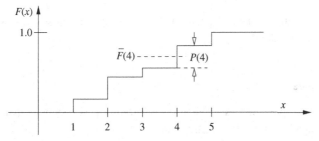

Figure 4.4 CDF graph for $K = 5$ letters.

If this error is also less than $P(x)/2$, then the approximation $\lfloor \overline{F}(x) \rfloor_{\ell(x)}$ still exceeds $F(x-1)$, so remains in the same rise. Therefore, we want the truncated length to satisfy

$$2^{-\ell(x)} < P(x)/2.$$

Solving for the smallest integer length $\ell(x)$, we determine that

$$\ell(x) = \lceil \log \frac{1}{P(x)} \rceil + 1, \tag{4.6}$$

where the logarithm is base 2, as usual. Therefore, we use the $\ell(x)$ binary digits of $\lfloor \overline{F}(x) \rfloor_\ell$ as the codeword for x. For example, suppose that $0.z_1, z_2, \ldots, z_\ell$ is the binary expansion of $\lfloor \overline{F}(x) \rfloor_\ell$. Then $z_1, z_2, \ldots, z_\ell$ is the codeword for x.

All numbers whose binary expansions have prefix $0.z_1, z_2, \ldots, z_{\ell(x)}$ lie in the interval $[0.z_1, z_2, \ldots, z_{\ell(x)}, 0.z_1, z_2, \ldots, z_{\ell(x)} + 2^{-\ell(x)})$, which is contained in the rise of the step at x. For $y \neq x$, its codeword must correspond to a number in a different interval, so must have a different prefix. Therefore, this code is prefix-free. The average length $\overline{\ell}$ of the SFE code can be bounded as follows:

$$\overline{\ell} = \sum_{x=1}^{K} P(x)\ell(x)$$

$$= \sum_{x=1}^{K} P(x)(\lceil \log \frac{1}{P(x)} \rceil + 1)$$

$$< \sum_{x=1}^{K} P(x)(\log \frac{1}{P(x)} + 2)$$

$$= H(X) + 2.$$

Clearly, $\overline{\ell} > \sum_{x=1}^{K} P(x) \log \frac{1}{P(x)} = H(X)$. Therefore, the average length is, at most, 2 bits per symbol larger than the entropy. Again, we have for i.i.d. sequences of length n,

$$nH(X) < \overline{\ell} < nH(X) + 2$$

$$H(X) < \frac{\overline{\ell}}{n} < H(X) + \frac{2}{n}. \tag{4.7}$$

The average length per letter approaches the entropy as closely as wanted for sequence length n sufficiently large.

4.3.1   SFE code examples

We present two examples. Tables 4.2 and 4.3 illustrate the SFE code construction and compare its average length to the Huffman code.

Table 4.2 Dyadic probabilities, $H(X) = 1.75$ bits

x	$P(x)$	$F(x)$	$\overline{F}(x)$	$\ell(x)$	$\lfloor \overline{F}(x) \rfloor_{\ell(x)}$	SFE codeword	Huffman codeword
1	0.25	0.25	0.125	3	0.001	001	10
2	0.50	0.75	0.500	2	0.10	10	0
3	0.125	0.875	0.8125	4	0.1101	1101	110
4	0.125	1.0	0.9375	4	0.1111	1111	111

4.3.1.1 Case of dyadic probabilities

The first case is a four-letter source with dyadic (integer powers of 2) probabilities 0.25, 0.50, 0.25, 0.125. Its entropy is 1.75 bits. The average length of the SFE code is 2.75 bits versus 1.75 bits for the Huffman code and the entropy.

4.3.1.2 Case of non-dyadic probabilities

The second case involves a five-letter source with non-dyadic probabilities 0.25, 0.40, 0.05, 0.15, 0.15. Its entropy is 2.066 bits. Notice that there is no need to order the letters by decreasing probabilities, as in Huffman coding. The average length of the SFE code is 3.45 bits versus 2.15 bits for the Huffman code and 2.066 bits for the entropy. Clearly, the SFE code requires a longer code length than the Huffman code for a scalar source. Note that the sixth bit from the $x = 3$ SFE codeword can be dropped. Like the Huffman code, one can approach the entropy by coding sequences of letters. It transpires that an extension of the SFE code to sequences, called the *arithmetic code,* is not only much easier to implement than the same for the Huffman code, but achieves an even closer approach to the entropy rate.

4.3.2 Decoding the SFE code

The decoder of an SFE code only needs to know the indexing of the letters with their probabilities. The decoder can therefore reconstruct beforehand the graph of $F(x)$ versus x and know that the midrise truncations $\lfloor \overline{F}(x) \rfloor_{\ell(x)}$ fall between $F(x - 1)$ and $F(x)$ for every x. When the decoder receives the $\ell(x)$ binary digits of $\lfloor \overline{F}(x) \rfloor_{\ell(x)}$, it can identify to which x it belongs.

Example 4.1 SFE code of Table 4.3

The decoder receives 1100, which translates to $0.1100_2 = 0.75_{10}$.[1] Since $F(3) < 0.75 < F(4)$, $x = 4$.

Table 4.3 Non-dyadic probabilities, $H(X) = 2.066$ bits

x	$P(x)$	$F(x)$	$\overline{F}(x)$	$\ell(x)$	$\lfloor \overline{F}(x) \rfloor_{\ell(x)}$	SFE codeword	Huffman codeword
1	0.25	0.25	0.125	3	0.001	001	10
2	0.40	0.65	0.45	3	0.011	011	0
3	0.05	0.70	0.6525	6	0.101001	101001	1111
4	0.15	0.85	0.775	4	0.1100	1100	1110
5	0.15	1.0	0.925	4	0.1110	1110	110

4.4   Arithmetic code

4.4.1   Preliminaries

Arithmetic coding is the name given to a particular method of applying SFE coding to sequences. The SFE codeword associated a number in a subinterval of real numbers between 0 and 1 ($[0,1)$) to a data sample. The codeword of an arithmetic code makes the same association, but to a sequence of samples. Again using the mapping of the letters to integers x, the subinterval of the SFE code for x has width $P(x)$, its lower boundary is $F(x-1)$ and its upper boundary is $F(x)$. The truncation of the midpoint of this interval to $\lceil \log_2 P(x) \rceil + 1$ binary digits identifies the subinterval and serves as the codeword. In order to apply this method to sequences $x^n = x_1 x_2 \cdots x_n$, we need to order the sequences to calculate the CDF $F(x^n)$. The order we adopt is the lexicographic order, which would be the numerical order of the sequences $x_1 x_2 \cdots x_n$ rewritten as fractions $0.x_1 x_2 \cdots x_n$ in a $(K+1)$-*ary* number system. Suppose that $K = 5$ as in our previous example. Then the elements of the sequences are the digits 1 to 5 (the digit 0 is not used). The number 0.423 exceeds 0.415, so 423 exceeds 415 and consequently $F(423) > F(415)$. Also, since 0.423 exceeds 0.415223, sequence 423 exceeds sequence 415223 and $F(423) > F(415223)$. We shall soon see the reason to order sequences with different numbers of elements in this way.

Now that the sequences are ordered, we wish to define a CDF, which we shall call a *sub-CDF*,

$$F^{\dagger}(x^n) \equiv \sum_{y^n < x^n} P(y^n).$$

We shall assume that sequences are i.i.d., so that

$$P(x_1 x_2 \cdots x_n) = P(x_1) P(x_2) \cdots P(x_n).$$

The sub-CDF excludes the probability $P(x^n)$ of the largest sequence from the regular CDF $F(x^n)$, i.e., $F(x^n) = F^{\dagger}(x^n) + P(x^n)$. Let us adopt a new notation for an interval. Normally, we denote $[\alpha, \beta)$ to mean the numbers starting with α at the lower end and ending infinitesimally short of β at the upper end. Instead, we shall use lower endpoint

α and width w to specify the interval and use the notation $|\alpha, w>$, $w = \beta - \alpha$. Now we are ready to describe the encoding process.

4.4.2 Arithmetic encoding

Let us use the sequence 423 as an example of how we use the lexicographic order to create the interval $|F^\dagger(423), P(423) >$ that contains the number associated with the codeword. First, we create the first-order CDF of a single letter sequence, as in Figure 4.4. In Figure 4.5, we label the coordinates of the rise intervals in the staircase. In parentheses, we write the numerical letter associated with its interval. Now when we encode the first letter 4 of 423, we are certain that 423 or any sequence starting with 4 must associate to the subinterval $|F^\dagger(4), P(4) >= |0.70, 0.15 >$. The sum of the probabilities of all possible sequences less than sequence 4 is $F^\dagger(4)$, i.e., $\sum_{y^n < 4} P(y^n) = P(1) + P(2) + P(3) = F^\dagger(4)$. Beyond the upper end of the interval is impossible, because those sequences begin with 5. Let us look at the second letter 2. Now we need to determine the interval $|F^\dagger(42), P(42) >$. If we scale the previous interval, in this case $|0, 1 >$, with its marked coordinates $F^\dagger(x), x = 1, 2, 3, 4, 5$ by $P(4)$ and map it into $|F^\dagger(4), P(4) >$, we see that the second subinterval is $|F^\dagger(42), P(42) >$. This scaling and mapping is shown in Figure 4.5, where the interval corresponding to 42 is $|0.7375, 0.06 >$. The lower end of this interval reflects the addition of sequence 41, the only 2-letter sequence less than 42, according to $F^\dagger(4) + P(41)$ and its width is $P(42) = P(4)P(2) = (0.25)(0.40) = 0.06$. All two-letter sequences less than sequence 4 have probability $F^\dagger(4)$. We proceed in the same way for encoding of the third letter 3. We scale and map $|0.7375, 0.06 >$ in the same way to identify the interval $|F^\dagger(423), P(423) >$ corresponding to 423. The lower end sub-CDF is found by addition of the $P(421)$ and $P(422)$ to $F^\dagger(42)$ and its width by $P(42)P(3)$. Notice that the lower end and width of the interval may be calculated recursively. For instance,

$$F^\dagger(423) = F^\dagger(42) + P(421) + P(422)$$
$$= F^\dagger(42) + P(42)(P(1) + P(2))$$
$$= F^\dagger(42) + P(42)F^\dagger(3).$$

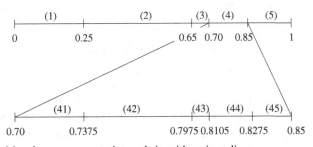

Figure 4.5 Mapping sequences to intervals in arithmetic coding.

The encoding of any length sequence proceeds in the same way as above. Seeing the recursive nature of these calculations for the codeword interval, we adopt the notation of an interval at the k-th stage as $|b_k, w_k >$, where b_k is the lower end or *base* and w_k is the width at stage k. Initially $|b_0, w_0 >= |0, 1 >$. So we see that the interval at stage k is

$$|b_k, w_k >= |b_{k-1} + F^\dagger(x_k)w_{k-1}, P(x_k)w_{k-1} >, k = 1, 2, \ldots, n. \qquad (4.8)$$

Clearly, from the above, $w_k = P(x_1)P(x_2) \cdots P(x_k)$. The width of the codeword interval upon encoding k letters equals the probability of these letters, which is the height of the k-th step in the CDF staircase, just as in the single letter case of the SFE code. Although it is harder to see, the base of the interval is the appropriate sub-CDF $F^\dagger(x^k)$. Due to the lexicographic ordering and the statistical independence of the sequence,

$$F^\dagger(x^k) = \sum_{y_1 y_2 \cdots y_k < x_1 x_2 \cdots x_k} P(y_1 y_2 \cdots y_k)$$

$$= Pr\{y_1 < x_1\} + Pr\{y_2 < x_2, y_1 = x_1\} + Pr\{y_3 < x_3, y_1 y_2 = x_1 x_2\}$$

$$+ \cdots + Pr\{y_k < x_k, y_1 y_2 \cdots y_{k-1} = x_1 x_2 \cdots x_{k-1}\}$$

$$= \sum_{y_1 < x_1} P(y_1) + \left(\sum_{y_2 < x_2} P(y_2)\right) P(x_1) + \left(\sum_{y_3 < x_3} P(y_3)\right) P(x_1)P(x_2)$$

$$+ \cdots + \left(\sum_{y_k < x_k} P(y_k)\right) P(x_1)P(x_2) \cdots P(x_{k-1})$$

$$= F^\dagger(x_1) + F^\dagger(x_2)P(x_1) + F^\dagger(x_3)P(x_1)P(x_2)$$

$$+ \cdots + F^\dagger(x_k)P(x_1)P(x_2) \cdots P(x_{k-1}),$$

which matches the recursive formula in (4.8).

After applying the recursion to the full sequence of n letters, the codeword interval is determined. The midpoint number of the interval, expressed as a binary fraction, is then truncated in precision to $\lceil \log_2(1/w_n) \rceil + 1$ bits, where $w_n = \prod_{k=1}^n P(x_k)$, to produce the digits of the codeword. The encoding algorithm is stated in detail in Algorithm 4.2.

Example 4.2 *Continuation of encoding 423.*

Continuing the example of encoding 423, where we found that 42 corresponds to $|0.7375, 0.06 >$, we use the recursive formula in Equation (4.8) to find the interval corresponding to 423.

$$b_3 = b_2 + F^\dagger(3) * w_2 = 0.7375 + 0.65 * 0.06 = 0.7765$$

$$w_3 = P(3)w_2 = 0.05 * 0.06 = 0.003.$$

The interval of 423 is calculated to be $|0.7765, 0.003 >$. The midpoint is $0.7780 = 0.11000111001 \cdots_2$, and when truncated to $\lceil \log_2(1/0.003) \rceil + 1 = 10$ binary digits, becomes 0.1100011100. The codeword for 423 is therefore 1100011100.

ALGORITHM 4.2 _____

Arithmetic encoding of input sequence x^n

1. Let $k = 1$.
2. Divide interval $[0, 1)$ into codeword intervals for the first letter x_1 of the source sequence x^n. These intervals are given by

$$I_1(x) = |b(x), w(x) >= |F^\dagger(x), P(x) >, x = 1, 2, \ldots, K.$$

3. Associate letter x_1 to interval $I_1(x_1)$.
4. Set $k = k + 1$. (Increment k by 1).
5. Divide the interval of the previous letter $I_{k-1}(x_{k-1})$ proportionately as before. The bases and widths of these intervals $I_k(x) = |b_k(x), w_k(x) >, k = 1, 2, \ldots, K$ are calculated recursively according to

$$b_k(x) = w_{k-1}(x_{k-1}) * F^\dagger(x) + b_{k-1}(x_{k-1}), \quad w_k(x) = w_{k-1}(x_{k-1}) * P(x).$$

6. Associate letter x_k to interval $I_k(x_k) = |b_k(x_k), w_k(x_k) >$.
7. If $k < n$, return to Step 4. Otherwise, continue.
8. Output binary codeword v, calculated from the last interval by

$$0.v = \lfloor (b_n(x_n) + 0.5 w_n(x_n))2 \rfloor \lceil \log_2(1/w_n(x_n)) \rceil + 1. \tag{4.9}$$

That is, the midpoint of the final interval, expressed as a binary fraction, is truncated to $\lceil \log_2(1/w_n(x_n)) \rceil + 1$ most significant digits.

_____

One obstacle to overcome in applying this procedure is the increase in required precision for each letter that is encoded. We wish to encode long sequences, since the longer the sequence, the closer is the approach of the rate to the optimal lower limit of the entropy. However, the precision requirement will eventually exceed the limit of the computer. Judicious scaling and rounding of the interval widths after encoding each letter allow the precision to remain fixed and the arithmetic operations to be fixed point. We shall not describe these operations here and suggest reading the excellent primer on arithmetic coding by Said [3].

4.4.3 Arithmetic decoding

The encoding process produces a codeword, which we denote as v, that associates the input sequence to a subinterval of the unit interval $[0, 1)$. This subinterval is the intersection of nested subintervals determined in each stage of the encoding. In other words, a subinterval at every stage contains its subsequent subinterval at the next stage. The decoding process follows this stage-by-stage procedure by successively determining the sequence x^k such that $0.v$ is contained in the subinterval $|F^\dagger(x^k), P(x^k) >$. Before writing down the recursion symbolically, let us illustrate again with the source in Table 4.3 and the example of the codeword $v = 1100011100$ for the input sequence 423.

Copying the encoder action, the decoder initializes the subintervals for the first letter as in Figure 4.5. Receiving $v = 1100011100$, it forms the associated number $0.v = 0.1100011100_2 = 0.77734375$ and sees that it is contained in the interval

[0.75, 0.85) associated with the letter 4. Therefore, 4 is the first letter. Now the decoder, again mimicking the encoder, divides interval [0.75, 0.85) into the same relative proportions, shown also in Figure 4.5, and locates 0.77734375 in subinterval [0.7375, 0.7975), which belongs to sequence 42. Similarly, we again divide the subinterval [0.7375, 0.7975) in the same proportions and find that 0.77734375 is in [0.7765, 0.7795), belonging to sequence 423.

We may know that 423 is the correct sequence, because we set it up that way, but the decoder does not know from receiving only v. Without further guidance, the decoder will continue to subdivide the current subinterval proportionally and decode more letters. Therefore, we must inform the decoder to stop, either by appending a special symbol or sending along with v a count of the number of letters in the source sequence.

The decoding is not actually done in the manner just described, because it requires receipt of the full codeword v, which can be quite long, perhaps millions of bits. We see that we need to receive only enough bits to locate v inside one of the code subintervals. Any extension of these bits determines a number inside this subinterval. For example, let us examine one bit at a time. Starting with $1 \rightarrow 0.1_2 = 0.5$ means that the codeword number is in $[1/2, 1)$,[2] which is not contained in any subinterval in Figure 4.5. Continuing, 11 corresponds to $0.11_2 = 0.75$ and the interval $[0.75, 1)$, likewise not contained in a codeword interval. Not until we examine 1100, corresponding to the interval [0.75, 0.8125) inside [0.70, 0.85), can we decode 4 as the first letter. Now, the interval [0.70, 0.85) is proportionally subdivided and the next 0 of v narrows the interval to [0.75, 0.78125), which is included in [0.7375, 0.7975), the subinterval of 42. The reader may check that it takes the receipt of the remaining bits of 11100 to identify the subinterval of 423.

In order to formalize this decoding procedure, we need to adopt some notation. Let v_i^j, $i < j$ denote a segment of the codeword v from digit i to digit j inclusive. Suppose that we have read the initial segment of j_k codeword digits $v_1^{j_k}$ to determine that $|0.v_1^{j_k}, 2^{-j_k} >$ lies within the codeword interval $|F^\dagger(x^k), P(x^k) >$. This means that

$$F^\dagger(x^k) \le 0.v_1^{j_k} < 0.v_1^{j_k} + 2^{-j_k} \le F^\dagger(x^k) + P(x^k).$$

The recursive calculation of the quantities $F^\dagger(x^k)$ and $P(x^k)$ will be shown shortly. Therefore, x^k, the first k letters of the input sequence are determined. Read $j_{k+1} - j_k$ more digits of v until the interval $|0.v_1^{j_{k+1}}, 2^{-j_{k+1}} >$ is contained in an interval $|F^\dagger(x^{k+1}), P(x^{k+1}) >$ for some input sequence x^{k+1}. The intervals for the possible sequences x^{k+1} are calculated recursively according to Equation (4.8). Continue in this way until a stop indicator is encountered or a known letter count is met. The decoding algorithm with recursive calculations is summarized in Algorithm 4.3.

This decoding algorithm does not include the mathematical tricks of scaling and truncation needed to avoid the precision overload. However, it does fulfill the objective of this section, which is to illuminate the fundamental principles. We recommend the excellent primer on arithmetic coding by Said [3] to explain these tricks to obtain fixed-point operations that avoid the precision overload.

ALGORITHM 4.3 ——————————————————————

Decoding an arithmetic code

1. Set $k = 1$.
2. Divide interval $[0, 1)$ into codeword intervals for the first letter x_1 of the source sequence x^n. These intervals are given by

$$|b(x), w(x) >= |F^\dagger(x), P(x) >, x = 1, 2, \ldots, K.$$

3. Read as many digits from the received codeword v until the interval implied by these digits fits into one of the codeword intervals in the previous step. Symbolically stated, read $v_1^{j_1}$ so that

$$F^\dagger(x) \leq 0.v_1^{j_1} < 0.v_1^{j_1} + 2^{-j_1} \leq F^\dagger(x) + P(x),$$

   for some $x = 1, 2, \ldots, K$, which is the decoded first letter. Denote this decoded first letter by $\hat{x}_1$ and this interval as $I_1(\hat{x}_1) = |b_1(\hat{x}_1), w_1(\hat{x}_1) >$.
4. Set $k = k + 1$. (Increment k by 1.)
5. Update the codeword intervals for decoding the k-th input letter by dividing the interval of the previous stage, $I_{k-1}(\hat{x}_{k-1}) = |b_{k-1}(\hat{x}_{k-1}), w_{k-1}(\hat{x}_{k-1})$, into the same proportions as before. The bases and widths of these intervals $|b_k(x), w_k(x) >, x = 1, 2, \ldots, K$ are calculated recursively according to

$$b_k(x) = w_{k-1}(\hat{x}_{k-1}) * F^\dagger(x) + b_{k-1}(\hat{x}_{k-1}), \quad w_k(x) = w_{k-1}(\hat{x}_{k-1}) * P(x).$$

6. Read $j_k - j_{k-1}$ more digits of v, so that the interval $|0.v_n 1^{j_k}, 2^{-j_k} >$ fits inside one of the k-th stage intervals $|b_k(x), w_k(x) >, x = 1, 2, \ldots, K$. Declare that $x = \hat{x}_k$ as the k-th decoded letter in the sequence.
7. Stop, if indicated by a marker or known count of the sequence length n. Otherwise, return to Step 4.

——————————————————————

4.5 Run-length codes

In many circumstances, consecutive repetitions of the same letter frequently occur. They are called *runs*. Instead of coding each letter in a run separately, one codes the letter and the length of the run as a pair. This method is called *run-length coding*. A common example is black and white text where 0 is black and 1 is white. Runs of 1 up to the width of a page are frequent, so it is economical to code the pair $\{1, r_1\}$, where r_1 is the white run length. Runs of black occur frequently enough that coding the pair $\{0, r_0\}$ also results in bit savings. Clearly the white and black pairs have different probability distributions. These pairs are usually coded with Huffman codes, although arithmetic codes can also be used.

When the alphabet consists of more than $K = 2$ letters, then the {letter, run-length} pairs generalize to $\{a_k, r_k\}$ for $k = 1, 2, \ldots, K$. One can continue to code pairs, say by a Huffman code, or more conveniently, code a_k separately and then r_k conditioned on a_k, since the probability distribution of r_k varies for each letter a_k. These two coding strategies are equivalent, because of the entropy relation

$$H(\mathcal{A}, \mathcal{R}) = H(\mathcal{A}) + H(\mathcal{R}/\mathcal{A}),$$

where $\mathcal{A}$ is the letter source ensemble and $\mathcal{R}$ is the run-length ensemble.

In the binary case, such as black and white text, the letters 0 and 1 do not have to be coded, because the end of a run signifies a change to the only other letter. A complication occurs because of the practical need to limit the run-lengths in a code. When a run-length exceeds this limit, then a zero-run length must be inserted. Consider the following example.

Example 4.3 The binary letter source emits the sequence:

$$11111111000111001111110000011111000.$$

The run-length alphabets are $r_1 \in \{0, 1, 2, 3, 4, 5, 6\}$ and $r_0 \in \{0, 1, 2, 3, 4\}$. Suppose that it is agreed that the first run belongs to letter 0. Then the run-length representation of the sequence is

$$0 \quad 6 \quad 0 \quad 2 \quad 3 \quad 3 \quad 2 \quad 6 \quad 4 \quad 0 \quad 1 \quad 5 \quad 3$$

The first 0 signifies a 0 run-length of letter 0, so that the following 6 belongs to a run of 6 1's. The second 0 signifies a 0 run-length of 0, so that the next 2 signifies a run of 2 1's. So the run of 8 1's is represented as 602. Likewise, the later run of 5 0's is represented as 401.

Let us continue with this example by encoding the run-length symbol representation. Suppose that the Huffman codes of the run-lengths r_1 and r_0 are those given in Table 4.4. Using this table, we find the sequence of codewords with their corresponding run-lengths and letters written directly below.

code:	10	01	10	00001	00	0001	011	01	11	10	010	11	00
run-lengths:	0	6	0	2	3	3	2	6	4	0	1	5	3
letter:	0	1	0	1	0	1	0	1	0	1	0	1	0

The spaces between codewords are shown for clarity and are not transmitted. As long as we know which code is applicable to a certain run-length, the codeword can be decoded instantaneously. Notice that this run-length coding example does not exhibit impressive performance, as the run-length code for the given data sequence of 35 bits consists of 33 bits, a savings of only 2 bits or 5.7%. We could save another two bits by changing the starting convention to letter 1. However, we wanted to show another use of the zero run-length symbol. Ordinarily, the data sequences are much longer, so that the start code length has negligible effect. Typically, run-length limits are much higher and run-length symbols are skewed more in probability toward larger values than in this example, whose purpose is just to illustrate the method.

Table 4.4 Run-length codes in Example 4.3

r_1 code		r_0 code	
length	codeword	length	codeword
0	10	0	10
1	00000	1	010
2	00001	2	011
3	0001	3	00
4	001	4	11
5	11		
6	01		

4.6 Alphabet partitioning: modified Huffman codes

4.6.1 Modified Huffman codes

One line in a standard width document consists of 1728 samples, so black and white runs would each have code alphabets of that size. Alphabets of such large size lead to practical difficulties of codebook design and memory. Large alphabets are especially troublesome when attempting adaptive coding or trying to exploit statistical dependence between samples in conditional coding. Large alphabets arise not only in run-length coding of documents, but also in transform coding, to be covered later, and when coding multiple symbols together that generate alphabet extensions.

One method that is used extensively to alleviate the problem of large alphabets is called *modified Huffman coding*. Suppose that the alphabet consists of integers with a large range, such as the run-lengths. We break the range into a series of intervals of length m, so that a number n is expressed as

$$n = qm + R.$$

The variable q is a quotient and R is a remainder in division of n by m. Given m, there is a one-to-one correspondence between n and the pair of numbers q and R. Encoding n is equivalent to coding the pair (q, R). If the pair members are coded separately and the size of the alphabet of n is N, then the size of the pair alphabet is $\lfloor N/m \rfloor + m$, which can amount to a significant reduction. For instance, consider the 1728 size alphabet for document coding. A typical $m = 64$ yields a size of 27 for the quotient q and 64 for the remainder R. Although the choice of the code is open, the quotient is usually coded with a Huffman code and the remainder either with a separate Huffman code or with natural binary (uncoded).

A special case of the modified Huffman code is often used when probabilities of values beyond a certain value are all very small. To illustrate, suppose that $0 \leq n \leq N$ and values $n > m$ have very small probability. Then, we code $0 \leq n \leq m$ with a Huffman code and the larger values with an ESCAPE codeword followed by $n - m$ in natural binary.

The loss of efficiency is often small when the statistics have certain properties and m is chosen judiciously. We adopt the notation that the corresponding calligraphic letter denotes the random variable or ensemble. We shall assume now that $n = 0, 1, 2, \ldots, N-1$ and let $Q = N/m$ be an integer. The increase in entropy due to independent coding of the $(Q, \mathcal{R})$ pair measures the theoretical performance loss. The original entropy may be expressed by

$$H(\mathcal{N}) = H((Q, R)) = H(Q) + H(\mathcal{R}/Q)$$

$$= H(Q) + \sum_{q=0}^{Q-1} P(q) H(\mathcal{R}/q), \tag{4.10}$$

where $P(q)$ denotes the probability of q. Coding $\mathcal{R}$ independently of Q means that

$$H(\mathcal{R}/Q) = H(\mathcal{R})$$

or, more specifically, that $H(\mathcal{R}/q)$ is the same for every q, which holds true when the probability distribution of the remainder is the same in every interval.

Suppose that we are able to choose m so that $\mathcal{R}$ has a nearly uniform distribution in every interval. Then we would not entropy code $\mathcal{R}$ and convey it with $\log_2 m$ natural bits. In such a circumstance, we would spend our efforts to entropy code the small alphabet Q with a Huffman or arithmetic code. The theoretical loss in efficiency or increase in bit rate would therefore be

$$\delta H = \log_2 m - \sum_{q=0}^{Q-1} P(q) H(\mathcal{R}/q) \tag{4.11}$$

Let us denote $Pr\{\mathcal{N} = n\}$, the probability that $\mathcal{N} = n$, as p_n, $n = 0, 1, 2, \ldots, N-1$. Then the probability of points in the q-th interval is

$$P(q) = \sum_{n=qm}^{(q+1)m-1} p_n$$

and the entropy increase in (4.11) can be manipulated algebraically to the expression

$$\delta H = \sum_{q=0}^{Q-1} \sum_{n=qm}^{(q+1)m-1} p_n \log\left(\frac{m p_n}{P(q)}\right). \tag{4.12}$$

We conclude from Equation (4.12) that δH can be quite small if $m p_n P(q) \approx 1$ or, equivalently, $p_n \approx P(q)/m$ for every q. The interpretation is that, when the probability distribution of $\mathcal{N}$ within every interval is nearly uniform, there is nearly zero rate penalty. Under these conditions, the coding complexity is therefore greatly reduced by this alphabet reduction strategy with almost no increase in bit rate.

4.6.2   Alphabet partitioning

Modified Huffman coding is a special case of a method called *alphabet partitioning*. In this general method, we partition the alphabet $\mathcal{A}$ into a number of (disjoint) subsets

$\mathcal{A}_1, \mathcal{A}_2, \ldots, \mathcal{A}_Q$, i.e., $\mathcal{A} = \cup_{q=1}^{Q} \mathcal{A}_q$. These subsets may be of different size and may be defined by range of probability rather than value. For example, uniform intervals of numbers as above may not be suitable, as they may not capture subsets of low or uniform probability. Let the number of points in subset $\mathcal{A}_q$ be m_q. Analogously, we use $\log m_q$ bits to code the points within every $\mathcal{A}_q$. We calculate the minimum combined rate, which is the entropy of the subset numbers plus the average of the fixed rates to code points within the subsets.

$$H_2 = \sum_{q=0}^{Q-1} P(q) \left[\log \left(\frac{1}{P(q)} \right) + \log m_q \right], \quad P(q) = \sum_{n \in \mathcal{A}_q} p_n. \tag{4.13}$$

Repeating the previous calculation with these general subsets, the increase in entropy is found to be

$$\delta H = \sum_{q=0}^{Q-1} \sum_{n \in \mathcal{A}_q} p_n \log \left(\frac{m_q p_n}{P(q)} \right). \tag{4.14}$$

Then, if $p_n \approx P(q)/m_q$ in $\mathcal{A}_q$ for all q, δH is small. So we reach the same conclusion that the points within the subsets should be approximately uniformly distributed in order to obtain nearly no rate increase. The ideal example to follow illustrates the significant reduction of complexity of alphabet partitioning.

Example 4.4 Consider a source with 80 letters having probabilities as follows:

$$P(a_k) = \begin{cases} 1/32, & k = 1, 2, \ldots, 16 \\ 1/64, & k = 17, \ldots, 32 \\ 1/128, & k = 33, \ldots, 48 \\ 1/256, & k = 49, \ldots, 80 \end{cases} \tag{4.15}$$

The obvious alphabet partitions are

$$\mathcal{A}_1 = \{a_k : k = 1, 2, \ldots, 16\}, \mathcal{A}_2 = \{a_k : k = 17, \ldots, 32\},$$

$$\mathcal{A}_3 = \{a_k : k = 33, \ldots, 48\}, \mathcal{A}_4 = \{a_k : k = 49, \ldots, 80\}$$

with probabilities

$$P(1) = 1/2, \ P(2) = 1/4, \ P(3) = 1/8, \ P(4) = 1/8.$$

From Equation (4.13), the minimal combined encoding rate is

$$H_2 = \frac{1}{2}\left(\log 2 + \log 16\right) + \frac{1}{4}\left(\log 4 + \log 16\right) + \frac{1}{8}\left(\log 8 + \log 16\right) + \frac{1}{8}\left(\log 8 + \log 32\right)$$

$$= \frac{7}{4} + \frac{33}{8} = 5\frac{7}{8},$$

which equals the sum of the subset entropy of 7/4 and the average fixed code length of 33/8. In fact, because of the uniform probabilities within the partitions, $\delta H = 0$, making H_2 equal to the original source entropy. Furthermore, the alphabet subsets have dyadic probabilities, so the Huffman code in Table 4.1 achieves the minimal codeword length equal to the subset entropy. The codeword for a_k is the joining of its partition codeword and the index within its partition. For example, the codeword c_{37} for a_{37} is

$$c_{37} = 110|0100.$$

(The first index in the partition is 0.) The separator | emphasizes the two parts and is not part of the codeword. The decoder needs only to search four codewords in the prefix to get the partition number and from that knows how many following bits represent the index. The Huffman code for the original source alphabet has 80 codewords, some of which are 8 bits long, so the decoding complexity has been greatly reduced, whether one uses a tree or table to store the code.

In this example, the partitions of the alphabet were obvious, because there were groups of letters with uniform probabilities. In general, groups of uniform probabilities are not present, so the grouping into partitions that minimize δH is not obvious. Said [4] has derived theorems and algorithms that achieve optimal symbol grouping to form alphabet partitions. This topic is beyond the scope of our purposes, so we refer the interested reader to Said's work.

4.7   Golomb code

The previous codes deal effectively with sources of random variables having a finite alphabet. However, there are circumstances where the alphabet does not have a finite limit. One such random variable is the number of successful attempts of some action until a failure occurs. Let the probability of success on a single attempt be θ, $0 < \theta < 1$. Then the probability of i successes in independent attempts until a failure is

$$P(i) = (1 - \theta)\theta^i, \quad i \geq 0.$$

This probability distribution is called the *geometric* distribution. It usually arises in run-length coding, where $P(i)$ would be the probability of a run of i zeros until a non-zero number occurs. Golomb [5] was the first to develop an optimal binary code for sources with the geometric probability distribution. His solution was to choose an integer m and encode i in two parts:

(a) unary code of quotient $i_q = \lfloor i/m \rfloor$;
(b) uniquely decodable code of remainder $i_r = i - mi_q$.

The unary code of an integer n is n 1's followed by 0 or n 0's followed by 1. We shall use the former convention. We shall explain shortly the details of the code of the

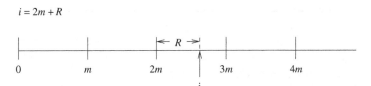

$i = 2m + R$

Figure 4.6   Illustration of number representation for Golomb coding.

second part. The order of the two parts is immaterial, except that the order above is more convenient for decoding. The idea of this code is illustrated in Figure 4.6. The interval of possible real numbers is broken into contiguous segments of length m. The unary code in the first part tells you the segment that contains i and the binary code in the second part tells you the location of i within that segment. Each part of the code is uniquely decodable, so the joined parts decoded separately yield a uniquely decodable code.

Selection of the best m, the one giving the shortest average code length, is an important element. Golomb derived optimal codes for integer m satisfying

$$\theta^m = 1/2,$$

but later, Gallager and Voorhis [6] proved that the optimal m satisfies

$$\theta^m + \theta^{m+1} \leq 1 < \theta^m + \theta^{m-1}.$$

In other words, m is the smallest integer such that $\theta^m \geq 1/2$.

Suppose that the number S of consecutive successes is currently n. Then the condition that $\theta^m = 1/2$ means that the event of an additional m successes halves the probability of S. ($\theta^{n+m}(1 - \theta) = \theta^m \theta^n(1 - \theta) = 1/2 * \theta^n(1 - \theta)$.) Therefore, an additional run of length m requires at least the entropy of 1 bit to encode. Therefore, a unary code for the number of length m segments is an optimal prefix-free code. The remainder i_r may be encoded with a prefix-free entropy code, but a natural binary code of fixed length $\lceil \log_2 m \rceil$ is usually just as efficient.[3] When m is not a power of 2, one can save bits using one less bit for certain smaller values of i_r. This procedure goes as follows:

1. Let $k = \lceil \log_2 m \rceil$. Code first $2^k - m$ values of i_r within the natural $k - 1$ bit representation.
2. Code remaining values of i_r with natural k bits of $i_r + 2^k - m$.

The full algorithm is detailed in Algorithm 4.4.

When m is a power of 2, the code takes a particularly simple form:

(a) unary code of number formed by higher order bits of i;
(b) $\log_2 m$ least significant bits of i.

Consider two examples, $m = 4$ and $m = 5$. For $m = 4$, each of the four remainder values is coded with its natural $k = 2$ bits. For $m = 5$, $k = 3$ and $2^k - m = 8 - 5 = 3$ lowest remainders (0,1,2) are coded with their natural 2 bits and the remaining values

ALGORITHM 4.4 _____

Golomb encoding

1. Initialization:
   (a) Given i to encode and integer m, do
      (i) compute quotient $i_q = \lfloor i/m \rfloor$;
      (ii) compute remainder $i_r = i - m * i_q$.
   (b) Compute $k = \lceil \log_2 m \rceil$ and $m_r = 2^k - m$.
2. Write i_q 1's followed by one 0.
3. If $m_r > 0$,
   (a) if $i_r < m_r$, write $k - 1$ bits of i_r; else, write k bits of $i_r + m_r$.
4. If $m_r = 0$, write k bits of i_r.

_____

Table 4.5 Golomb codes for $m = 4$ and $m = 5$

i	$m = 4$ codeword	$m = 5$ codeword
0	0 00	0 00
1	0 01	0 01
2	0 10	0 10
3	0 11	0 110
4	10 00	0 111
5	10 01	10 00
6	10 10	10 01
7	10 11	10 10
8	110 00	10 110
9	110 01	10 111
10	110 10	110 00

of 3 and 4 are represented with the 3 bits of the numbers 6 and 7. We present these two Golomb codes for values of i from 0 to 10 in Table 4.5.

There are also two parts in decoding a Golomb codeword. The decoder knows m and therefore the $k - 1$ bits representing low remainders $0, 1, \ldots, 2^k - m - 1$. The decoder counts the number of 1's preceding the first 0 to get i_q and then decodes the remaining bits of the codeword to a number i_r. The final decoded number is $i = mi_q + i_r$. The decoding differs slightly for the cases of $m = 2^k$ and $m \neq 2^k$. When $m = 2^k$, the receiver counts the number of 1's preceding the first 0 to determine i_q. It then reads the k bit number following this first 0 and sets it to i_r. Then it computes the result $i = mi_q + i_r$. When m is not a power of 2, it still counts the number of 1's before the first 0 to get i_q. However, the number of bits belonging to i_r could be k or $k - 1$ ($k = \lceil \log_2 m \rceil$). After the first 0, if the next $k - 1$ bits match one of the low remainders, then i_r is the number represented by these bits. The remainder is the actual value and the decoder declares $i = mi_q + i_r$. If there is no match, the k bits following the first 0 convey the value of i_r. The translation of the actual value of i_r by $2^k - m$ precludes occurrence of any of these $k - 1$ bit prefixes. Suppose that the numerical value of these

ALGORITHM 4.5 _____

Golomb decoding

1. Initialization:
   (a) Given m, calculate $k = \lceil \log_2 m \rceil$ and $m_r = 2^k - m$.
   (b) Store $k - 1$ bit patterns of remainders $0, 1, \ldots, m_r - 1$.
2. Count number i_q of 1's preceding first 0 in received word.
3. Determine whether the $k - 1$ bit pattern (following the first 0) matches any of the stored remainder patterns.
   (a) If there is a match, set numerical value of match to i_r.
   (b) If not, set $i_r = R - m_r$, where R is the value of the k bit pattern following the first 0.
4. Decode number $i = i_q * m + i_r$.

_____

k bits is R. Then, $i_r = R - (2^k - m)$ and the decoder declares that $i = m i_q + i_r$. The decoding algorithm is summarized in Algorithm 4.5.

_____

Example 4.5 Example of Golomb decoding
Suppose that the operative code is Golomb, $m = 10$, and the received word is 101110.... The decoder knows beforehand that: (a) $k = 4$ and $2^k - m = 6$; (b) remainders 0, 1, 2, 3, 4, 5 are coded with 3 bits; and (c) remainders 6, 7, 8, 9 are coded as 12, 13, 14, 15 with 4 bits. The initial 10 of the received word says that $i_q = 1$. The three 1's following the 0 do not match the numbers 0, 1, 2, 3, 4, or 5. Therefore, the four bits 1110 equalling decimal 14 is the translated remainder. Therefore, $i = 10 * 1 + 14 - 6 = 18$ is the decoded number.

_____

4.8 Dictionary coding

The codes studied thus far are efficient when the statistical model is known or estimated accurately. A mismatch between the actual model and the model used for creating the code results in deterioration of efficiency. A code that works efficiently without any modeling assumption or calculation is called a *universal code*. Such a code is desirable, because statistics usually vary for most data sources and adapting the code accordingly usually consumes most of the computation in the encoding algorithm.

The universal codes in use today have their origins in two compression algorithms described in landmark papers by Ziv and Lempel [7, 8]. These papers were more concerned with formal proofs that the compression efficiencies of these two algorithms approach closely the optimal performance of algorithms based on the actual statistics of the source. The earlier 1977 algorithm, known as LZ77 (for Lempel–Ziv 1977), gave rise to the popular file compression algorithms of PKZip, Zip or WinZip, gzip, and PNG, to name but a few. The later 1978 algorithm, LZ78 (for Lempel–Ziv 1978), in a modified form named LZW due to Welch [9], gave rise to UNIX compress and GIF.

Both these algorithms build a dictionary of strings of symbols from the input data sequence. The principle behind such a strategy is simple. Consider that we have been given a fixed, indexed dictionary of all possible symbol strings to a long length. Starting from the first data symbol, the input sequence is searched to find the longest segment that matches a string in the dictionary. The code of the index of this matching string is written to the output. The next portion of the input sequence is then searched to find the longest string that matches a string in the dictionary and the code of its index is written to the output. The process continues until all the symbols in the input sequence have been coded.

As an artificial example, consider the data symbols to be the four letters a, b, c, and d. Suppose that the dictionary consists of all strings of length 8 of these four letters. This dictionary contains $4^8 = 65\,536$ strings, so it requires 16 bits in natural binary to represent every index. Let the input sequence consist of 100 000 data symbols. We must find 12 500 matches in the dictionary to the data sequence, so writing 16 natural bits for each match yields an output code of 200 000 bits. So we have not saved anything at all from the simple scheme of writing 2 bits for each letter in the sequence. Any savings have to come from entropy coding of the indices in the dictionary. In natural language, certain strings of letters will seldom occur or not occur at all and others will occur quite frequently. Those not occurring at all should be removed from the dictionary. In other words, the entropy of non-random strings in a dictionary is typically far smaller than the base-2 logarithm of the size of its string population. This example points out the possible inefficiency of using a fixed dictionary containing all possible strings. Methods that build an adaptive dictionary are potentially more efficient, because they will contain the strings that actually occur in the text and none that does not occur. The LZ77 and LZ78 universal codes build dictionaries adaptively and take advantage of disparateness in frequency of occurrence of strings in order to compress an input data sequence.

4.8.1   The LZ78 code

The LZ78 code is taken first, as it is the simpler to explain. The idea is to parse the input string by the uniquely occurring substrings of any length and to build and index the dictionary with the unique substrings that have previously occurred. An example is the best way to describe the method.

Consider the input string to be *cabbcabbacbcababcac....* Parsing the string according to the uniquely occurring substrings yields:

$$|c|a|b|bc|ab|ba|cb|ca|bab|cac|....$$

We must explain that we parse once we have not seen a substring (including a single letter) that has not occurred previously. In order to encode this parsing, we must build a dictionary of the phrases that have not occurred previously and convey them through their indices in the dictionary. The procedure is that for the next parse, look into the dictionary to find the longest phrase that matches that part of the input string. For example, consider the next to last (9-th) parse *bab*. *ba* matched the *ba* in the 6-th parse and

Table 4.6 LZ78 encoding of string $cabbcabbacbcababbcac \dots$

Phrase	c	a	b	bc	ab	ba	cb	ca	bab	cac
Index	1	2	3	4	5	6	7	8	9	10
Output	$(0, C(c))$	$(0, C(a))$	$(0, C(b))$	$(3, 1)$	$(2, 3)$	$(3, 2)$	$(1, 3)$	$(1, 2)$	$(6, 3)$	$(8, 1)$

parsed string:$|c|a|b|bc|ab|ba|cb|ca|bab|cac| \dots$

together with the next letter b gives a new phrase to be added to the dictionary. To convey this parse, we need to send two numbers, the index of ba, which is 6, and the index of the following letter b, which is 3. This doublet of numbers must be sent for every parse. We shall illustrate the steps of this algorithm through the current example.

The dictionary starts as empty. Each of the first three letters c, a, and b, has no previous match. They are conveyed by 0 for no match and the code for the letter, denoted by $C(c)$, $C(a)$, or $C(b)$. Therefore, the first three doublets and dictionary entries are $(0, C(c))$, $(0, C(a))$, and $(0, C(b))$, and c, a, and b, respectively, as shown in Table 4.6. The next input letter b is in the dictionary, so we look to its succeeding letter c to see whether bc is in the dictionary. It is not, so bc goes into the dictionary as entry 4 and its code doublet is $(3, 1)$, the first component being the index of the matching phrase and the second the index of the following letter. Repeating this procedure through the 8-th parse of ca, we now come to two letters of the input string, ba, that match entry 6 in the dictionary. Together with the next letter b, the phrase bab does not match any previous entry in the dictionary, so bab is added as entry 9 with the doublet code of $(6, 3)$ for this parse. Table 4.6 shows the development of the dictionary and coding for the given fragment of the input string.

As the input string grows, longer and longer phrases start to enter the dictionary. Eventually, the dictionary growth must stop to limit the memory usage. The remainder of the input string can be encoded with the phrases of the dictionary thus created.

The decoding procedure is straightforward. Each doublet represents a parse that progresses from beginning to the end of the input string. The decoder sees only the codewords in the order of the third column in Table 4.6. It builds the dictionary in the same way as does the encoder. The first three codewords $(0, C(c))$, $(0, C(a))$, and $(0, C(b))$ give c, a, and b, respectively, as dictionary entries and output phrases. The next codeword $(3, 1)$ decodes as bc, which becomes entry 4 in the dictionary. And so it goes until all the codewords have been read.

4.8.2 The LZW algorithm

Welch [9] published a modification of LZ78 that encoded the input string with a series of single indices instead of doublets. This modification has been named the Lempel–Ziv–Welch or LZW algorithm. In this method, the dictionary is initialized with the different single letters. Then the first letter in the input string corresponds to an index in the dictionary, and with the next letter appended forms a new phrase to enter into the dictionary. Only the index of the match is encoded. The appended letter becomes the first letter of the next phrase for which to find a match. Let us illustrate this procedure with the same input string:

$$cabbcabbacbcababcac\ldots$$

The initial dictionary is:

Index	Dictionary Phrase
1	a
2	b
3	c

The first letter c is in the dictionary, but ca is not, so ca is placed as entry 4 in the dictionary and 3, the index of the match to c, is encoded. The last letter, a, of the current entry becomes the first letter of the next candidate phrase for a match. Appending the next letter produces ab, which becomes entry 5 in the dictionary and 2, the index of a, is encoded. Since the last letter of a new phrase is encoded as the first letter in the next match, the current index and the next index together determine the current dictionary entry. This stratagem is crucial to correct decoding. The same procedure continues for the rest of the input string. The full procedure is illustrated below with arrows pointing to the letters that end a new phrase entry and begin a new phrase to search for a match. At the tails of the arrows are the new phrases entering the dictionary, which can be read from the string starting at the previous arrow. Every phrase minus its final letter matches a previous phrase in the dictionary and its index is the encoder output. The output index is written directly below its corresponding dictionary phrase.

LZW encoding

Input	c	a	b	b	ca	bb	a	c	bc	ab	ab	ca	$c\ldots$
		↑	↑	↑	↑	↑	↑	↑	↑	↑	↑	↑	↑
Phrase		ca	ab	bb	bc	cab	bba	ac	cb	bca	aba	abc	cac
Output		3	1	2	2	4	6	1	3	7	5	5	4

The output of the encoder, here 312246137554 for the fragment shown, is sent to the decoder. Let us see how the decoder acts upon this input to reconstruct the input string. The initial dictionary containing the three letter indices is assumed to be known at the decoder. The decoder receives the indices and builds upon this dictionary using the same rules as in the encoder. The first received index of 3 says that c is the first matching phrase. Then c will be the first letter of the matching phrase corresponding to the next index. The next index of 1 means that a is the next letter and ca is the new dictionary phrase with index of 4. Now a becomes the first letter of the next matching phrase, indicated by the next index of 2 signifying b. Therefore b is the next letter and ab is the new phrase inserted into the dictionary with index of 5. Let us skip ahead to the decoder inputs of 4 and 6. These indices indicate successive matches to the phrases ca and bb.

The first b in bb is appended to the previous matched phrase of ca to form the new phrase cab to enter into the dictionary. The following index input of 1 indicates a match to a, so bba becomes the next new entry to the dictionary. The following diagram shows the progression of decoding the sequence of indices at the decoder input and building the dictionary.

						LZW decoding						
Input	3	1	2	2	4	6	1	3	7	5	5	4
	↓	↓	↓	↓	↓	↓	↓	↓	↓	↓	↓	↓
Output	c	a	b	b	ca	bb	a	c	bc	ab	ab	ca
Phrase		ca	ab	bb	bc	cab	bba	ac	cb	bca	aba	abc ca ...
Index		4	5	6	7	8	9	10	11	12	13	14 15

Notice that the end of the encoder input string contained a final c, which was not coded, but could form the new phrase cac that was added to the dictionary. Since the decoder has no knowledge of this c, all that is known of the last dictionary phrase is that it begins with the last match of ca. Sometimes we run into the situation of receiving an input whose dictionary entry is incomplete, so cannot be decoded immediately. For example, suppose that the next decoder input is 15, which is the incomplete last entry in the dictionary. Nevertheless, we know that entry 15 begins with ca, so we append the first letter c to ca to get cac, which, being a new phrase, must be the complete entry 15. Therefore, the input of 15 decodes to cac. The encoding and decoding steps of the LZW coding algorithm are stated in general in Algorithm 4.6.

Various coding systems that use LZW differ in the way that the indices are coded. One can use an adaptive entropy code, Huffman or arithmetic, where the binary codeword length of an index is approximately the base-2 logarithm of its inverse probability. A simpler way is to expand the dictionary in octave stages up to a certain size limit. The indices in each octave may be represented by a fixed number of bits, which increases by 1 bit in the next octave. Therefore by sending the octave number along with the bits for an index, one can achieve a shorter code than that from a single large dictionary. For example, suppose that the dictionary builds to 256 entries in the first stage and then to 512 in the second stage and so forth until it reaches a prescribed limit of 2048. In the first- and second-stage dictionaries, there are 256 entries, so that 8 bits can represent their indices. In the third stage there are 512 entries and in the fourth and final stage there are 1024 entries. These indices require 9 and 10 bits, respectively. Along with the bits for a given index, 2 bits must be added to indicate its octave. With one large dictionary of 2048 entries, every index requires 11 bits in natural binary representation. There is clearly a potential for bit savings in the expanding dictionary scheme, especially when one entropy codes the two bits that indicate the containing octave.

4.8.3 The LZ77 coding method

The LZ77 method does not extract unique phrases to enter into an indexed dictionary in memory. The dictionary consists of phrases in the recently encoded portion of the input

ALGORITHM 4.6 _____

The Lempel–Ziv–Welch (LZW) coding algorithm

Encoder

1. Index letters of source alphabet and put as entries to initially empty dictionary.
2. Read letters from input until phrase matches one in the dictionary. (The first match is a single letter.)
3. Output index of matched phrase.
4. Write matched phrase appended by next input letter to dictionary as next entry with index incremented by 1.
5. If there are more input letters, return to Step 2. If not, stop.

Decoder

1. Index source letters and enter into dictionary in same order as in the encoder.
2. Read first index from input and write its associated phrase in the dictionary to the output.
3. Read next index from input; decode its associated phrase by lookup in the dictionary; write next entry to dictionary as the previous decoded phrase appended by the first letter of the current decoded phrase. If current index is one (1) plus the highest index in the dictionary, its dictionary entry begins with the previous matched phrase. Decode this partial phrase. The previous matched phrase appended by the first letter of this partial phrase is the sought dictionary entry and decoder output.
4. If there are more input indices, return to Step 3. Otherwise, stop.

string that match phrases in the portion about to be encoded. The location of the match is indicated by a pointer to the phrase in the encoded portion. Figure 4.7 shows a buffer containing letters of the input string. To the left of the vertical bar are letters or symbols that have already been encoded. The terms letters and symbols will be used interchangeably here. This portion of the buffer is called the *search buffer*. The portion that is to the right of the vertical bar contains letters not yet encoded and is called the *look-ahead buffer*. One looks into the search buffer to find the longest phrase that matches the phrase starting at the first letter in the look-ahead buffer. The pointer to the matching phrase in the search buffer is represented by two constants: the *offset* from the first letter in the look-ahead buffer to the beginning of the matching phrase in the search buffer; and the *length* of the matched phrase. Figure 4.7 portrays a match by the shaded pattern in the two buffer portions with its offset and length. Besides the offset and length of the match, the encoder also encodes the next symbol in the look-ahead buffer that follows the matched phrase, so the encoder emits a series of triplets (offset, length, next) to represent the input string. We adopt the notation o for offset, ℓ for length, and $C(x)$ for the code of the next symbol x.

Figure 4.7   Buffer and pointer definitions in LZ77 coding.

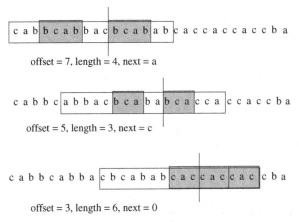

offset = 7, length = 4, next = a

offset = 5, length = 3, next = c

offset = 3, length = 6, next = 0

Figure 4.8 Example of first three stages of LZ77 coding.

Again, an example teaches best the LZ77 coding method. We consider the same input string as before extended by several more letters: *cabbcabbacbcababcaccaccaccba*. The input buffer has a total length $w = 16$ with the search buffer length of $L_s = 10$. Figure 4.8 shows the first few stages of the buffer contents and the coding.

Referring to the figure, we observe that the longest match in the search buffer (to the left of the vertical line) to a phrase in the look-ahead buffer is *bcab*, which has an offset of 7 and a length of 4. The next letter in the look-ahead buffer is *a*, so the triplet of $(7, 4, C(a))$ is the output. Then the buffer window slides five symbols to the right up to the next symbol not yet coded. For the next stage portrayed in the middle buffer in the figure, the longest match in the search buffer to a phrase starting at the first symbol in the look-ahead buffer is *bca*, having an offset of 5 and a length of 3. The next look-ahead symbol is *c*, so the triplet output is $(5, 3, C(c))$. The window now slides four symbols to the right for the next stage. Referring to the bottom buffer in the figure, the matching phrase of *caccac* extends from the search buffer into the look-ahead buffer. Here, the length of 6 exceeds the offset of 3. We display it as two overlapped shaded rectangles. Once we examine the decoding procedure, the reason that we can extend the match into the look-ahead buffer will become evident. Also, the match extends to the end of the look-ahead buffer, which is a rare occurrence, since in practice these buffers are thousands of symbols long. In this rare case, we send a 0 for the next symbol and shift the window only 6, so that the look-ahead buffer begins at the next symbol to be coded.

That raises the question: why must we encode the symbol following the match? The reason is that the symbol may not yet have been encountered in the input string and thus may not match any symbol in the search buffer. In this case, the triplet code is $(0, 0, C(x))$, where x denotes the symbol. Since sending this next symbol wastes bits, we shall later discuss a way to avoid this practice.

The decoding follows the course of the encoding. First, the search buffer is filled with known symbols agreed by both encoder and decoder. The look-ahead buffer starts as empty. When a triplet $(o, \ell, C(x))$ is received, the decoder moves to an offset o in

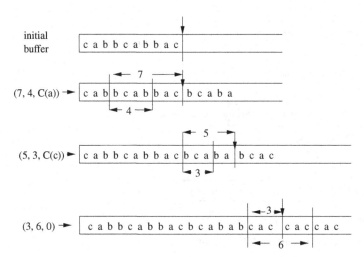

Figure 4.9   Example of first three stages of LZ77 decoding.

the search buffer and copies ℓ symbols to the beginning of the look-ahead buffer. It then decodes $C(x)$ and puts x in the next location in the look-ahead buffer. The start of the look-ahead buffer is moved $\ell + 1$ positions to the right for receiving the next triplet, whereby this process is repeated. Actually, there is no need to maintain a look-ahead buffer, only a pointer to the start position after the last decoded symbol.

We shall illustrate the decoding for our example. The decoder knows the initial search buffer contents and receives the three triplets $(7, 4, C(a))$, $(5, 3, C(c))$, and $(3, 6, 0)$. In Figure 4.9, we illustrate the initial search buffer contents and the start pointer after the last symbol. Upon receipt of $(7, 4, C(a))$, the decoder copies the $\ell = 4$ symbols *bcaba* offset from the pointer by $o = 7$ positions and writes them to the empty positions after the pointer followed by symbol a, decoded from $C(a)$. The pointer then moves to the right by five positions. The reception of the next triplet $(5, 3, C(c))$ causes the decoder to copy the $\ell = 3$ symbols *bca* that are offset to the left by $o = 5$ positions to the empty positions after the pointer, followed by c. The next triplet $(3, 6, 0)$ requires slightly different actions, because the match length exceeds the offset. Starting from the offset $o = 3$, we copy the first three symbols *cac* to the empty positions following the pointer and then continue copying three more after the pointer into the next empty positions in the buffer. The $\ell = 6$ symbols decoded are therefore *caccac*. No symbol follows, because of 0 in the third coordinate. The input string up to this point has been reproduced at the decoder. The pointer shifts six positions to the right if there are any more triplets to decode. The encoding and decoding steps are stated in general in Algorithm 4.7.

The number of bits needed to encode a triplet depends on the sizes of the buffers and source alphabet. The offset cannot exceed L_s, the size of the search buffer, so it requires $\lceil log_2 L_s \rceil$ natural binary bits. The match length can be as large as w, the total length of the buffer (search plus look-ahead), so it requires $\lceil log_2 w \rceil$ natural binary bits. Assuming that the size of the source alphabet is A, the next symbol code $C(x)$ requires $\lceil log_2 A \rceil$ natural binary bits, so the number of natural bits n_{tp} needed to encode a triplet is

ALGORITHM 4.7 ——

The Ziv–Lempel 1977 (LZ77) coding algorithm

Encoder

1. Initialization (see Figure 4.7):
    (a) Set sizes of search and look-ahead buffers.
    (b) Fill search buffer with default symbols known at decoder. Fill look-ahead buffer with input symbols to be coded.
    (c) Set reference point to beginning of look-ahead buffer.
2. Search phrases anchored at reference point in look-ahead buffer that match phrases in search buffer and select the longest phrase match.
3. Record position of longest match in search buffer (offset from reference point) and the length of the match in number of symbols. The length can extend into the look-ahead buffer, in which case the length ℓ exceeds the offset o.
4. Encode offset, length of longest match, and symbol in look-ahead buffer following the match as the triplet $(o, \ell, C(x))$.
5. Move reference point $\ell + 1$ positions to the right if $C(x) \neq 0$. Otherwise move ℓ positions.
6. If there are more input symbols in the look-ahead buffer, return to Step 2. Otherwise, stop.

Decoder

1. Fill search buffer portion of output buffer with default symbols known at the encoder. Leave rest of buffer empty. Set reference point after last default symbol.
2. Read triplet $(o, \ell, C(x))$ from input. Copy ℓ symbols starting at o positions to left of reference point and write them starting at first empty position to right of reference point.
3. If $C(x) \neq 0$, decode $C(x)$ to x and write it to next empty position.
4. Move reference point to next empty position. If $C(x) \neq 0$, move reference point $\ell + 1$ positions to right; otherwise move ℓ positions.
5. If there are more triplets to receive, return to Step 1; otherwise, stop.

——

$$n_{tp} = \lceil log_2 L_s \rceil + \lceil log_2 w \rceil + \lceil log_2 A \rceil.$$

For our toy example, $n_{tp} = 4 + 4 + 2 = 10$ bits per triplet for a total of 30 bits to encode 15 letters, which could be straightforwardly encoded naturally with 30 bits. Our choices of buffer sizes, besides being too short for practical application, were far from ideal. Ziv and Lempel recommend the following approximate rule for buffer sizes:

$$w \approx L_s A^{hL_s}, \quad 0 < h < 1.$$

One of the disadvantages of the LZ77 method compared to the LZ78 method is that it relies on finding phrases to match from the recent past. The buffer should therefore be long enough to be able to match long periodically occurring phrases. One of the advantages of the LZ77 method is that it does not need to maintain a dictionary in storage, so uses less memory than LZ78. Variants of LZ77, such as gzip, PKZip, and WinZip, use different forms of entropy coding of the triplets to achieve better efficiency.

    One variant of LZ77, called LZSS [10], dispenses with the third member of the triplet, the next symbol. As mentioned previously, coding of the next symbol is needed only to convey the occurrence of a symbol not yet encountered. One can do the same more efficiently by sending a flag bit signifying whether or not a new symbol is encountered. If yes, send a 1 and the new symbol following the offset and length; if no, just send 0 following the offset and length. So almost all of the time, only a doublet of (o, ℓ) followed by 0 is encoded.

4.9   Summary remarks

This chapter taught lossless coding methods, some of which have the potential to achieve the entropy limit on rate. As the reader saw, practical considerations did not allow the achievement of the rate of entropy, but at least they came close in many circumstances. These methods are called entropy coding methods, as are other lossless methods, the rates of which cannot theoretically achieve entropy. We tried to present the entropy coding methods that are most commonly used, which include Huffman coding, arithmetic coding, run-length coding, Golomb coding, and Lempel–Ziv coding. Our approach was tutorial with many illustrative examples and algorithm routines. Also contained in this chapter is the important method of alphabet partitioning, which does not appear in other textbooks, and which greatly simplifies entropy coding. For readers who wish to learn more lossless coding methods, they are referred to the textbooks by Sayood [11] and Salomon [12].

Problems

4.1 (Prob. 9.9 [14]) Suppose that an i.i.d. binary source has letter probabilities $p(0) = 1/3$ and $p(1) = 2/3$.

(a) Construct a Huffman code for the triples (three consecutive letters) produced by this source.

(b) Find the average length of this Huffman code and compare it to the entropy of the source.

4.2 (Prob. 9.8 [14]) Suppose that a memoryless binary source has probabilities $P(0) = 0.6$, $P(1) = 0.4$ and is coded with a SFE binary code. Construct by hand a table associating the eight possible input sequences of length 3 with their SFE codewords.

4.3 We shall use the SKIERS image (available in path datasets/images/skiers.pgm in this book's website[4]) to generate a random sequence of symbols $\{0, 1, 2, 3, 4, 5, 6, 7\}$.

Take each pixel of integer value x ($0 \geq x \leq 255$) and map it to a symbol i in the above symbol set according to the rule:

$$i = \left\lfloor \frac{x}{32} \right\rfloor.$$

($\lfloor z \rfloor \equiv$ integer part of z.)

(a) Estimate the probabilities of the symbols 0 to 7 generated from the SKIERS image. Use these probabilities in a fixed model to arithmetic code the sequence of symbols generated in this way.

(b) Now use an adaptive arithmetic code to encode the same symbol sequence.

(c) Compare the compression ratios or bits per symbol of the compressed files of Parts (a) and (b) above.

- You may use the arithmetic code software written by A. Said in the course website under the folder software http://www.cambridge.org/pearlman.

- Source code for the arithmetic code may also be found in http://www.cipr.rpi.edu/~pearlman/arithcode and is exactly that chronicled in the article by Witten *et al.* [13]. This article explains the use of the code and the building of the probability models.

- Another source for arithmetic coding software: http://www.cipr.rpi.edu/students/~wheeler

4.4 The Golomb code can be used to encode the integer values of an image. (One can pretend that the values are run-lengths.) Let the source be again the sequence of integers generated in Problem 4.9.

(a) Encode this source with a Golomb code.

(b) Determine the compression ratio of your code.

4.5 A source has six letters with the probabilities of 0.20, 0.20, 0.10, 0.08, 0.08, 0.34.

(a) By hand, construct a Huffman code for this source.

(b) Determine the average codeword length of this code.

(c) What is the reduction factor from a fixed length code for this source? Comment on whether you think that using a Huffman code is worth the effort.

4.6 Repeat Problem 4.9, substituting construction of an SFE code for a Huffman code.

4.7 Suppose that an i.i.d. binary source has letter probabilities $p(0) = 1/3$ and $p(1) = 2/3$.

(a) Construct a Huffman code for the triples (three consecutive letters) produced by this source.

(b) Find the average length of this Huffman code and compare it to the entropy of the source.

(c) Describe the SFE code for the triplet source and calculate its average length.

4.8 (Prob. 9.14 [14]) In this problem, we consider noiseless source coding for a source with an infinite alphabet. Suppose that $\{X_n\}$ is an i.i.d. source with probability mass function

$$p_X(k) = C2^{-k}, k = 1, 2, \ldots$$

(a) Find C and the $H(X_0)$, the entropy of the random variable X_0.

(b) Assume that there is no constraint on the length of binary noiseless code-words. Construct a noiseless source code using something like a Huffman code. It is not an obvious Huffman code, because you can not start with the merging of the two least probable source symbols. It should, however, be a prefix code and have the smallest possible average length. What is the average codeword length? Is the code optimal?

(c) Suppose now that you have a maximum codeword length of 100 binary symbols. Hence, it is clearly no longer possible to encode the sequence noise-lessly. Suppose that you simply truncate the code of the previous part by encoding input symbols 1–100 using that code, and all other symbols get mapped to the same binary codeword as 100. What is the resulting probability of error and what is the resulting average codeword length in bits?

4.9 Run-length coding is a popular method of encoding binary data when the proba-bilities of the binary symbols are fairly disparate. Assume a memoryless binary source with probabilities $p(0) = 0.9, \; p(1) = 0.1$. The encoding of this source is done in two stages, first counting the number of zeros between successive ones in the source output and then encoding these run-lengths into binary codewords by a Huffman code. The first stage of encoding maps the source segments into a sequence of decimal run-lengths of zeros by the following table.

Source segments	Run lengths
1	0
01	1
001	2
0001	3
00001	4
000001	5
000000	6

Thus, the sequence 100100000000110001 is mapped as

```
100   100000   00   1   1000   1
 0      2       6   1    0     3
```

We will assume that all source sequences end in 1. Note that all possible source segments may be composed from those occurring in the table.

(a) Calculate the average number of source digits per run length digit.

(b) Determine the Huffman code for the run lengths.

(c) Calculate the average length of the Huffman code for the run lengths.

(d) What is the average compression ratio, that is, the ratio of the average number of source digits to average number of (binary) codeword digits?

4.10 Repeat all parts of Problem 4.9 for a Golomb code for the run-lengths.

4.11 Consider the memoryless four-letter source $\{0,1,2,3\}$ with corresponding prob-abilities $\{0.20, 0.36, 0.18, 0.26\}$.

(a) Give the minimum number of bits per source letter for encoding this source without error.

(b) Construct an optimal code for encoding letters from this source one at a time. Give the average number of bits per source letter for this code.

(c) Construct an optimal code for encoding letters from this source two at a time.

Give the number of bits per source letter for this code. Is it an improvement over your code in Part (b).

Comment on whether you think entropy coding is worth doing for this source.

Notes

1. When it is not clear from the context, trailing subscripts indicate the base of the number system.
2. Any addition of bits beyond the k-th position adds less than 2^{-k}.
3. In run-length coding, the segment length (m) is hardly ever large enough that the entropy for this m-letter geometric distribution is significantly smaller than $\lceil \log_2 m \rceil$.
4. http://www.cambridge.org/pearlman.

References

1. D. A. Huffman, "A method for the construction of minimum redundancy codes," *Proc. IRE*, vol. 40, pp. 1098–1101, Sept. 1952.
2. F. Jelinek, *Probabilistic Information Theory*. New York: McGraw-Hill, 1968.
3. A. Said, "Arithmetic coding," in *Lossless Compression Handbook*, ed. K. Sayood. San Diego, CA: Academic Press, 2003, ch. 5.
4. A. Said, "On the reduction of entropy coding complexity via symbol grouping: I redundancy analysis and optimal alphabet partition," HP Laboratories, Palo Alto, CA, Report HPL-2004-145, Aug. 2004.
5. S. W. Golomb, "Run-length encodings," *IEEE Trans. Inf. Theory*, vol. IT-12, no. 3, pp. 399–401, July 1966.
6. R. G. Gallager and D. van Voorhis, "Optimal source codes for geometrically distributed integer alphabets," *IEEE Trans. Inf. Theory*, vol. 21, no. 2, pp. 228–230, Mar. 1975.
7. J. Ziv and A. Lempel, "A universal algorithm for data compression," *IEEE Trans. Inf. Theory*, vol. IT-23, no. 3, pp. 337–343, May 1977.
8. J. Ziv and A. Lempel, "Compression of individual sequences via variable-rate coding," *IEEE Trans. Inf. Theory*, vol. IT-24, no. 5, pp. 530–536, Sept. 1978.
9. T. A. Welch, "A technique for high-performance data compression," *IEEE Computer*, vol. 17, no. 6, pp. 8–19, June 1984.
10. T. C. Bell, "Better OPM/L text compression," *IEEE Trans. Commun.*, vol. COM-34, no. 12, pp. 1176–1182, Dec. 1986.
11. K. Sayood, *Introduction to Data Compression*, 3rd edn. San Francisco, CA: Morgan Kaufmann Publishers (Elsevier), 2006.
12. D. Salomon, *Data Compression: The Complete Reference*, 3rd edn. New York: Springer, 2004.
13. H. Witten, R. M. Neel, and J. G. Cleary, "Arithmetic coding for data compression," Commun. AcM, vol. 30, no. 6, June 1987. pp. 520–540.
14. A. Gersho and R. M. Gray, *Vector Quantization and Signal Compression*, Boston, Dordrecht, London: Kluwer Academic Publishers, 1992.

Further reading

T. M. Cover and J. A. Thomas, *Elements of Information Theory*. New York, NY: John Wiley & Sons, 1991, 2006.

A. Said and W. Pearlman, "Low-complexity waveform coding via alphabet and sample-set partitioning," in *Proceedings of the SPIE: Visual Communications and Image Processing*, vol. 3024, San Jose, CA, Feb. 1997, pp. 25–37.

Y. Q. Shi and H. Sun, *Image and Video Compression for Multimedia Engineering*. Boca Raton: CRC Press, Taylor & Francis Group, 2008.

5 Lossy compression of scalar sources

5.1 Introduction

In normal circumstances, lossless compression reduces file sizes in the range of a factor of 2, sometimes a little more and sometimes a little less. Often it is acceptable and even necessary to tolerate some loss or distortion between the original and its reproduction. In such cases, much greater compression becomes possible. For example, the highest quality JPEG-compressed images and MP3 audio are compressed about 6 or 7 to 1. The objective is to minimize the distortion, as measured by some criterion, for a given rate in bits per sample or equivalently, minimize the rate for a given level of distortion.

In this chapter, we make a modest start toward the understanding of how to compress realistic sources by presenting the theory and practice of quantization and coding of sources of independent and identically distributed random variables. Later in the chapter, we shall explain some aspects of optimal lossy compression, so that we can assess how well our methods perform compared to what is theoretically possible.

5.2 Quantization

The sources of data that we recognize as *digital* are discrete in value or amplitude and these values are represented by a finite number of bits. The set of these discrete values is a reduction from a much larger set of possible values, because of the limitations of our computers and systems in precision, storage, and transmission speed. We therefore accept the general model of our data source as continuous in value. The discretization process is called *quantization*. The number of continuous values in a finite amplitude interval, no matter how small, is uncountably infinite. Quantization is therefore absolutely necessary to reduce the infinite set of possible values to a finite one and therefore the representation of a value by an infinite number of bits to a finite number. In this chapter, we shall study the most efficient ways to effect this outcome.

5.2.1 Scalar quantization

Consider a stationary, memoryless, continuous-amplitude source X with probability density function $q(x)$.[1] We call X a source letter and wish to approximate it by one of the so-called reproduction letters (values) $\{y_1, y_2, \ldots, y_K\}$. These reproduction letters are often called *quantizer values* and the method of associating X with a quantizer value is called *quantization*. One should bear in mind that the actual picture is a sequence of source letters of the same probability distribution being emitted independently from the source and each letter being quantized, i.e., associated with a quantizer value, individually. Therefore, we call this process *scalar quantization*.

In order to measure the loss of accuracy due to the quantization, we need a measure of the distortion created by replacing the true source value $X = x$ with its reproduction y_k, for any $k = 1, 2, \ldots, K$. We denote this distortion measure by $d(x, y_k)$. We need not specify this distortion measure now, but must insist that it be a non-negative, non-decreasing function of $|x - y_k|$, the absolute error or Euclidean distance between x and y_k. Given $X = x$ from the source and armed with this measure of distortion, we choose the quantizer value that minimizes the distortion. Stating this formally, choose y_k such that

$$d(x, y_k) \leq d(x, y_\ell), \quad \text{for all } \ell \neq k.$$

The index k of the best match conveys y_k, because the set of quantizer values is known and agreed upon. Therefore, at most, $\log_2 K$ bits in the binary expansion of k are needed to convey y_k. This procedure of quantization and binary encoding of k is called pulse code modulation or PCM and is illustrated in Figure 5.1.

Assuming no corruption in transmission, the receiver reads the index k and just looks up the reconstruction value y_k in a table. The total distortion incurred is exactly $d(x, y_k)$, as measured in the quantization of x to y_k.

Because we have defined a distortion measure and a set of reproduction symbols beforehand, we really do not need to compute the distortions $d(x, y_k)$ after $X = x$ is emitted from the source. One can specify beforehand intervals

$$I_k = \{x : d(x, y_k) \leq d(x, y_\ell), \quad \text{all } \ell \neq k\} \quad k = 1, 2, \ldots, K, \quad (5.1)$$

which is the set of x's which are closest in distortion measure to y_k for every $k = 1, 2, \ldots, K$. For a distortion measure which is a monotone, non-decreasing function f of the distance between x and y_k, i.e.,

$$d(x, y_k) = f(|x - y_k|),$$

the I_k's are non-overlapping intervals. The endpoints of these intervals we denote as x_{k-1} and x_k:

$$I_k = [x_{k-1}, x_k) = \{x : x_{k-1} \leq x < x_k\}$$

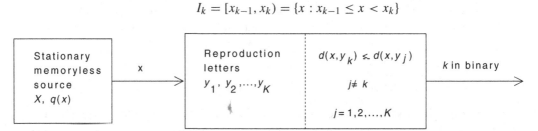

Figure 5.1 Model of scalar quantization.

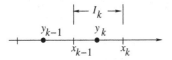

Figure 5.2   Threshold and reproduction points and intervals.

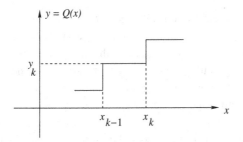

Figure 5.3   Graphical form of quantizer function.

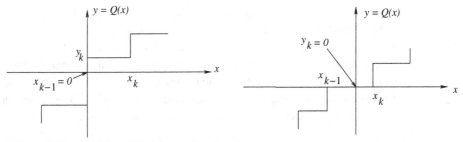

Figure 5.4   Midrise (left) and midtread (right) quantizer functions.

The definitions of these points and intervals are depicted in Figure 5.2.

Upon emitting $X = x$, we accept as our reproduction of x the value y_k if $x \in I_k$.

If x is in $I_k = [x_{k-1}, x_k)$, then x is *mapped* or quantized to y_k. Using functional notation

$$y_k = Q(x) \quad \text{if} \quad x \in I_k \quad (\text{or } x_{k-1} \le x < x_k). \tag{5.2}$$

As already mentioned, $\{y_1, y_2, \ldots, y_K\}$ are called *quantizer levels*. The interval endpoints $\{x_0, x_1, x_2, \ldots, x_{K-1}, x_K\}$ are called *decision thresholds*. A graph of the function of quantization is a staircase function with the level of the k-th step y_k with ends at x_{k-1} and x_k, as shown in Figure 5.3. If there is a k such that $y_k = 0$, the quantizer is called *midtread*. If there is no k such that $y_k = 0$ and a k such that $x_k = 0$, the quantizer is called *midrise*. Figure 5.4 clearly illustrates the origin of these names.

The quantization error (noise) is measured by the average distortion, which is the expected value of $d(x, Q(x))$, given by

$$D = E[d(x, Q(x))] = \int_{-\infty}^{\infty} d(x, Q(x))q(x)dx,$$

where $E[]$ is the expectation operator. Since $Q(x) = y_k$ if $x_{k-1} \leq x < x_k$,

$$D = \sum_{k=1}^{K} \int_{x_{k-1}}^{x_k} d(x, y_k) q(x) dx \qquad (5.3)$$

When $x_0 = -\infty$ and $x_K = +\infty$, the portions of D in the two end intervals $(-\infty, x_1)$ and $[x_{K-1}, +\infty)$ are called *overload error* or *distortion*, while the contributions to D in the other regions are called *granular distortion*. Note that since the errors in different regions add, the overload and granular errors add. The thresholds x_1 and x_{K-1} that mark the finite boundaries of the two overload regions are chosen to be on the tails of the probability function. The granular and overload regions are depicted with respect to a typical probability density function in Figure 5.5. The actual error $d(x, y_k)$ in the overload regions, below x_1 and above x_{K-1}, is unbounded, but, when weighted by the probability density function, should contribute very little to D. The actual errors $d(x, y_k)$ in the interior regions are bounded and usually small, but when weighted by the probability density function and combined contribute almost all of the average distortion D.

The overall design objective of a quantizer with a given number of quantization levels K is to choose the decision thresholds and quantization levels to minimize the quantization noise D in Equation 5.3. Being a multi-dimensional optimization problem solvable only numerically, simpler, constrained solutions are sought that are more practical for implementation. In the following sections, we shall examine some of these simpler, more practical solutions before tackling the general optimization.

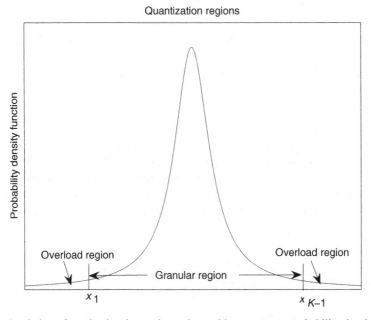

Figure 5.5   Depiction of overload and granular regions with respect to a probability density function.

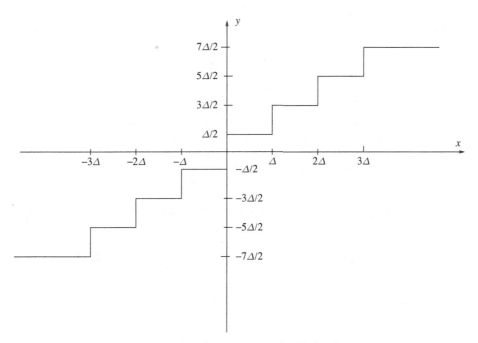

Figure 5.6 Midrise uniform quantizer characteristic for $K = 8$ quantization levels.

5.2.2 Uniform quantization

A significant reduction in the number of solution parameters may be accomplished by restricting the decision thresholds to be uniformly spaced. Such a design is called a *uniform quantizer*. The quantizer levels can either be uniformly or non-uniformly spaced. Besides the simplified solution afforded by a uniform quantizer, it can be implemented readily in hardware and very simply in software.

5.2.2.1 Uniform thresholds and levels

Let us begin with the common case of an even probability density function, $q(x) = q(-x)$, of the stationary source X. Then the finite decision thresholds among $x_0, x_1, \ldots, x_K$ are uniformly spaced by distance Δ and are placed symmetrically about the origin $x = 0$. When the number of decision intervals K is even, there is a threshold at $x = 0$ and the quantizer is midrise. Otherwise, when K is odd, the quantizer is midtread. A midrise quantizer characteristic for $K = 8$ is depicted with its thresholds and levels in Figure 5.6. For K even, the decision thresholds are therefore

$$x_{\min}, -(K/2 - 1)\Delta, -(K/2 - 2)\Delta, \ldots, -\Delta, 0, \Delta, 2\Delta, \ldots, (K/2 - 1)\Delta, x_{\max}, \tag{5.4}$$

where $x_{\min} = -\infty$ or a very small number and $x_{\max} = +\infty$ or a very large number. We assume always that $x_{\min} = -x_{\max}$ for an even probability density function.

For the example in Figure 5.6 where $K = 8$, the thresholds are

$$-\infty, -3\Delta, -2\Delta, -\Delta, 0, \Delta, 2\Delta, 3\Delta, +\infty.$$

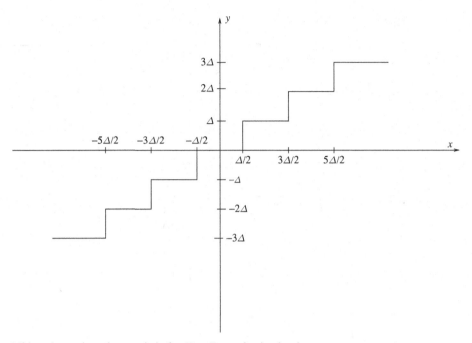

Figure 5.7   Midtread quantizer characteristic for $K = 7$ quantization levels.

The quantization levels are chosen to be the midpoints of the interior intervals, so are uniformly spaced by Δ reaching outward from the origin. Therefore, for K levels they are

$$-\frac{K-1}{2}\Delta, -\frac{K-3}{2}\Delta, \ldots, -\frac{\Delta}{2}, \frac{\Delta}{2}, \ldots, \frac{K-3}{2}\Delta, \frac{K-1}{2}\Delta. \qquad (5.5)$$

For K odd, the decision thresholds are

$$x_{\min}, -\frac{K-2}{2}\Delta, -\frac{K-4}{2}\Delta, \ldots, -\frac{\Delta}{2}, \frac{\Delta}{2}, \frac{3\Delta}{2}, \ldots, \frac{K-2}{2}\Delta, x_{\max}. \qquad (5.6)$$

For the example in Figure 5.7 where $K = 7$, the thresholds are

$$-\infty, -5\Delta/2, -3\Delta/2, -\Delta/2, \Delta/2, 3\Delta/2, 5\Delta/2, +\infty.$$

The quantization levels are again chosen to be uniformly spaced by Δ as the midpoints of the interior intervals. Therefore, for odd K, they are

$$-\frac{K-1}{2}\Delta, -\frac{K-3}{2}\Delta, \ldots, -2\Delta, -\Delta, 0, \Delta, \ldots, \frac{K-3}{2}\Delta, \frac{K-1}{2}\Delta. \qquad (5.7)$$

For $K = 7$, the quantizer levels are $-3\Delta, -2\Delta, -\Delta, 0, \Delta, 2\Delta, 3\Delta$, as shown in Figure 5.7.

5.2.2.2   Implementation of uniform quantization

As mentioned previously, realization of a uniform quantizer is particularly easy in software or hardware. The index of the quantization decison interval or bin is particularly easy to generate in uniform quantization. Given the input x, the following mathematical operation produces an index q of the quantization bin (decision interval) of a uniform midrise quantizer:

$$q = \lfloor x/\Delta \rfloor \tag{5.8}$$

The index q is encoded for transmission or storage. The reconstruction (often called *dequantization*) y at the midpoint of the bin is simply

$$y = (q + 1/2)\Delta. \tag{5.9}$$

Uniform midtread quantization is no more complicated. Given x, the index q is given by

$$q = \left\lfloor \frac{x}{\Delta} + 1/2 \right\rfloor ; \tag{5.10}$$

and the midpoint reconstruction value y, given q, is simply

$$y = q\Delta. \tag{5.11}$$

Midtread quantization distinguishes itself from midrise quantization, because it has a reconstruction value of 0. In fact,

$$q = 0, \text{ if } -\Delta/2 \le x < +\Delta/2.$$

Therefore, it acts as a threshold quantizer in that all inputs whose magnitudes are less than the threshold of $\Delta/2$ are set to 0.

5.2.2.3   Optimum step size

The distance between successive decision thresholds Δ is usually called the *step size* of a uniform quantizer. The step size is the only unknown parameter for a uniform quantizer with a fixed number of levels of a source with a given *even* probability density function. If Δ is too small, source values in the overload region will likely cause large errors too frequently. On the other hand, if Δ is too large, the more probable occurrence of values in the granular region will be too coarsely quantized and cause too much distortion. The graph of mean squared error distortion D versus Δ in Figure 5.8 for the Gaussian (normal) probability density (see Equation (5.48)) illustrates this behavior. We seek to solve for the intermediate value of Δ that minimizes the average distortion D for an even probability density function $q(x)$.

First, we need to specify a distortion measure $d(x, y_k)$. The choice is always controversial, because it always depends on how one uses the data. The most common measure by far is squared error,

$$d(x, y_k) = (x - y_k)^2. \tag{5.12}$$

Mean squared error (MSE) is the expected value of this distortion measure. Substituting the squared error in (5.12) into the average distortion expression in (5.3), the MSE is expressed mathematically as

$$D = \sum_{k=1}^{K} \int_{x_{k-1}}^{x_k} (x - y_k)^2 q(x) dx. \tag{5.13}$$

Squared error is the most analytically tractable distortion measure and leads to closed-form solutions in some cases. Closed-form and analytically-based solutions, even for simplistic error measures and source models, bring us valuable insight into solutions for more complicated and realistic measures and models. Furthermore, different methods of quantization and signal processing are usually rated according to their MSE performance. Therefore, we adopt squared error as our distortion measure.

We now derive the average distortion, which is MSE, for uniform quantizers. To do so, we substitute the thresholds and quantizer levels into the average distortion formula in Equation (5.13). For K odd, substitution of the decision thresholds $x_k = -\frac{K-2k}{2}\Delta$, quantization levels $y_k = -\frac{K-(2k-1)}{2}\Delta$, and $d(x, y_k) = (x - y_k)^2$ for $k = 1, 2, \ldots, K$ yields the expression

$$D = \int_{x_{\min}}^{-\frac{K-2}{2}\Delta} \left(x + \frac{K-1}{2}\Delta \right)^2 q(x) dx \tag{5.14}$$

$$+ \sum_{k=2}^{K-1} \int_{-\frac{K-2(k-1)}{2}\Delta}^{-\frac{K-2k}{2}\Delta} \left(x + \frac{K-(2k-1)}{2}\Delta \right)^2 q(x) dx$$

$$+ \int_{\frac{K-2}{2}\Delta}^{x_{\max}} \left(x - \frac{K-1}{2}\Delta \right)^2 q(x) dx$$

This expression simplifies somewhat when we account for the even symmetry of the probability density function $q(x)$ and the distortion measure. Using these properties, the average distortion becomes

$$D = 2 \int_{x_{\min}}^{-\frac{K-2}{2}\Delta} \left(x + \frac{K-1}{2}\Delta \right)^2 q(x) dx \tag{5.15}$$

$$+ 2 \sum_{k=2}^{(K-1)/2} \int_{-\frac{K-2(k-1)}{2}\Delta}^{-\frac{K-2k}{2}\Delta} \left(x + \frac{K-(2k-1)}{2}\Delta \right)^2 q(x) dx + 2 \int_{-\Delta/2}^{0} x^2 q(x) dx$$

A similar expression for D can be derived in the same way for K even.

Figure 5.8 was drawn by substituting the normal (Gaussian) probability density function (mean 0, variance 1) into Equation (5.15) using $K = 9$ quantizer levels and calculating the resulting distortions for a set of step sizes Δ. The minimum distortion D of 0.0308 is obtained at $\Delta = 0.52$.

A necessary condition for the minimum D is that the derivative with respect to Δ be zero. Clearly, this solution is intractable analytically and must be approached numerically. Usually, a Newton–Raphson algorithm that iteratively seeks the root of $\frac{dD}{d\Delta}$ is employed to determine the minimizing Δ for a given value of K. Such an algorithm was used to generate Table 5.1 in the Appendix that lists the optimum step sizes and consequent MSE and entropy for $K = 1$ to $K = 64$ output levels.

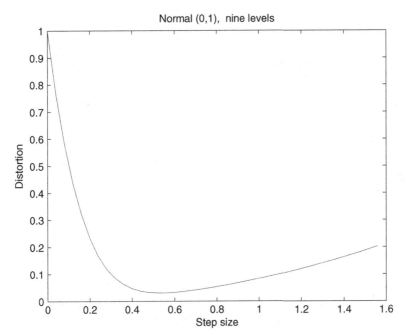

Normal (0,1), nine levels

Figure 5.8 MSE versus step size for uniform quantization of a normal (0,1) source with $K = 9$ quantization levels.

5.2.2.4 Uniform threshold quantizers

When the decision thresholds are fixed, the midpoints between successive thresholds are generally not the reconstruction points that minimize the distortion. Once thresholds are fixed and known, the rate to represent the index of the decision interval is also fixed. So, for a known source probability distribution, the distortion can generally be lowered by better choices of the reconstruction points. We shall now derive these optimum reconstruction points. The decoder has prior knowledge of the set of decision thresholds and receives an index that locates the decision interval into which the source value falls. Let the k-th decision interval be denoted by $\mathcal{I}_k = [x_{k-1}, x_k)$. Given the probability density function $q(x)$ of the source and the distortion measure $d(x, y_k)$ between the source value x and its reconstruction y_k in the corresponding decision interval, the average distortion caused by the quantization of source values in this known k-th interval can be expressed as

$$D_k = \frac{\int_{x_{k-1}}^{x_k} d(x, y_k) q(x) dx}{\int_{x_{k-1}}^{x_k} q(x) dx}$$

The point y_k that minimizes D_k is the solution to $\frac{dD_k}{dy_k} = 0$. Applying this solution to the squared error distortion measure, $d(x, y_k) = (x - y_k)^2$, we obtain

$$0 = \frac{\int_{x_{k-1}}^{x_k} (x - y_k) q(x) dx}{\int_{x_{k-1}}^{x_k} q(x) dx}. \tag{5.16}$$

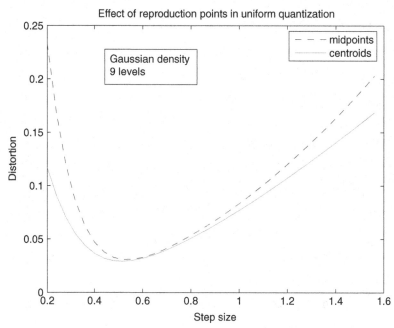

Figure 5.9   Distortion of midpoint versus centroid reproduction points for quantization of a normal (0,1) source with $K = 9$ quantization levels.

Solving for y_k,

$$y_k = \frac{\int_{x_{k-1}}^{x_k} xq(x)dx}{\int_{x_{k-1}}^{x_k} q(x)dx},$$ (5.17)

the centroid of $q(x)$ in the interval $[x_{k-1}, x_k)$. Therefore, the set of centroids in the intervals minimizes the MSE. A quantizer with uniform decision intervals and with centroids of these intervals as reconstruction points is called a *uniform threshold quantizer*.

We repeat the same experiment as before to illustrate the effect of using the centroids of the intervals as the quantization points. Again, we quantize the Gaussian source with $K = 9$ levels, varying the step size and calculating the resulting MSE. In Figure 5.9 we show the results from both uniform and centroid reconstructions. As expected, the distortion for centroid reconstruction is always smaller than that for midpoint reconstruction. The distortions of the two curves are farther apart for low values of step size, because the overload (outer) region predominates and the centroid and midpoint reproduction points are relatively far apart. As the intervals get larger, the granular (inner) region starts to dominate the distortion, where the centroids and midpoint distances are fairly close together, no larger than the step size. In the important range of step sizes that give the smallest distortions, centroid reconstruction yields only slightly smaller distortion. In the figure, the minimum distortion of the centroid reconstruction is only 0.0016 below the 0.03082 minimum of the midpoint reconstruction. The difference in minimum distortion will grow even smaller with a larger number of levels K. As the step size grows even larger, the two curves diverge because the centroids and midpoints within the decision intervals in the increasingly dominant granular region grow farther apart.

5.3 Non-uniform quantization

The general framework for scalar quantization was introduced in Section 5.2.1, before specializing to uniform quantization. Now we explain the optimum choices of the threshold and reproduction values for non-uniform quantization. For a given number of quantizer levels K, we must solve jointly for the decision intervals $I_k = [x_{k-1}, x_k)$ and quantizer levels y_k, $k = 1, 2, \ldots, K$, that minimize the expected distortion D in Equation (5.3). Necessary conditions are

$$\frac{\partial D}{\partial x_k} = 0 \text{ and} \tag{5.18}$$

$$\frac{\partial D}{\partial y_k} = 0, \quad k = 1, 2, \ldots, K$$

That is, we are seeking the stationary points of

$$D = \sum_{k=1}^{K} \int_{x_{k-1}}^{x_k} d(x, y_k) q(x) dx$$

To seek a stationary point, we can first fix the $\{y_k\}_{k=1}^{K}$ set and find the best set of decision thresholds. When $d(x, y_k) = f(|x - y_k|)$ with f a monotone and non-decreasing function of its argument, the search for $I_k = [x_{k-1}, x_k)$ such that $d(x, y_k) \leq d(x, y_\ell)$ for all $\ell \neq k$ must place x_k halfway between y_k and y_{k+1}, i.e.,

$$x_k = \frac{y_{k+1} + y_k}{2} \quad k = 1, 2, \ldots, K - 1 \tag{5.19}$$

Then I_k is the set of points which are closest to y_k in distance. You may check this result by $\frac{\partial D}{\partial x_k} = 0$ (holding everything else constant). To find the best $\{y_k\}$ a necessary condition is that $\frac{\partial D}{\partial y_k} = 0$ $k = 1, 2, \ldots, K$

$$\frac{\partial D}{\partial y_k} = 0 = \int_{x_{k-1}}^{x_k} \frac{\partial}{\partial y_k} (d(x, y_k)) q(x) dx \tag{5.20}$$

We have already seen the solution in (5.17) when the distortion measure is squared-error: $d(x, y_k) = (x - y_k)^2$. Substituting into (5.20), we obtained

$$\int_{x_{k-1}}^{x_k} \frac{\partial}{\partial y_k} (x - y_k)^2 q(x) dx = 0$$

$$-2 \int_{x_{k-1}}^{x_k} (x - y_k) q(x) dx = 0$$

$$y_k = \frac{\int_{x_{k-1}}^{x_k} x q(x) dx}{\int_{x_{k-1}}^{x_k} q(x) dx} \quad k = 1, 2, \ldots, K. \tag{5.21}$$

As mentioned previously, y_k is the centroid of the probability density in the decision interval between x_{k-1} and x_k for every k. It may also be interpreted as a conditional mean, i.e.,

$$y_k = E[X/X \in [x_{k-1}, x_k)] \qquad k = 1, 2, \ldots, K.$$

This result is not unexpected when one looks at the possible actions of the decoder. Assuming uncorrupted reception of the codeword, the decoder knows only the range interval I_k, so sets the quantization point to the centroid of the region, which is the conditional mean above, in order to minimize the MSE. Are these conditions sufficient for a minimum? Not always. They are if the Hessian matrix

$$\mathcal{H} = \left\{ \frac{\partial^2 D}{\partial y_k \partial y_\ell} \right\}_{k,\ell=1}^{K}$$

is positive definite. A simpler condition by Fleischer [1] is that if $\log q(x)$ is concave $\left(\frac{\partial^2 (\log q(x))}{\partial x^2} < 0 \right)$ and distortion is squared error, then the minimum above is a global minimum. The distortion function for this condition was extended to a convex and increasing weighting function of absolute error by Trushkin [2].

The following algorithms compute $\{y_k\}$ and $\{x_k\}$ through iteration of the conditions in Equations (5.20) or (5.21) and (5.19).

I. Lloyd (Method I) algorithm [3]:
  1. Choose set $\{x_1, \ldots, x_{K-1}\}$ arbitrarily $(x_o = x_{\min}, x_K = x_{\max})$.
  2. Compute centroids y_k by Equation (5.20) or (5.21).
  3. Compute new set of thresholds by Equation (5.19).
  4. Set stopping criterion on decrease of D.
    If not met, go back to 2.
    If met, stop.
II. Lloyd–Max algorithm [3, 4]:
  1. Choose y_1, arbitrarily. $(x_o = x_{\min})$. Set $k = 1$.
  2. Calculate x_k by $y_k =$ centroid of $q(x)$ in $[x_{k-1}, x_k)$.
  3. Calculate y_{k+1} by $x_k = (y_{k+1} + y_k)/2$.
  4. If $k < K$, set $k = k + 1$ and go back to 2.
    If $k = K$, compute $y_c =$ centroid of $q(x)$ in $[x_{K-1}, x_K)$.
    If $|y_K - y_c| > \epsilon$ (error tolerance), return to 1 for a new choice of y_1. Otherwise stop.

Non-uniform quantization should achieve lower distortion than uniform quantization for the same number of levels K. We have compared the distortion minima of non-uniform and uniform quantization versus K for the Gaussian probability density with variance $\sigma^2 = 1$ and the squared-error criterion. Figure 5.10 plots the minimum distortions D in dB from $K = 1$ to 128. Tables 5.1 and 5.2 in the Appendix present the exact distortions in this plot from $K = 1$ to 64.

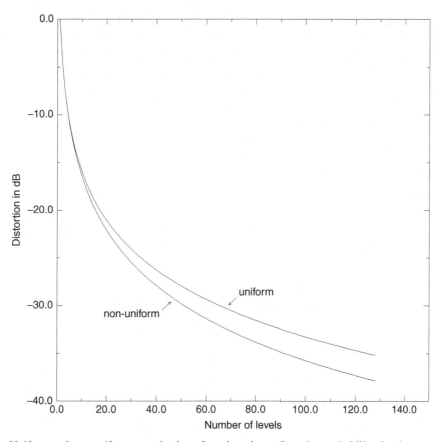

Figure 5.10   Uniform and non-uniform quantization of a unit variance Gaussian probability density.

As expected from the additional restriction of uniform spacing of the decision thresholds, the uniform quantizer has more distortion for every K than that of the non-uniform quantizer. The trend of these results is true for any probability density and any reasonable distortion measure.

5.3.1   High rate approximations

Numerical solutions, while valuable, do not yield intuition into general characteristics of quantization performance. Therefore, we develop some approximate analytical solutions. Consider a quantizer, uniform or non-uniform, with a large number of quantization levels K. We set the interior boundaries of the overload regions, so that the average distortion contributed by these regions is so small that it can be neglected. For example, when the probability density function is symmetric, the distortion in the overload regions can safely be neglected when the overload regions are $x < -4\sigma$ and $x > +4\sigma$, where σ^2 is the variance. Then the granular region is divided into $K - 2$ decision intervals $I_k = [x_{k-1}, x_k)$ of width $\Delta_k = x_k - x_{k-1}$. The decision intervals are assumed to be small enough for large K that the probability density function within each interval is nearly uniform. Therefore the values of the probability density function within the interval are approximately the same as the value at the reconstruction point y_k, i.e., $q(x) \approx q(y_k)$. The MSE D is then calculated according to

$$
\begin{aligned}
D &= \sum_{k=1}^{K} \int_{x_{k-1}}^{x_k} (x - y_k)^2 q(x)dx \\
&\approx \sum_{k=2}^{K-1} q(y_k) \int_{x_{k-1}}^{x_k} (x - y_k)^2 dx \\
&= \frac{1}{3} \sum_{k=2}^{K-1} [(x_k - y_k)^3 - (x_{k-1} - y_k)^3]
\end{aligned}
\tag{5.22}
$$

The optimal reproduction point is the centroid, which is approximately the midpoint of the interval, i.e.,

$$
y_k \approx (x_{k-1} + x_k)/2
$$

Substituting the approximate y_k into (5.22) and simplifying, we obtain

$$
D \approx \frac{1}{12} \sum_{k=2}^{K-1} q(y_k)(x_k - x_{k-1})^3
\tag{5.23}
$$

$$
= \frac{1}{12} \sum_{k=2}^{K-1} q(y_k)\Delta_k^3
\tag{5.24}
$$

$$
\approx \frac{1}{12} \sum_{k=2}^{K-1} P(y_k)\Delta_k^2
\tag{5.25}
$$

where $P(y_k)$ is the probability of y_k. The substitution of $P(y_k)$ in the last step relies on K being very large, so that Δ_k is always very small. Therefore, D is approximately $1/12$ of the average of the Δ_k^2 for I_k in the granular region.

Most often, the above approximation for D is applied to uniform quantization of a symmetric probability distribution. In this case, $\Delta_k = \Delta$, a constant for the applicable k's. Therefore, for this case,

$$
D = \frac{\Delta^2}{12} \sum_{k=2}^{K-1} P(y_k) \approx \frac{\Delta^2}{12}
\tag{5.26}
$$

This MSE equals the variance of a uniform distribution with width of support Δ. It is no accident, because every decision interval is of width Δ and the MSE in that interval is exactly the second moment or variance. In fact, if we wanted to derive just this last formula, we could have started by stating that the error within every decision interval in the granular region is approximately uniformly distributed from $-\Delta/2$ to $+\Delta/2$ and calculated its mean squared value of $\Delta^2/12$. However, we shall make use later of the approximate distortion formulas for the case of non-uniform quantization in (5.23) to (5.25). It is useful to remember that the variance of a uniform distribution of support Δ is $\Delta^2/12$.

Example 5.1 Consider uniform quantization of a source with a symmetric probability distribution of variance σ^2. We set the boundaries of the overloaod regions to $x_1 = -4\sigma$ and $x_{K-1} = 4\sigma$, so we have $K - 2$ decision intervals in the granular region of width

$$\Delta = 8\sigma/(K - 2) \approx 8\sigma/K$$

for large K. Substituting the approximate value of Δ into (5.26), we obtain

$$D = \frac{(8\sigma/K)^2}{12} = \frac{16\,\sigma^2}{3\,K^2} \tag{5.27}$$

Notice the multiplicative factor of σ^2. MSE always scales by this factor. Notice also that MSE is inversely proportional to K^2. In fact, as we show later, declining by the square of the number of levels is the fastest possible. In order to put 5.27 in terms of rate, let us assume that K is an integer power of 2, i.e., $K = 2^n$. So $n = \log_2 K$ is the rate in bits per sample. The MSE decreases exponentially with rate as shown in the following expression

$$D = \frac{16}{3}\sigma^2 2^{-2n}. \tag{5.28}$$

The above formula is often given in terms of signal to quantization noise ratio, defined as

$$SQNR = \frac{\sigma^2}{D} \tag{5.29}$$

Often SQNR is expressed in units of decibels (dB), in which case we designate it as $SQNR_{db} \equiv 10\log_{10} SQNR$. Thus, the signal to quantization noise ratio in dB of a uniform quantizer with a large number of levels becomes

$$SQNR_{dB} \approx 6.02n - 7.27 \text{ dB} \tag{5.30}$$

This signal-to-noise ratio gains 6 dB for each additional bit. The –7.27 dB seems to be the penalty for using non-optimal coding. Scalar quantization is a non-optimal coding method.

5.4 Companding

Uniform quantization has the advantages of simplicity and efficiency, a subject that we address more specifically later. However, with non-uniform quantization, for the same number of quantization levels, one obtains lower distortion, because one can set larger decision intervals when the probability is large and smaller ones when the probability is small. A method of quantization that incorporates the positive attributes of both kinds of quantization is called *companding*. We shall explain this method.

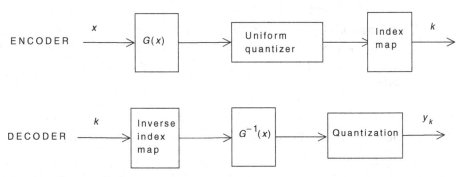

ENCODER $\xrightarrow{\quad x \quad}$ $G(x)$ $\longrightarrow$ Uniform quantizer $\longrightarrow$ Index map $\xrightarrow{\quad k \quad}$

DECODER $\xrightarrow{\quad k \quad}$ Inverse index map $\longrightarrow$ $G^{-1}(x)$ $\longrightarrow$ Quantization $\xrightarrow{\quad y_k \quad}$

Figure 5.11   A companding quantization system.

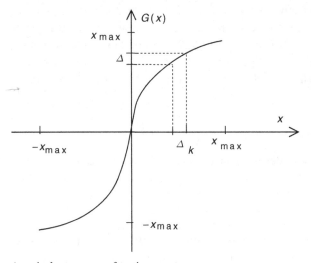

Figure 5.12   A typical compressor function.

The idea is to transform or map the input with a monotone, non-linear function and quantize the output of the mapping with a uniform step size quantizer. This non-linear mapping expands the range of the small input values and contracts the range of the large values. The final reproduction point is obtained through the inverse of this non-linear function. A block diagram of such a system is shown in Figure 5.11. The quantization of the output of the non-linear mapping produces an index of the uniform decision interval. At the decoder, the inverse of this mapping associates this index with the interval of source values corresponding to the uniform interval found at the encoder. The reconstruction point is usually the midpoint of the source interval. The non-linear mapping of the input is called the *compressor* and the inverse mapping producing the decoder output the *expandor*. These names are merged together to name the overall system a *compandor* or companding system.

A typical non-linear function $G(x)$ is depicted in Figure 5.12 for $0 \leq x \leq x_{\max}$, where we define $x_{\max} \equiv x_{K-1}$, the inner boundary of the positive overload region. This function is monotone increasing from $G(0) = 0$ to $G(x_{\max}) = x_{\max}$. It starts at a slope larger than one, so that it expands a range of small values. The decreasing slope acts to compress the range of the higher values of x. In fact, the compressor function $G(x)$ for $x \geq 0$ has all the characteristics of a scaled cumulative probability function, such as $x_{\max} F_X(x/x_{\max})$, where $F_X(x)$ is the cumulative probability distribution function of the source. For negative values of x, we insist that $G(x)$ be an odd function, i.e., $G(x) = -G(-x)$.

Companding quantizers have been used in audio and speech coding systems. Standard compressor functions are the so-called μ-law and A-law functions given below for positive values of $s = x/x_{\max}$.

μ-Law Compressor

$$G_\mu(s) = \frac{\log_e(1 + \mu s)}{\log_e(1 + \mu)}, \quad 0 \leq s \leq 1 \tag{5.31}$$

A-Law Compressor

$$G_A(s) = \begin{cases} \frac{As}{1 + \log_e A}, & 0 \leq s \leq 1/A \\ \frac{1 + \log_e(As)}{1 + \log_e A}, & 1/A \leq 1 \end{cases} \tag{5.32}$$

In many realistic scenarios, the variance of the input signal is known only within some range. For uniform quantization, one sets the step size according to an assumed signal variance. When the actual signal variance varies from the assumed one, then the chosen step size is either too large or too small for good performance. On the other hand, these companding quantizers display a much broader range of input variances giving acceptable quality than the uniform quantizer with the same number of levels. For example, the μ-law quantizer with $\mu = 255$ using 128 levels achieves SQNR's exceeding 25 dB for a 40 dB range of input variances compared to 10 dB for the uniform quantizer with 128 levels. (See Problem 5.7.)

5.4.1 Distortion at high rates

We shall now develop an approximation to MSE distortion for general companding quantizers applicable to high rates. First, we consider a quantization interval of width Δ that corresponds to some interval of width Δ_k belonging to interval $I_k = [x_{k-1}, x_k)$ of the input to the compressor. For small Δ, the following relationships hold:

$$\Delta = G(x_{k+1}) - G(x_k) \tag{5.33}$$
$$\approx G'(y_k)(x_k - x_{k-1})$$
$$= G'(y_k)\Delta_k,$$

$$G'(y_k) \approx \frac{\Delta}{\Delta_k} \tag{5.34}$$

where $G'(y_k)$ denotes the derivative of $G(x)$ evaluated at $x = y_k$. $G'(0)$ is often called the *companding gain*. Assuming a symmetric probability density,

$$\Delta = 2x_{\max}/K,$$

so that

$$\Delta_k \approx \frac{2x_{max}}{KG'(y_k)}. \tag{5.35}$$

Substituting this approximation to Δ_k into the approximate distortion formula in Equation (5.24) and simplifying, we obtain

$$D \approx \frac{x_{max}^2}{3K^2} \sum_{k=2}^{K-2} \frac{q(y_k)\Delta_k}{(G'(y_k))^2}. \tag{5.36}$$

We recognize in the limit of large K that the sum above tends to an integral.

$$D \approx \frac{x_{max}^2}{3K^2} \int_{-x_{max}}^{x_{max}} \frac{q(y)}{(G'(y))^2} dy. \tag{5.37}$$

This formula for D, known as *Bennett's formula* [5], holds for an arbitrary symmetric $q(x)$ and any suitable compressor function $G(x)$ when the number of levels K is large.

It is of interest to determine the optimum compressor function to see what is the minimum possible distortion. In order to do so, one must set a constraint on $G'(y)$ that appears in the integral in Equation (5.37). Without going into detail, the constraint must be that the integral of $G'(y)$ in the granular region must be a constant, because it is proportional to a point density function having $K - 2$ total points. Through calculus of variations, the solution for the minimizing $G'(y)$ under this constraint is

$$G'_{opt}(y) = C_1(q(y))^{1/3},$$
$$C_1 = x_{max} \left[\int_0^{x_{max}} (q(y))^{1/3} dy \right]^{-1}.$$

Now integrating $G'_{opt}(y)$ between 0 and x and evaluating the constant through $G(x_{max}) = x_{max}$, the optimum compressor function is

$$G_{opt}(x) = x_{max} \left[\int_0^{x_{max}} (q(y))^{1/3} dy \right]^{-1} \int_0^x (q(y))^{1/3} dy \tag{5.38}$$

Combining Equations (5.38) and (5.37) and using the even symmetry, we obtain the minimum MSE distortion

$$\min D \approx \frac{2}{3K^2} \left[\int_0^{x_{max}} (q(y))^{1/3} dy \right]^3. \tag{5.39}$$

Example 5.2 A common model for realistic signals is the Laplacian probability distribution. The Laplacian probability density with mean zero and variance σ^2 is expressed as

$$q(x) = \frac{1}{\sqrt{2}\sigma} \exp(-\sqrt{2}|x|/\sigma), \quad -\infty < x < \infty.$$

The "heavier tails" compared to the Gaussian density mean that large magnitude signal values are more likely to occur. Let us calculate the minimum distortion of the optimum compandor in Equation (5.39) using this probability distribution. With $x_{max} = 4\sigma$, the result of the calculation is

$$\min D \approx 2.747\sigma^2/K^2.$$

The signal-to-noise quantization ratio is therefore

$$SQNR = 0.3641K^2$$

Assuming that $K = 2^n$, n being the number of bits per sample, and expressing SQNR in units of decibels, we find that

$$SQNR_{dB} = 6.02n - 4.39.$$

Therefore, in the high rate regime, we have gained almost 3 dB (7.27–4.39) in SQNR by use of optimal companding.

5.5   Entropy coding of quantizer outputs

The code rate for the previous quantization schemes is $R = \log_2 K$ bits per sample. Let us now consider the block of samples $(x_1, x_2, \ldots, x_N)$ emitted from the memoryless source X with probability density function $q(X)$ to be the input to the quantizer. The quantizer independently quantizes each source sample in turn and outputs the corresponding sequence of decision interval indices $(q_1, q_2, \ldots, q_N)$. Each q_n, $n = 1, 2, \ldots, N$ corresponds uniquely to one of the quantizer levels $\{y_1, y_2, \ldots, y_K\}$. The system schematic is shown in Figure 5.13.

The output of the quantizer can be considered as a memoryless, discrete-alphabet source with letter probabilities

$$P(y_1), P(y_2), \ldots, P(y_K)$$

$$P(y_k) = P(q_k) = \Pr\{x_{k-1} \leq X < x_k\} = \int_{x_{k-1}}^{x_k} q(x)dx \quad k = 1, 2, \ldots, K$$

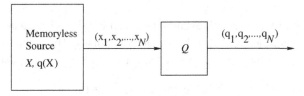

Figure 5.13   Quantization of a sequence from the source.

This new source can therefore be encoded without error at a rate arbitrarily close to its entropy of

$$H = \sum_{k=1}^{K} P(y_k) \log_2 \frac{1}{P(y_k)} \quad \text{bits per sample,}$$

say, by a Huffman code on blocks of N outputs for N sufficiently large. This quantizer output entropy is often called the *index entropy*, since the probabilities of the quantizer levels equal the probabilities of the interval indices, i.e., $P(y_k) = P(q_k)$, for all k. Such further encoding of the quantized source is called *entropy coding*. The motivation for entropy coding is savings of rate, because $H \leq \log_2 K$, the previous rate without entropy coding. As no additional error or distortion is introduced, the error is that in the quantization alone. Therefore, with entropy coding, the distortion D and the entropy H are the important quantities of performance in the quantization. The number of quantizer levels K is no longer a determining factor.

So, with a view to entropy coding the quantizer outputs, we should look for a different measure of quantizer performance. Instead of constraining or fixing the number of levels, we should constrain the entropy H to be no greater than some value H_0 and adjust the quantizer levels to minimize D. Equivalently, we can fix $D \leq D_0$ for some distortion D_0 and adjust the quantizer levels to minimize H.

The probabilities of the quantizer outputs are almost never equal, so that H is almost always strictly less than $\log_2 K$. However, what is really surprising is that for a given distortion, the output entropy of the uniform quantizer is less than that of the non-uniform quantizer for the same distortion. Another way of stating this fact is that the distortion of the uniform quantizer is less than that of the non-uniform quantizer for the same entropy. In Figure 5.14 is plotted the MSE D against the output entropy H for entropy-coded and uncoded quantizer outputs for the Gaussian density. The "uncoded" curves have a rate of $\log_2 K$ whereas the "coded" curves have a rate of

$$H = -\sum_{k=1}^{K} P(y_k) \log_2 P(y_k) \text{ bits per sample.}$$

For a given rate, the coded curves give the lower distortions, whether the quantizer is uniform or non-uniform. However, as the rate grows larger, the uniform quantizer outperforms the non-uniform quantizer when the outputs are entropy-coded. This latter result seems to be true for any density and squared-error distortion measure.

Example 5.3 We wish to achieve a MSE no larger than 0.003146 in quantization of a source with a Gaussian probability distribution of unit variance. Using the tables in the Appendix, determine the smallest rate required under the following conditions.

1. The quantization is uniform and the outputs are not coded.
2. The quantization is non-uniform and the outputs are not coded.

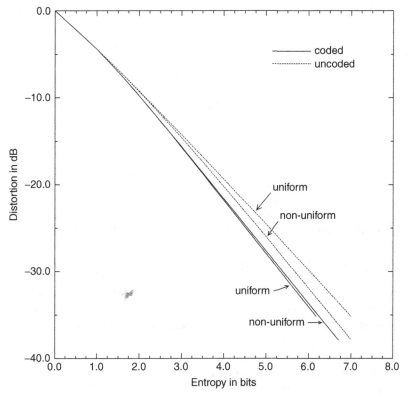

Figure 5.14 Comparison of uncoded and entropy-coded quantization for non-uniform and uniform quantizer levels. The source is the unit variance Gaussian and the distortion is squared error.

3. The quantization is uniform and the outputs are entropy coded.
4. The quantization non-uniform and the outputs are entropy coded.

Solution:

1. Looking at Table 5.1 for uniform quantization, MSE = 0.003146 can be achieved with 34 output levels. The uncoded rate is $\lceil \log_2 K \rceil = 6$ bits per sample. We need to round up to the nearest integer, because the number of bits per sample must be an integer for uncoded outputs.
2. Looking at Table 5.2 for non-uniform quantization, 29 output levels are needed to reach MSE just under 0.003146. The uncoded rate is $\lceil \log_2 K \rceil = 5$ bits per sample.
3. For entropy coded outputs the rate is the entropy. The entropy in the table entry of Item (1) above is 4.523380 bits. Rounding is not needed for coded outputs, because the rate is the average of variable length codewords.
4. Likewise, looking at the same table entry as in Item (2), the minimum rate is the entropy of 4.591663 bits, less than that of its uncoded output, but greater than that of the uniform quantizer's coded output.

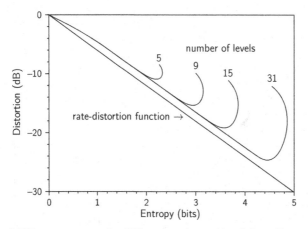

Figure 5.15   MSE versus entropy for different numbers of levels in uniform quantization.

5.5.1   Entropy coded quantizer characteristics

Since the uniform quantizer is superior in entropy coding, we need only to consider a uniform quantizer and vary the level spacing Δ (for symmetric densities). Here is the method of Goblick and Holsinger [6] to accomplish the above.

1. Choose K, the number of levels.
2. Choose Δ, the uniform quantizer spacing.
3. Calculate D and H.
4. Return to step 2 (to choose a new Δ).
   If there are enough points to trace a curve, go to 5.
5. Return to 1 (to choose a new K).

   Goblick and Holsinger plotted a figure similar to that of Figure 5.15 for a Gaussian density with squared-error distortion. For each K a curve is traced where the lower part is a straight line when H is plotted versus D on a log scale. Every curve eventually coincides with the same straight line as distortion D increases and entropy H decreases. Higher rates and lower distortion are obtained for the higher values of K. The lower envelope of these curves, which is that straight line, is the minimum entropy for a given distortion. This straight line envelope is well described by the following equation for rates above 1.75 bits:

$$R^*(D) = 1/4 + R(D) \quad \text{bits per sample} \tag{5.40}$$

where $R(D) = \frac{1}{2}\log_2 \frac{\sigma^2}{D}$ is the rate-distortion function of the Gaussian source with squared-error for $\sigma^2 > D$. The coordinates (D, R) of the rate-distortion function of a source with a given distortion criterion give the lowest rate R theoretically achievable for average distortion not exceeding D. Therefore, there exists a uniform quantizer, whose rate for a given distortion is only 1/4 bit larger than that of the best possible encoder (for this source and distortion measure). In another paper, Gish and Pierce [7] proved that for any source density and squared error, a uniform quantizer is asymptotically optimal (for large rates) and its entropy is only 0.255 bits per sample larger than the corresponding point of the rate-distortion function $R(D)$. The rate-distortion function $R(D)$ and the Gish–Pierce bound will be treated in a later section of this chapter.

There are more sophisticated algorithms in the literature than the one just described for minimizing D with respect to the entropy constraint $H \leq H_o$. For those who want to delve further into the subject and see results for other memoryless sources, see [8] and [9]. We shall describe such an algorithm later in the next chapter on vector quantization.

5.5.2 Null-zone quantization

Uniform quantization can be modified slightly to improve performance without noticeable decrease of implementation simplicity. An important and useful modification is to provide flexibility in the region of source values that quantize to zero. This region is called the *null zone* or *dead zone* of the quantizer. The most commonly used null zone is $-\Delta \leq x \leq \Delta$. This interval that quantizes to 0 is then double the length of the other intervals. The numerical operation to implement this quantizer is particularly simple.

Given the input x, the following mathematical operation produces an index q of the quantization bin (decision interval):

$$q = \text{sign}(x)\lfloor |x|/\Delta \rfloor \tag{5.41}$$

The index q is encoded for transmission or storage. The reconstruction (often called *dequantization*) y is given by

$$y = \begin{cases} (q + \xi)\Delta, & q > 0 \\ (q - \xi)\Delta, & q < 0 \\ 0, & q = 0 \end{cases} \tag{5.42}$$

where $0 \leq \xi < 1$. Heretofore, the bin index q will be called the *quantizer level*. The parameter ξ allows flexibility in setting the position of the reconstruction point within the interval. The value $\xi = 0.5$ places the reconstruction at the interval's midpoint. ξ may be set to place the reconstruction value at the centroid of the quantization interval. It has been derived through a model and confirmed in practice that $\xi \approx 0.38$ usually works well. Note that this is a midtread quantizer, because when $-\Delta \leq x \leq \Delta$, the quantizer level and reconstruction value are both 0.

Double the step size has been shown to be a nearly optimal choice of the best width for the null zone. However, it is only true for some probability density functions and certain bit rates. In order to include other widths for the null zone in our considerations, let us choose the null zone to be $-t\Delta/2 \leq x \leq t\Delta/2$, where $1 \leq t \leq 2$. The associated quantizer operation is

$$q = \begin{cases} \lceil \frac{x}{\Delta} - t/2 \rceil, & x \geq 0 \\ \lfloor \frac{x}{\Delta} + t/2 \rfloor, & x \leq 0 \end{cases} \tag{5.43}$$

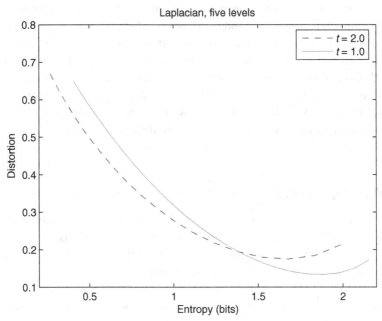

Figure 5.16 Double and normal null zone widths in 5-level uniform quantization of a Laplacian probability density with zero mean and unit variance.

Note that the formula yields $q = 0$ for $-t\Delta/2 \le x \le t\Delta/2$. Therefore, $t = 2$ gives the double width null zone and $t = 1$ gives the normal width.

Dequantization follows analogously.

$$y = \begin{cases} (q + t/2 + \xi)\Delta, & q > 0 \\ (q - t/2 - \xi)\Delta, & q < 0 \\ 0, & q = 0 \end{cases} \tag{5.44}$$

In what sense is the double step size preferable to the normal step size for the null zone? To answer this question, the objective criterion must be minimum MSE subject to a limit on entropy. The case for illustration is midpoint reconstruction ($\xi = 0.5$) in uniform step size quantization. The formula for the MSE distortion is a simple modification of the formula in Equation (5.15) for the null zone width. We proceed by calculating distortion D versus entropy H as the step size Δ varies for $t = 1.0$ and $t = 2.0$. We use the zero mean Laplacian probability density,

$$q(x) = \frac{1}{\sqrt{2}\sigma} \exp(-\sqrt{2}|x|/\sigma), \quad -\infty < x < \infty,$$

where σ^2 is the variance of this probability distribution. The graph in Figure 5.16 displays the results with $K = 5$ quantizer levels and $\sigma^2 = 1$.

From this figure, we see that the double width zone ($t = 2.0$) gives lower distortion at rates up to 1.4 bits, where the $t = 1$ and $t = 2$ curves cross. This distortion advantage of $t = 2$ decreases gradually with rate until it disappears at 1.4 bits. The gradual decline

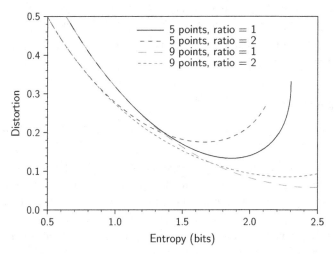

Figure 5.17   MSE versus entropy for 5 and 9 quantization levels with normal (ratio = 1) and double (ratio = 2) null zone. The probability density function is Laplacian with zero mean and unit variance. The reproduction points are the midpoints of the uniform width decision intervals (quantizer bins).

of the double width null zone advantage up to a rate where it disappears entirely and becomes a disadvantage is characteristic of any number of quantization levels.

Use of a larger number of quantization levels allows the double width advantage to extend to higher bit rates. In fact, the graphs for the same zone width for different numbers of quantization levels converge at lower rates, similar to the normal width zone case in Figure 5.15. Figure 5.17 illustrates this phenomenon for five and nine quantization levels with normal and double width zones for midpoint reproduction points and the Laplacian density. Not shown in this figure are the step sizes that produced the various points. These step sizes decrease in the counterclockwise direction from the vertical axis. If one wishes to operate at 1 bit where the graphs of five and nine levels coincide, then one needs only five quantizer levels to achieve a (normalized) distortion of 0.275. The corresponding step size is $\Delta = 1.19$ and null zone width $2\Delta = 2.38$. Recall that attaining this entropy of 1 bit requires entropy coding of the quantizer outputs.

The advantage of a wider null zone seems to be applicable only to probability densities that are highly peaked at zero. For example, there does not seem to be an advantage to a wider null zone for the Gaussian distribution. Furthermore, centroid reproduction points provide no advantage for a wider than normal null zone.

5.6   Bounds on optimal performance

Having theoretical limits on performance is extremely valuable in assessing the efficiency of a coding method. Knowing these limits can determine whether or not to spend further effort to seek better performance. We can determine optimal performance limits, however, only for limited classes of sources and distortion measures. By the necessity to avoid undue complexity in computation and implementation, most compression methods are designed for these limited classes. In this section, we shall present some of these theoretical limits without formal proofs. For more details, we refer the reader to a good textbook on information theory [10, 11].

5.6.1   Rate-distortion theory

The performance limits are expressed as a function called the rate-distortion function, denoted by $R(D)$. The value R of this function evaluated at average distortion D is the minimum rate of a code that can achieve an average distortion not exceeding D. Conversely, if the code rate is less than R, then the average distortion must be greater than D. The proof of these statements requires proving a coding theorem of information theory and is therefore beyond the scope of this discussion. However, definition and calculation of this function will allow the necessary comparison of some practical methods to theoretical limits.

Consider a source that emits a sequence of independent realizations of a random variable X with probability density $q_X(x)$. (Such a sequence is called *i.i.d.*) The reproduction Y of each source element X is assumed to be produced by a fictional memoryless "channel" or *test channel* governed by the conditional probability density function $p_{Y/X}(y/x)$. This test channel with its input and output is depicted in Figure 5.18.

This test channel description relating output to input, being probabilistic, is not to be confused with the action of coding, in which the equivalent channel is entirely deterministic. A coding theorem establishes the relationship to coding. Let us define a measure of distortion $d(x, y)$ when the source $X = x$ and its reproduction $Y = y$. Such measure is often called a single-letter distortion measure, because it does not change for different elements of the source and reproduction sequences. The average distortion per source letter can now be calculated using the joint density function $q_X(x)p_{Y/X}(y/x)$ of the input and output of the test channel.

$$D = E[d(X, Y)] = \int \int d(x, y)q_X(x)p_{Y/X}(y/x)dxdy, \qquad (5.45)$$

where $E[]$ is the expectation operator and the limits of integration are understood to be $-\infty$ to ∞. The (average) mutual information between the input and output is defined as

$$I(p) \equiv I(X; Y) = E\left[\log \frac{p_{Y/X}(y/x)}{w_Y(y)}\right] \qquad (5.46)$$

$$= \int \int q_X(x)p_{Y/X}(y/x) \log \frac{p_{Y/X}(y/x)}{w_Y(y)} dxdy$$

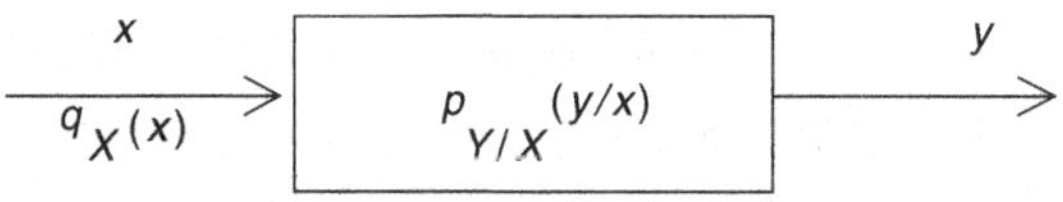

Figure 5.18   Test channel.

where $w_Y(y) = \int q_X(x)p_{Y/X}(y/x)\,dx$ is the probability density function of the output Y. The source probability distribution $q_X(x)$ is considered to be known, so that the mutual information $I(X; Y)$ is a function of the test channel and is hence denoted as $I(p)$ above.

Now we are ready to define the rate-distortion function. The rate-distortion function $R(D)$ is the minimum of the (average) mutual information over all test channels $p_{Y/X}(y/x)$ giving an average distortion no greater than some fixed value D. The mathematical statement is

$$R(D) = \min_{p_{Y/X}} \{I(p) \ : \ E[d(X, Y)] \le D\} \tag{5.47}$$

We state now the theorem that establishes the connection of $R(D)$ to optimal coding.

THEOREM 5.1 (rate-distortion theorem and converse) *For stationary and ergodic sources (under certain technical conditions) and a single-letter-based distortion measure $d(x, y)$, given $D \ge 0$, $\epsilon > 0$, and a sequence length N sufficiently large, there exists a code of rate R such that*

$$R < R(D) + \epsilon.$$

Furthermore, there exists no code of rate

$$R < R(D)$$

with average distortion D or less.

The rate-distortion function $R(D)$ is established now as an unbeatable lower bound on rate for a code with average distortion D. The theorem is stated for the general stationary, ergodic case. Certainly, these conditions are met for the simpler i.i.d. case discussed here. The rate-distortion theorem is likewise valid for discrete-alphabet sources.

In some cases $R(D)$ can be evaluated or bounded analytically. Otherwise there is a computational algorithm due to Blahut [12] for evaluating the points of $R(D)$. $R(D)$ is a monotone, non-increasing, convex $\bigcup$ (concave) function of D. Some important special cases, where exact formulas or analytical bounds are known, follow.

Case 1: $R(D)$ for a Gaussian source X with zero mean and variance σ^2 and squared error distortion measure.

$$q_X(x) = \frac{1}{\sqrt{2\pi\sigma^2}} \exp\left\{-\frac{x^2}{2\sigma^2}\right\}, \quad -\infty < x < +\infty \tag{5.48}$$

$$d(x, y) = (x - y)^2 \tag{5.49}$$

$$R(D) = \max\left\{0, \frac{1}{2} \log \frac{\sigma^2}{D}\right\} \quad \text{bits per source letter or} \tag{5.50}$$

$$D(R) = \sigma^2 2^{-2R} \quad, \quad R \geq 0 \tag{5.51}$$

Note the two extremal points $D = 0$, $R = \infty$ and $R = 0$, $D = \sigma^2$. The first one says that infinite rate is required for zero mean squared error, as expected. The second says that a mean squared error of σ^2 can be achieved for zero rate. This is done by setting any input x to zero, the mean value. The reception of all zeros is certain, so no information rate is required to convey it.

Case 2: An upper bound to $R(D)$.

Given a source of unknown density and variance σ^2 and squared error distortion measure, the rate-distortion function is overbounded as:

$$R(D) \leq \max\left\{0, \frac{1}{2}\log\frac{\sigma^2}{D}\right\} \tag{5.52}$$

The upper bound is the rate-distortion function of the Gaussian source. This means that the Gaussian source of the same variance is the most difficult to encode, since it requires the largest rate to obtain the same distortion.

Case 3: A lower bound to $R(D)$.

A lower bound to $R(D)$ for squared error distortion is

$$R(D) \geq h(X) - \frac{1}{2}\log(2\pi e D) \tag{5.53}$$

where $h(X)$ is the differential entropy of the source, defined by

$$h(X) = -\int_{-\infty}^{+\infty} q_X(x)\log q_X(x)dx.$$

This lower bound is the *Shannon lower bound* for the squared-error distortion measure.

5.6.2   The Gish–Pierce bound

The previous section dealt with optimal coding, which achieves the minimum possible distortion for a given limit on entropy (rate). The tenets of the coding theorems seem to dictate that optimal performance can be approached only through coding of arbitrarily long sequences of random variables and use of unlimited computational resources. It is useful to ask the question of what are the performance limits within the restricted framework of scalar quantization. As usual, the question can only be answered in general in the realm of fine quantization.

Gish and Pierce [7] proved that, in the limit of a large number of quantization levels (fine quantization), minimum MSE distortion for a given entropy can be realized with uniform step size quantization. Therefore all we have to do is derive the distortion and entropy in this limit to find the optimal performance limit.

Let us assume uniform quantization of source with probability density $q(x)$ using a small step size δ and a large number of levels K. There is no need to specify whether the reproduction point is the midpoint or the centroid, because they coincide under the small step size approximation. We have already derived Equation (5.26) for the approximate MSE distortion of

$$D \approx \frac{\delta^2}{12},$$ (5.54)

which neglects the contribution of the overload regions. Continuing with these assumptions, the entropy may be expressed by

$$H = - \sum_{k=1}^{K} P(y_k) \log P(y_k),$$

where the probability

$$P(y_k) = \int_{x_k}^{x_k+\delta} q(x)dx \approx q(x_k)\delta.$$

Therefore, in the limit of large K, the entropy approximation becomes

$$H \approx - \sum_{k=1}^{K} q(x_k) \log(q(x_k)\delta)\delta$$ (5.55)

$$\approx - \int_{-\infty}^{\infty} q(x) \log q(x)dx - \log \delta \int_{-\infty}^{\infty} q(x)dx$$

$$= h(q) - \log \delta,$$

where

$$h(q) = - \int_{-\infty}^{\infty} q(x) \log q(x)dx$$

is the differential entropy of the probability density $q(x)$. Using Equations (5.54) and (5.56), we obtain the final form of the optimal rate-distortion characteristic of a scalar quantizer in the fine quantization limit:

$$H = h(q) - \frac{1}{2} \log 12D.$$ (5.56)

This result is usually referred to as the Gish–Pierce bound for high rate scalar quantization. Notice that it depends on the differential entropy of the probability distribution and that the distortion is MSE. Gish and Pierce also found that uniform quantization is still optimal in extending this result to other distortion measures.

Example 5.4 Let the scalar source X be Gaussian with mean μ and variance σ^2. This probability density is

$$q(x) = \frac{1}{\sqrt{2\pi\sigma^2}} \exp(-(x - \mu)^2/2\sigma^2).$$

Calculating its differential entropy using base e logarithms,

$$h(q) = - \int_{-\infty}^{\infty} q(x) \log q(x) dx$$

$$= - \int_{-\infty}^{\infty} q(x) \left[-\log(\sqrt{2\pi\sigma^2}) - (x-\mu)^2/2\sigma^2 \right] dx$$

$$= \frac{1}{2} \log(2\pi\sigma^2) \int_{-\infty}^{\infty} q(x) dx + \frac{1}{2\sigma^2} \int_{-\infty}^{\infty} (x-\mu)^2 q(x) dx \quad (5.57)$$

$$= \frac{1}{2} \log(2\pi e \sigma^2) \quad (5.58)$$

The last step Equation (5.58) follows from noticing in Equation (5.57) that the integral over the whole range of the density equals 1 and that the integral in the second term of Equation (5.57) is exactly the variance σ^2. The base of the logarithm can be either e for units of nats or 2 for units of bits.

Now substituting this Gaussian differential entropy into Equation (5.56), the optimal rate-distortion characteristic is found to be

$$H = \frac{1}{2} \log(2\pi e \sigma^2) - \frac{1}{2} \log 12D \quad (5.59)$$

$$= \frac{1}{2} \log \frac{\sigma^2}{D} + \frac{1}{2} \log \frac{\pi e}{6}.$$

The first term above is the rate-distortion function of the Gaussian source with squared error criterion when $\sigma^2 > D$, certainly the case here. The second term evaluates to 0.255 bits and is consequently the penalty incurred for use of quantization instead of optimal coding. The derived result in Equation (5.60) matches the empirical formula of Goblick and Holsinger in Equation (5.40) inferred from the graph in Figure 5.15.

In the case of high rate or optimal quantization, with squared error as the distortion measure, the distortion seems to decrease exponentially at a rate of twice the number of bits per sample (2^{-2R}). Therefore, the following ad hoc formula has been accepted as a suitable description of the distortion versus rate characteristic of a minimum MSE quantizer:

$$D = \sigma^2 g(R) 2^{-2R}, \quad (5.60)$$

where σ^2 is the source variance, R is the rate in bits per sample, and $g(R)$ is a slowly varying, algebraic function of R which depends on the source probability distribution [12]. For a Gaussian source, the function $g(R)$ must exceed 1 or it would violate the lower bound of the corresponding distortion-rate function in Equation (5.51).

5.7 Concluding remarks

The research and number of articles on quantization exploded after Lloyd [3] and Max [4] wrote their papers. The literature on quantization is so vast that no single chapter can do it justice. So as not to distract the reader with a plethora of citations, many direct references were omitted in the text, but instead were included in the chapter's reference. Even our choice here to divide into two chapters, one for scalar quantization just concluded, and the next one for vector quantization, cannot embrace all this work. We have tried to present the essential material as concisely and clearly as possible from an engineer's point of view: keeping mathematical abstraction to a minimum and using plenty of figures and examples to teach the concepts and how to apply them. Some of the material in the section on null-zone quantization cannot be found elsewhere. Other subjects, such as entropy coding of quantizer ouputs and choice of reproduction points, are treated differently than in other textbooks or articles. We concluded the chapter with a section on attainable theoretical bounds on performance, so that we can measure how well our practical techniques are doing. We hope that the readers find this chapter beneficial to their capability to comprehend and use scalar quantization in their work or studies.

5.8 Appendix: quantization tables

Table 5.1 Optimum uniform quantization of Gaussian source: 1 to 64 output levels

Output levels	Step size	Distortion (MSE)	Entropy (bits)	Output levels	Step size	Distortion (MSE)	Entropy (bits)
1	0.000000	1.000000	0.000000	33	0.183310	0.003313	4.487124
2	1.595769	0.363380	1.000000	34	0.178739	0.003146	4.523880
3	1.224006	0.190174	1.535789	35	0.174405	0.002991	4.559592
4	0.995687	0.118846	1.903730	36	0.170290	0.002848	4.594320
5	0.842986	0.082178	2.183127	37	0.166377	0.002715	4.628122
6	0.733433	0.060657	2.408506	38	0.162652	0.002592	4.661037
7	0.650770	0.046860	2.597576	39	0.159100	0.002477	4.693120
8	0.586019	0.037440	2.760570	40	0.155710	0.002370	4.724409
9	0.533822	0.030696	2.903918	41	0.152471	0.002270	4.754943
10	0.490778	0.025687	3.031926	42	0.149372	0.002177	4.784758
11	0.454623	0.021856	3.147615	43	0.146404	0.002089	4.813888
12	0.423790	0.018853	3.253191	44	0.143560	0.002007	4.842364
13	0.397159	0.016453	3.350309	45	0.140831	0.001930	4.870215
14	0.373906	0.014500	3.440248	46	0.138210	0.001857	4.897468
15	0.353411	0.012889	3.524015	47	0.135692	0.001788	4.924148
16	0.335201	0.011543	3.602417	48	0.133269	0.001724	4.950281
17	0.318903	0.010405	3.676112	49	0.130936	0.001663	4.975887
18	0.304225	0.009434	3.745642	50	0.128689	0.001605	5.000988

Table 5.1 (cont.)

Output levels	Step size	Distortion (MSE)	Entropy (bits)	Output levels	Step size	Distortion (MSE)	Entropy (bits)
19	0.290931	0.008598	3.811462	51	0.126522	0.001550	5.025604
20	0.278830	0.007873	3.873952	52	0.124432	0.001498	5.049753
21	0.267763	0.007239	3.933440	53	0.122414	0.001449	5.073453
22	0.257601	0.006682	3.990205	54	0.120464	0.001402	5.096720
23	0.248234	0.006189	4.044490	55	0.118579	0.001358	5.119571
24	0.239570	0.005751	4.096505	56	0.116756	0.001315	5.142020
25	0.231531	0.005360	4.146436	57	0.114991	0.001275	5.164081
26	0.224050	0.005008	4.194446	58	0.113282	0.001237	5.185767
27	0.217070	0.004692	4.240679	59	0.111627	0.001200	5.207092
28	0.210540	0.004406	4.285264	60	0.110022	0.001165	5.228066
29	0.204418	0.004146	4.328317	61	0.108465	0.001132	5.248702
30	0.198665	0.003909	4.369941	62	0.106954	0.001100	5.269011
31	0.193248	0.003693	4.410228	63	0.105488	0.001069	5.289002
32	0.188139	0.003495	4.449264	64	0.104063	0.001040	5.308686

Table 5.2 Optimum non-uniform quantization of Gaussian source: 1 to 64 output levels

Output levels	Distortion (MSE)	Entropy (bits)	Output levels	Distortion (MSE)	Entropy (bits)
1	1.000000	0.000000	33	0.002359	4.773025
2	0.363380	1.000000	34	0.002226	4.815004
3	0.190174	1.535789	35	0.002104	4.855793
4	0.117482	1.911099	36	0.001991	4.895457
5	0.079941	2.202916	37	0.001888	4.934058
6	0.057978	2.442789	38	0.001792	4.971650
7	0.044000	2.646931	39	0.001703	5.008284
8	0.034548	2.824865	40	0.001621	5.044010
9	0.027853	2.982695	41	0.001545	5.078870
10	0.022937	3.124583	42	0.001474	5.112906
11	0.019220	3.253506	43	0.001407	5.146156
12	0.016340	3.371666	44	0.001345	5.178655
13	0.014063	3.480744	45	0.001287	5.210437
14	0.012232	3.582051	46	0.001233	5.241533
15	0.010737	3.676630	47	0.001182	5.271972
16	0.009501	3.765328	48	0.001134	5.301780
17	0.008467	3.848840	49	0.001089	5.330984
18	0.007593	3.927741	50	0.001047	5.359608
19	0.006848	4.002518	51	0.001007	5.387673
20	0.006208	4.073584	52	0.000969	5.415203
21	0.005653	4.141290	53	0.000934	5.442216
22	0.005170	4.205942	54	0.000900	5.468732
23	0.004746	4.267806	55	0.000868	5.494769

Table 5.2 (cont.)

Output levels	Distortion (MSE)	Entropy (bits)	Output levels	Distortion (MSE)	Entropy (bits)
24	0.004372	4.327112	56	0.000838	5.520344
25	0.004041	4.384064	57	0.000809	5.545473
26	0.003746	4.438843	58	0.000782	5.570170
27	0.003483	4.491610	59	0.000756	5.594452
28	0.003246	4.542507	60	0.000732	5.618331
29	0.003032	4.591663	61	0.000708	5.641821
30	0.002839	4.639193	62	0.000686	5.664934
31	0.002664	4.685202	63	0.000665	5.687683
32	0.002505	4.729784	64	0.000644	5.710078

Problems

5.1 Show that the optimal (Lloyd–Max) quantizer for a distortion measure defined by $d(x, y) = |x - y|$ satisfies the equations:

$$x_k = \frac{1}{2}(y_k + y_{k+1}), k = 1, 2, \ldots, K - 1 \quad (x_0 = -\infty, x_K = +\infty)$$
$$y_k = \text{median}\{x \mid x \in (x_{k-1}, x_k]\}, \quad k = 1, 2, \ldots, K$$

where $\{x_k\}$ are the decision thresholds and $\{y_k\}$ are the quantizer (reproduction) levels.

What is the intuitive explanation for Equation (5.61) holding for both squared error and absolute error measures?

5.2 (a) Prove that the optimal (Lloyd–Max) quantizer $Y = Q(X)$, $E = X - Y$ for a given probability density function (zero-mean) and squared-error distortion measure satisfies the following equations.

$$\rho_{EX} = \sigma_E/\sigma_X,$$
$$\rho_{XY} = \sqrt{1 - (\sigma_E^2/\sigma_X^2)},$$
$$\rho_{EY} = 0,$$

where $\rho_{UV} \equiv E[UV]/\sqrt{E[U^2]E[V^2]}$ is the correlation coefficient between zero mean random variables U and V.

(b) Calculate ρ_{XY} and ρ_{EX} for the optimal k-level quantizer for uniform density (and squared-error).

5.3 Let the source probability density function $p_X(x)$ be an even function of x ($p_X(x) = p_X(-x)$). Prove that the output entropy H of a symmetric quantizer with an even number of levels K is no smaller than 1 bit per sample. (For what value of K is the minimum always achieved?)

5.4 It is interesting to study the behavior of optimal entropy-coded quantizers at low bit rates. Let the source probability density function $p_X(x)$ be Laplacian with variance $\sigma_X^2 = 2$ $(p_X(x) = \frac{1}{2}\exp -|x|, -\infty < x < +\infty)$ and let the distortion measure be squared error. We shall derive the characteristic of a three-level quantizer with finite thresholds at $+a$ and $-a$ versus a.

   (a) Noting that the fixed thresholds, the output entropy, is determined, derive that the optimal quantizer levels are: $-(1 + a), 0, (1 + a)$.
   (b) Derive equations that give the minimum rate (output entropy) and normalized average distortion D/σ_X^2 in terms of the parameter $\theta = e^{-a}$.
   (c) Plot the minimal rate R versus D/σ_X^2 by varying the value of θ. Identify values of a and H that give the minimum possible D. (These are the parameters of the $K = 3$ Lloyd–Max quantizer.) Compare the value of $H = 1$ bit per sample to that of the optimal $K = 2$ quantizer of rate $R = 1$ bit per sample. In fact, calculate that this distortion for $K = 2$ is achieved at the rate of 0.4867 bits per sample with the $K = 3$ optimal quantizer.

5.5 Design an optimal (MMSE) $K = 4$ quantizer for the one-sided exponential density function

$$q(x) = \begin{cases} e^{-x}, & x \geq 0 \\ 0, & x < 0. \end{cases}$$

Use either an analytical method or the Lloyd I algorithm. Aim for at least two decimal places accuracy. (Outputs are not entropy-coded.)

5.6 Suppose a memoryless source emits random variables X with the same probability density function as Problem 5.5. Suppose that the source values are now quantized with a uniform quantizer of step size Δ with an infinite number of levels, i.e., for all k

$$y_k = Q(x) = (k - 1/2)\Delta, \text{ if } x \in ((k - 1)\Delta, k\Delta].$$

   (a) Find the MSE, $E[(X - Q(X))^2]$.
   (b) Find the entropy of the quantized outputs, $H(Q(X))$.
   (c) Find the differential entropy $h(X) = -\int q(x) \log q(x)dx$ of the source.
   (d) Compare $H(Q(X))$ to the Shannon Lower Bound. What happens as $\Delta \to 0$?

5.7 Compute and plot SNR in dB versus loading fraction (reciprocal of loading factor) V/σ, $V = x_{K-1}$) in dB from –40 to +5 dB for a 128-level quantizer when the signal input is Laplacian (two-sided exponential probability density function) for both a uniform and a μ-law quantizer*. In each case specify the dynamic range achievable in dB for a 25 dB SNR requirement and include overload effects. Plot both curves on a single graph with linear/linear axes. Show all formulas used. Outputs are not entropy coded.
   *Compandor characteristic: $G(x) = V\frac{\ln(1+\mu|x|/V)}{\ln(1+\mu)}\text{sgn}(x), |x| \leq V$
   (If needed, assume that $\mu = 255$.)

5.8   Consider a random variable X that assumes values $\pm a$ each with probability P_1, and a value of 0 with probability P_0. For the case of $P_0 = 0.5$ and $P_1 = 0.25$,
   (a) show that the symmetric 1-bit MMSE quantizer is defined by output levels $0.5a$ and a MSE of $a^2/4$;
   (b) show that the non-symmetric 1-bit quantizer with output levels $-a/3$ (for $x \leq 0$) and $+a$ (for $x > 0$) has a smaller MSE, equal to $a^2/6$.

5.9   The MSE distortion of an even-symmetric quantizer with an odd number of quantization levels was derived in Equation (5.15). Derive a similar equation for the distortion for an even number of quantization levels (midrise quantizer).

5.10   (Prob. 6.14 [13]) Suppose that a random variable X has a two-sided exponential (Laplacian) probability density function

$$f_X(x) = \frac{\lambda}{2}e^{-\lambda|x|}.$$

A three-level, MMSE quantizer Q for X has the following form

$$Q(x) = \begin{cases} +b, & \text{if} \quad x > a \\ 0, & \text{if} \quad -a \leq x \leq a \\ -b, & \text{if} \quad x < -a \end{cases}$$

   (a) Find an expression for b as a function of a so that the centroid condition is met.
   (b) For what value of a will the quantizer using b chosen as above satisfy both the Lloyd conditions for optimality? What is the resulting MSE?
   (c) Specialize your answer to the case $\lambda = 1$.
   (d) If you optimally entropy coded the outputs of this quantizer, what would be its bit rate?

5.11   We wish to study centroid and midpoint reconstruction in quantizing the positive random variable X with the exponential probability density function

$$q(x) = \begin{cases} \frac{1}{T}e^{-x/T}, & x > 0 \\ 0, & x \leq 0. \end{cases}$$

Let the decision interval be $a \leq x \leq b$, where $a = T$ and $b = (1 + 2^{-k})T$, $k = 0, 1, 2, 3$.
   (a) Calculate the MSE for quantizing X to midpoint reconstruction points in the decision intervals $[T, (1 + 2^{-k})T]$, for $k = 0, 1, 2$, and 3.
   (b) Calculate the centroids of the decision intervals $[T, (1 + 2^{-k})T]$, for $k = 0, 1, 2$, and 3.
   (c) Calculate the MSE that results when quantizing X to the centroid reconstruction point for each of the four decision intervals.
   (d) Construct a table or graph comparing the MSE of midpoint and centroid reconstructions for the different size decision intervals. State conclusions about the trend of the results.

5.12 (Prob. 5.5 [13]) The MSE of a quantizer with $M = 128$ levels, overload point $x_{max} = 5$, and compressor slope of 100 at the origin is estimated by Bennett's integral. The quantizer was designed for a given normalized probability density and for the nominal input power level of $\sigma^2 = 4.0$. Explain why the integral becomes invalid as the input level becomes small. What is the correct asymptotic formula for the SNR as σ approaches 0? Consider separately two cases, first when σ is in the neighborhood of 0.003, and second when σ is less than 0.0005. Assume a smooth and well-behaved compressor curve.

5.13 We wish to quantize the output of a stationary, memoryless Gaussian source for subsequent transmission over a channel of limited capacity. The source has zero mean and variance one and the distortion measure in the system is squared error $(d(u, v) = (u - v)^2)$. The quantizer is a uniform, even-symmetric $N = 4$-level quantizer with midpoint quantization levels as shown in the figure.

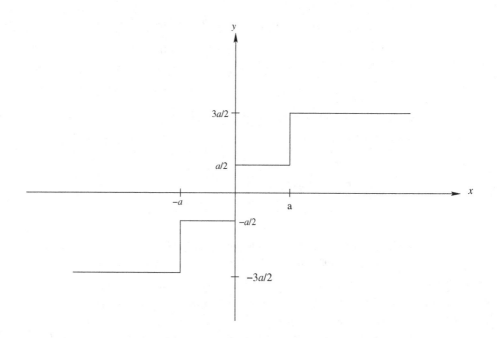

Solve by computer for the step size a that gives the MMSE D_{min}.

5.14 The requirement in a certain communications system is to obtain an average distortion (squared error) per source symbol $\bar{d} \le 0.007$. The source is Gaussian with zero mean and unit variance and the distortion measure is squared error $(d(u, v) = (u - v)^2)$.

(a) There are various options for encoding the quantizer outputs V for achieving this distortion, but they may require different rates. Using the tables in the Appendix, ascertain in each case below the minimum rate R in *bits per symbol* required to meet the distortion requirement.

(i) Each quantizer output v is indexed by a fixed number of binary digits;

(ii) each quantizer output v is indexed by a single N-ary symbol;

(iii) long blocks of quantizer outputs are encoded into fixed length sequences of binary digits;

(iv) each quantizer output v is encoded into a number of binary digits, where the number may vary depending on the particular output; and

(v) blocks of 100 quantized outputs v are encoded into binary sequences of different lengths dependent on the particular block of v's.

(vi) Why might the scheme in (v) be superior to the one in (iv)?

(b) All of the schemes above seek to encode V, the quantizer output. Quantizing the source U places a limitation on the minimum rate required to achieve $\bar{d} \leq$ 0.007. Determine the rate R for the optimal code for U with this distortion objective.

5.15 We wish to compare different encoding schemes for a memoryless Gaussian source ($\mathcal{N}(0, 1)$) with MSE distortion.

Using the table for the MMSE Gaussian uniform quantizer, plot on one graph the following rates R versus logarithm of the MSE.

(a) The data rate R in bits per sample when quantizer outputs are converted directly to binary digits.

(b) The data rate R in bits per sample when quantized outputs are converted to N-ary symbols, where N is the number of output levels.

(c) The data rate R in bits per sample when quantizer outputs are entropy coded.

(d) The data rate R in bits per sample of an optimal code for this source.

Note

1. Hereafter, the subscript X in the density $q_X(x)$ will be dropped when the random variable is obvious.

References

1. P. E. Fleischer, "Sufficient conditions for achieving minimum distortion in a quantizer," in *IEEE International Convention Record – Part I*, vol. Part I, 1964, pp. 104–111.
2. A. V. Trushkin, "Sufficient conditions for uniqueness of a locally optimal quantizer for a class of convex error weighting functions," *IEEE Trans. Inf. Theory*, vol. 28, no. 2, pp. 187–198, Mar. 1982.
3. S. P. Lloyd, "Least squares quantization in PCM," *IEEE Trans. Inf. Theory*, vol. 28, no. 2, pp. 129–137, Mar. 1982.
4. J. Max, "Quantizing for minimum distortion," *IRE Trans. Inform. Theory*, vol. 6, no. 1, pp. 7–12, Mar. 1960.
5. W. R. Bennett, "Spectra of quantized signals," *Bell Syst. Tech. J.*, vol. 27, pp. 446–472, July 1948.
6. T. J. Goblick Jr. and J. L. Holsinger, "Analog source digitization: a comparison of theory and practice," *IEEE Trans. Inf. Theory*, vol. 13, no. 2, pp. 323–326, Apr. 1967.
7. H. Gish and J. N. Pierce, "Asymptotically efficient quantizing," *IEEE Trans. Inf. Theory*, vol. 14, no. 5, pp. 676–683, Sept. 1968.
8. A. N. Netravali and R. Saigal, "Optimum quantizer design using a fixed-point algorithm," *Bell Syst. Tech. J.*, vol. 55, pp. 1423–1435, Nov. 1976.
9. N. Farvardin and J. W. Modestino, "Optimum quantizer performance for a class of non-gaussian memoryless sources," *IEEE Trans. Inf. Theory*, vol. 30, no. 3, pp. 485–497, May 1984.
10. T. M. Cover and J. A. Thomas, *Elements of Information Theory*. New York, NY: John Wiley & Sons, 1991, 2006.
11. R. G. Gallager, *Information Theory and Reliable Communication*. New York, NY: John Wiley & Sons, 1968.
12. R. E. Blahut, "Computation of channel capacity and rate-distortion functions," *IEEE Trans. Inf. Theory*, vol. 18, no. 4, pp. 460–473, July 1972.
13. A. Gersho and R. M. Gray, *Vector Quantization and Signal Compression*. Boston, Dordrecht, London: Kluwer Academic Publishers, 1992.

Further reading

A. Alecu, A. Munteanu, J. Cornelis, S. Dewitte, and P. Schelkens, "On the optimality of embedded deadzone scalar-quantizers for wavelet-based l-infinite-constrained image coding," *IEEE Signal Process. Lett.*, vol. 11, no. 3, pp. 367–370, Mar. 2004.

D. A. Huffman, "A method for the construction of minimum redundancy codes," *Proc. IRE*, vol. 40, pp. 1098–1101, Sept. 1952.

N. S. Jayant and P. Noll, *Digital Coding of Waveforms*. Englewood Cliffs, NJ: Prentice Hall, 1984.

M. Oger, S. Ragot, and M. Antonini, "Model-based deadzone optimization for stack-run audio coding with uniform scalar quantization," in *Proceedings of the IEEE International Conference on Acoustics Speech Signal Process.*, Las Vegas, NV, Mar. 2008, pp. 4761–4764.

W. A. Pearlman, "Polar quantization of a complex gaussian random variable," in *Quantization*, ser. Benchmark Papers in Electrical Engineering and Computer Science, ed. P. Swaszek Van Nostrand Reinhold, 1985, vol. 29.

W. Pearlman, "Polar quantization of a complex Gaussian random variable," *IEEE Trans. Commun.*, vol. COM-27, no. 6, pp. 101–112, June 1979.

W. Pearlman and G. Senge, "Optimal quantization of the Rayleigh probability distribution," *IEEE Trans. Commun.*, vol. COM-27, no. 1, pp. 101–112, Jan. 1979.

C. E. Shannon, "A mathematical theory of communication," *Bell Syst. Tech. J.*, vol. 27, pp. 379–423 and 632–656, July and Oct. 1948.

C. E. Shannon and W. Weaver, *The Mathematical Theory of Communication*. Urbana, IL: University of Illinois Press, 1949.

G. J. Sullivan, "Efficient scalar quantization of exponential and Laplacian random variables," *IEEE Trans. Inf. Theory*, vol. 42, no. 5, pp. 1365–1374, Sept. 1996.

6 Coding of sources with memory

6.1 Introduction

Independent scalar quantization of a sequence of samples from a source is especially inefficient when there is dependence or memory among the samples. Even if there is no memory, where the samples are truly statistically independent, scalar quantization of each sample independently, although practical, is a suboptimal method. When neighboring samples provide information about a sample to be quantized, one can make use of this information to reduce the rate needed to represent a quantized sample. In this chapter, we shall describe various methods that exploit the memory of the source sequence in order to reduce the rate needed to represent the sequence with a given distortion or reduce the distortion needed to meet a given rate target. Such methods include predictive coding, vector coding, and tree- and trellis-based coding. This chapter presents detailed explanations of these methods.

6.2 Predictive coding

The first approach toward coding of sources with memory is using scalar quantization, because of its simplicity and effectiveness. We resort to a simple principle to motivate our approach. Consider the scenario in Figure 6.1, where a quantity u_n is subtracted from the source sample x_n at time n prior to scalar quantization (Q). The same quantity u_n is added to the quantized difference $\tilde{e}_n$ to yield the reconstructed output y_n. What is remarkable is that the reconstruction error, $x_n - y_n$, equals the quantization error, $e_n - \tilde{e}_n$. The following equations prove this fact.

$$
\begin{aligned}
e_n &= x_n - u_n \\
y_n &= \tilde{e}_n + u_n \\
x_n - y_n &= e_n + u_n - (\tilde{e}_n + u_n) \\
&= e_n - \tilde{e}_n
\end{aligned}
\tag{6.1}
$$

We know from the previous chapter that samples with a small range or alphabet can be represented with the same accuracy using fewer bits than those with a large range or alphabet. Therefore, the choice of u_n should produce a difference e_n that tends to be small with high probability. The best choice, therefore, is the *maximum*

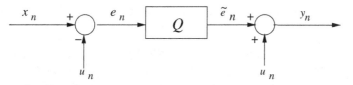

Figure 6.1 Depiction of subtraction and addition of same quantity before and after quantization (Q), whereby reconstruction error equals quantization error.

a priori probability (MAP) estimate of x_n from a group of neighboring samples. Without specifying the members of this neighborhood group, let us denote it by $\mathcal{G}_n$. The MAP estimate $\widehat{x}_n$ of x_n is given by

$$\widehat{x}_n = \arg\max_x p(x/\mathcal{G}_n), \tag{6.2}$$

where $p(x/\mathcal{G}_n)$ is the conditional probability density function of x given the neighborbood $\mathcal{G}_n$. Here it is assumed that the source is stationary, so that this probability function is independent of time shifts. Clearly, setting $u_n = \widehat{x}_n$ produces the difference e_n with the highest probability of being small.

MAP estimation involves a too slow operation and too complex computation for most purposes, so one seeks a suboptimal solution for u_n. When adjacent samples in the source sequence are highly dependent (strongly correlated), the simplest choice for u_n is the previous sample x_{n-1}. Under these circumstances the difference $e_n = x_n - x_{n-1}$ is highly likely to be small. This choice of the previous sample is common and often quite effective.

The most common estimate of x_n to use for u_n is the linear minimum mean squared estimate (LMMSE). This estimate is a linear combination of samples in the neighborhood group $\mathcal{G}_n$. The coefficients in the linear combination are chosen to minimize the MSE,

$$E[x_n - \widehat{x}_n]^2 = E[e_n^2], \tag{6.3}$$

which is the variance of the e_n, the input to the quantizer. When the source sequence is Gaussian, the MAP and LMMSE estimators are the same.

We have been considering the data source to be a time sequence of samples. When that is the case, the neighborhood group of samples should be comprised of some number of the most recent past samples. Then the estimator is called a *predictor*, for obvious reasons. Sometimes, the data sequence does not need to be ordered in time, as in the case of images. The neighborhood can then consist of samples that surround the sample to be estimated. Nonetheless, the following section will develop the solution for optimal linear prediction in a time sequence of samples.

6.2.1 Optimal linear prediction

Let $\ldots, x_{-1}, x_0, x_1 \ldots, x_n$ be a stationary, zero mean source sequence. The linear prediction of x_n from the past p values is

$$\widehat{x}_n = \sum_{k=1}^{p} a_k x_{n-k} \tag{6.4}$$

where $a_1, a_2, \ldots, a_p$ are the prediction coefficients. The well-known solution for the vector of prediction coefficients $\mathbf{a}^o = (a_1^o, a_2^o, \ldots, a_p^o)^t$ that minimizes the MSE $E[|\widehat{x}_n - x_n|^2]$ follows the projection principle that the error $x_n - \widehat{x}_n$ be orthogonal to the data or past values, i.e.,

$$E[(x_n - \widehat{x}_n) x_{n-k}] = 0, \quad k = 1, 2, \ldots, p \tag{6.5}$$

which results in the matrix (Yule–Walker) equation

$$\Phi_x^{(p)} \mathbf{a}^o = \phi_x \tag{6.6}$$

where $\phi_x = (\phi_x(1), \phi_x(2), \ldots, \phi_x(p))^t$; $\phi_x(\ell) = E[x_n x_{n-\ell}]$ is the autocorrelation function of the source with lag ℓ, $\ell = 1, 2, \ldots, p$; and $\Phi_x^{(p)} = \{\phi(|k - \ell|)\}_{k,\ell=0}^{p-1}$ is the p-th order autocorrelation matrix of the source. Re-writing the Yule–Walker equation in terms of its individual elements, it is expressed more explicitly as

$$\begin{pmatrix} \phi_x(0) & \phi_x(1) & \cdots & \phi_x(p-1) \\ \phi_x(1) & \phi_x(0) & \cdots & \phi_x(p-2) \\ \vdots & \vdots & \vdots & \vdots \\ \phi_x(p-1) & \phi_x(p-2) & \cdots & \phi_x(0) \end{pmatrix} \begin{pmatrix} a_1^o \\ a_2^o \\ \vdots \\ a_p^o \end{pmatrix} = \begin{pmatrix} \phi_x(1) \\ \phi_x(2) \\ \vdots \\ \phi_x(p) \end{pmatrix} \tag{6.7}$$

The MMSE can be expressed as

$$\begin{aligned} e_{\text{min}}^2 = \min E[|x_n - \widehat{x}_n|^2] &= \phi_x(0) - \phi_x^t \mathbf{a}^o \\ &= \phi_x(0) - \phi_x^t (\Phi_x^{(p)})^{-1} \phi_x \tag{6.8} \end{aligned}$$

The autocorrelation matrix Φ_p is always non-negative definite. Assuming that it is strictly positive definite assures that its inverse $(\Phi_x^{(p)})^{-1}$ is also positive definite, so that $\phi_x^t (\Phi_x^{(p)})^{-1} \phi_x > 0$ for any vector ϕ_x (by definition of positive definite). Therefore, the minimum error variance e_{min}^2 is always less than $\phi_x(0)$, the variance of the source.

Example 6.1 Consider the first-order, autoregressive (AR(1)) source model, given by

$$x_n = \rho x_{n-1} + z_n \tag{6.9}$$

$$= \sum_{j=0}^{\infty} \rho^j z_{n-j} \tag{6.10}$$

The probability distribution of this source is first-order Markov (Markov-1), because, given the previous sample x_{n-1}, the current random variable x_n depends only on the current noise z_n and no other past members of the sequence where ρ is a constant such that $0 < \rho < 1$, and z_n is i.i.d. with mean zero and variance σ_z^2. The expression for x_n

in Equation (6.10) assumes that in the infinite past the sequence was initially zero. That makes the source x_n stationary with zero mean and covariance function

$$\phi_x(k) = E[x_n x_{n-k}]$$

$$= E[\sum_{i=0}^{\infty} \rho^i z_{n-i} \sum_{j=0}^{\infty} \rho^j z_{n-k-j}]$$

$$= \sum_{i=0}^{\infty} \sum_{j=0}^{\infty} \rho^{i+j} E[z_{n-i} z_{n-k-j}] \tag{6.11}$$

for $k \geq 0$. Because the noise sequence is i.i.d.,

$$E[z_{n-i} z_{n-k-j}] = \begin{cases} \sigma_z^2, & \text{if } i = k+j \\ 0, & \text{otherwise.} \end{cases} \tag{6.12}$$

Therefore, substitution of (6.12) into (6.11) reveals the source covariance function to be

$$\phi_x(k) = \sum_{j=0}^{\infty} \rho^{k+2j} \sigma_z^2$$

$$= \sigma_z^2 \rho^k \sum_{j=0}^{\infty} \rho^{2j}$$

$$= \sigma_z^2 \frac{\rho^k}{1 - \rho^2}, \quad k \geq 0. \tag{6.13}$$

Evaluating $\phi_x(k)$ at $k = 0$ reveals the variance σ_x^2 of the source to be related to the noise variance according to

$$\sigma_x^2 = \frac{\sigma_z^2}{1 - \rho^2} \tag{6.14}$$

The covariance of the Markov-1, AR(1) source is thus often expressed as

$$\phi_x(k) = \sigma_x^2 \rho^k, \quad k \geq 0. \tag{6.15}$$

Because the covariance (or autocorrelation) function is always an even function, we conclude that

$$\phi_x(k) = \sigma_x^2 \rho^{|k|}, \quad k = 0, \pm 1, \pm 2, \ldots. \tag{6.16}$$

Solving for the optimum linear predictor for a Markov-1, AR(1) source is reasonably straightforward. We illustrate a solution for an order-2 prediction in the following example.

Example 6.2 Again, we consider the AR(1) source in (6.9) and an order-2 predictor $\hat{x}_n = a_1 x_{n-1} + a_2 x_{n-2}$. The Yule–Walker equations in this case are

$$\sigma_x^2 \begin{pmatrix} 1 & \rho \\ \rho & 1 \end{pmatrix} \begin{pmatrix} a_1^o \\ a_2^o \end{pmatrix} = \sigma_x^2 \begin{pmatrix} \rho \\ \rho^2 \end{pmatrix}$$

Solving, we obtain

$$a_1^o = \rho \text{ and } a_2^o = 0.$$

The first-order predictor is optimum, as expected, since the value to be predicted depends explicitly only on the single, most recent, past value, so there is no need to consider a predictor of higher order for Markov-1 sequences. Analogously, the optimum predictor for a Markov-m (AR(m)) sequence is of order m.

6.2.2   DPCM system description

The fact that the variance of the prediction error, usually called the *prediction residual*, is smaller than that of the source motivates its use in a workable coding system. Therefore, we would want to take

$$u_n = \widehat{x}_n = \sum_{k=1}^{p} a_k x_{n-k}.$$

The obstacle is that a prediction is made from past values of the source and that these values are not available at the destination, since any reconstruction of past values must contain some error if the source has continuous amplitude. If one does proceed with prediction from true values at the source by always holding the past p values in memory and attempts reconstruction at the destination, the reconstruction error will accumulate as decoding progresses. One must therefore predict using reconstructed values at the source which are the same as those at the destination. The method that actualizes the quantization and prediction from quantized values is called *Differential Pulse Code Modulation* or, more commonly, *DPCM*. A DPCM system is illustrated in Figure 6.2.

In this system the prediction residual e_n, which is the difference between the current source value x_n and a prediction $\widehat{x}_n$, is quantized to $\widetilde{e}_n$, whose binary code is transmitted to the destination. The prediction $\widehat{x}_n$ has been formed from a linear combination of the past p reconstructed values of the source $\widetilde{x}_{n-1}, \ldots, \widetilde{x}_{n-p}$, which in turn can only be derived from past $\widetilde{e}_n$'s. Therefore, the current reconstruction $\widetilde{x}_n$ is the sum of its prediction $\widehat{x}_n$ and the quantized prediction error $\widetilde{e}_n$. With this arrangement of feeding the prediction value $\widehat{x}_n$ back to add to the quantized prediction error (feedback around the quantizer), the error in the reconstruction $\widetilde{x}_n$ is equal to the error in the quantization of e_n. That is,

$$x_n - \widetilde{x}_n = (x_n - \widehat{x}_n) - (\widetilde{x}_n - \widehat{x}_n)$$
$$= e_n - \widetilde{e}_n$$

The decoder can form the reconstructions $\widetilde{x}_n$ from $\widetilde{e}_n$ as does the encoder. The operation is shown also in Figure 6.2.

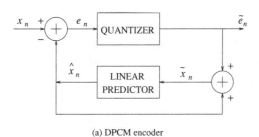

(a) DPCM encoder

(b) DPCM decoder

Figure 6.2 DPCM encoder and decoder.

6.2.3 DPCM coding error and gain

The sequence of prediction residuals is a stationary scalar source whose values are independently quantized for MMSE. The distortion versus rate function is proportional to the source variance, which is the mean squared value of the prediction residual. It is not $e_{\min}^2$ in Equation (6.8), because in DPCM the prediction takes place on reconstructed past values, not actual past values. If we make the approximation that the predictor coefficients are designed for actual past values using the source covariances $E[x_n x_{n-\ell}]$, $\ell = 0, 1, 2, \ldots, p$ and that the prediction $\widehat{x}_n$ from the quantized reconstructions is approximately the same as that from actual past values, then $e_{\min}^2$ in Equation (6.8) is the variance of the prediction residual. This approximation becomes more realistic at high rates when the quantization error is small and the reconstructions become close to their actual values. Therefore, the approximation is

$$E[\widehat{e}_n^2] \approx \phi_x(0) - \phi_x^t (\Phi_x^{(p)})^{-1} \phi_x$$

The gain in DPCM coding over PCM coding may be defined as

$$G_{\text{DPCM/PCM}} = \frac{D_{\text{PCM}}}{D_{\text{DPCM}}}$$

Since the distortions (MSEs) of both PCM and DPCM have the same unit variance distortion versus rate function, such as that in Equation (5.60), the gain is just the ratio of their respective source variances, which is

$$G_{\text{DPCM/PCM}} = \frac{\phi_x(0)}{\phi_x(0) - \phi_x^t (\Phi_x^{(p)})^{-1} \phi_x} \tag{6.17}$$

This gain exceeds one for a source with memory and is always optimistic due to the approximation of assuming prediction from non-quantized values.

Example 6.3 Let us now calculate the gain of DPCM when coding the Markov-1 source using Equation (6.17) for $m = 2$ order prediction. Although second-order prediction is one more order than necessary, Equation (6.17) still gives correct results. The covariance matrix and its inverse are, respectively,

$$\Phi_x^{(2)} = \sigma_x^2 \begin{pmatrix} 1 & \rho \\ \rho & 1 \end{pmatrix} \text{ and } (\Phi_x^{(2)})^{-1} = \frac{\sigma_x^2}{1 - \rho^2} \begin{pmatrix} 1 & -\rho \\ -\rho & 1 \end{pmatrix}$$

using $\phi_x(k) = \sigma_x^2 \rho^k$, $k = 0, 1, 2$. Substituting into Equation (6.17) and simplifying, we obtain

$$G_{\text{DPCM/PCM}} = \frac{\sigma_x^2}{\sigma_x^2(1 - \rho^2)} = \frac{1}{1 - \rho^2}.$$

If we used Equation (6.17) with first-order prediction, the relevant substitutions are

$$\phi_x(1) = \sigma_x^2 \rho \text{ and } (\Phi_x^{(1)})^{-1} = \frac{1}{\sigma_x^2}$$

and we discover the same result. The higher the correlation parameter $\rho < 1$, the larger is the coding gain, as expected. Notice also that the source variance always cancels out in the gain formulas, so one could always use normalized covariance functions or assume that $\sigma_x^2 = 1$ for calculating these gains.

6.3   Vector coding

6.3.1   Optimal performance bounds

6.3.1.1   Source coding with a fidelity criterion

When the real-valued source letters are not discrete but continuous amplitude, the entropy rate is generally infinite. Perfect reconstruction is therefore impossible with a finite code rate. In order to reproduce the source vector of length N at a remote point, a certain amount of distortion must be accepted. First, a measure of distortion is defined between the source vector realization $\mathbf{x}$ and its corresponding reproduction $\mathbf{y}$ and denoted by $d_N(\mathbf{x}, \mathbf{y})$. At the source is a list of K possible reproduction vectors $\{\mathbf{y}_1, \mathbf{y}_2, \dots, \mathbf{y}_K\}$ of length N called the codebook or dictionary C. When $\mathbf{x}$ is emitted from the source, the codebook C is searched for the reproduction vector $\mathbf{y}_m$ that has the least distortion, i.e.,

$$d_N(\mathbf{x}, \mathbf{y}_m) \leq d_N(\mathbf{x}, \mathbf{y}_k) \text{ for all } k \neq m$$

and the binary index of m is transmitted to the destination. The rate R of the code in bits per source letter or bits per dimension is therefore

$$R = \frac{1}{N} \log_2 K.$$

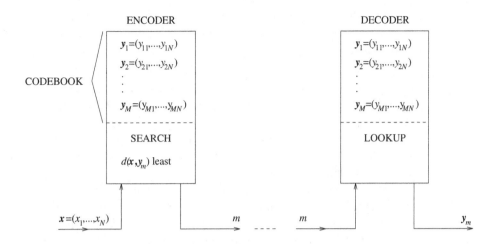

Transmission of m : $\leq \log_2 M$ bits per vector

or $\log_2 M/N$ bits per sample

Figure 6.3   Block or vector coding system model.

As the same codebook and indexing are stored at the destination in a table, the lookup in the table produces the same reproduction vector and same distortion found at the source through search if the channel is assumed to be distortion-free. A diagram of the system model is shown in Figure 6.3. Since the source vector is a random variable (of N dimensions), the average distortion obtained for a given codebook C can in principle be calculated with a known source probability distribution. Through the search of the codebook, the minimum distortion reproduction vector $\mathbf{y}_m$ is a function of the source vector $\mathbf{x}$ and is also a random variable.

The average distortion per source letter of the code C is therefore

$$E\left[\frac{1}{N} d_N(\mathbf{X}, \mathbf{y}_m(\mathbf{X}))/C\right]$$

where the expectation is calculated for the given source distribution $q_{\mathbf{X}}(\mathbf{x})$. The optimal code C is the one that produces the least average distortion for the same rate R, i.e.,

$$\min_{C} E\left[\frac{1}{N} d_N(\mathbf{X}, \mathbf{y}_m(\mathbf{X}))/C\right]$$

where the minimization is over all codebooks C of rate R.

To find an optimal code through an exhaustive search through an infinite number of codes of size $K = 2^{RN}$ is truly an impossible task. Furthermore, optimal performance is obtained in the limit of large N, as it can be proved that as N approaches infinity, $E\left[\frac{1}{N} d_N(\mathbf{X}, \mathbf{y}_m(\mathbf{X}))/C\right]$ approaches a limit in a monotone and non-increasing fashion. We must therefore resort to another approach to find the minimum average distortion of a code of rate R. This approach is through a theory called rate-distortion theory and actually yields a method of code construction which is "highly probable" to achieve optimal performance. We shall now introduce this theory to make the above statements more precise.

6.3.1.2 Rate-distortion theory

Consider a continuous-amplitude vector source ensemble $\mathbf{X}$ and a reproduction vector ensemble $\mathbf{Y}$ and a stochastic relationship between them governed by the conditional probability density function $p_{\mathbf{Y}/\mathbf{X}}(\mathbf{y}/\mathbf{x})$ of a reproduction vector $\mathbf{Y} = \mathbf{y}$ given a source vector $\mathbf{X} = \mathbf{x}$. The conditional probability density function $p_{\mathbf{Y}/\mathbf{X}}(\mathbf{y}/\mathbf{x})$ is called a test channel, which is depicted in Figure 6.4.

This test channel description relating output to input, being probabilistic, is not coding, in which the equivalent channel is entirely deterministic. A theorem to be stated later will establish the relationship to coding. The average distortion per source letter can be calculated for the joint density function $q_{\mathbf{X}}(\mathbf{x})\, p_{\mathbf{Y}/\mathbf{X}}/(\mathbf{y}/\mathbf{x})$ for the test channel joint input–output ensemble $\mathbf{XY}$ as

$$E\left[\frac{1}{N}d_N(\mathbf{X}, \mathbf{Y})\right] = \int\int \frac{1}{N}d_N(\mathbf{x}, \mathbf{y})q_{\mathbf{X}}(\mathbf{x})p_{\mathbf{Y}/\mathbf{X}}(\mathbf{y}/\mathbf{x})\, d\mathbf{x}d\mathbf{y} \tag{6.18}$$

The (average mutual) information per letter between the input and output ensembles is

$$\frac{1}{N}I_N(\mathbf{X}; \mathbf{Y}) = E\left[\frac{1}{N}\,\log\,\frac{p_{\mathbf{Y}/\mathbf{X}}(\mathbf{y}/\mathbf{x})}{w_{\mathbf{Y}}(\mathbf{y})}\right]$$

$$= \frac{1}{N}\int\int q_{\mathbf{X}}(\mathbf{x})p_{\mathbf{Y}/\mathbf{X}}(\mathbf{y}/\mathbf{x})\,\log\,\frac{p_{\mathbf{Y}/\mathbf{X}}(\mathbf{y}/\mathbf{x})}{w_{\mathbf{Y}}(\mathbf{y})}\, d\mathbf{x}d\mathbf{y} \tag{6.19}$$

where $w_{\mathbf{Y}}(\mathbf{y}) = \int q_{\mathbf{X}}(\mathbf{x})p_{\mathbf{Y}/\mathbf{X}}(\mathbf{y}/\mathbf{x})\, d\mathbf{x}$ is the probability density function of the output vector random variable $\mathbf{Y}$. The source vector probability distribution $q_{\mathbf{X}}(\mathbf{x})$ is considered to be known, so that the information $I_N(\mathbf{X}; \mathbf{Y})$ is a function of the test channel and is hence denoted as $I_N(p)$.

Consider now the problem of minimizing the average distortion per letter over all test channels $p_{\mathbf{Y}/\mathbf{X}}(\mathbf{y}/\mathbf{x})$ giving an average mutual information per letter no greater than some fixed rate $R \geq 0$. The mathematical statement is

$$D_N(R) = \inf_p\left\{E\left[\frac{1}{N}\,d_N(\mathbf{X}, \mathbf{Y})\right] : \frac{1}{N}I(p) \leq R\right\} \tag{6.20}$$

The result is an average distortion $D_N(R)$ which is a function of R and depends on N. The function $D_N(R)$ is called the N-tuple distortion-rate function. The distortion-rate function $D(R)$ is

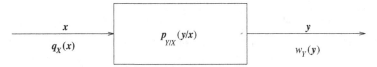

Figure 6.4   Test channel.

$$D(R) = \inf_N D_N(R) = \lim_{N \to \infty} D_N(R) \qquad (6.21)$$

The corresponding inverse functions, $R_N(D)$ and $R(D)$, are stated mathematically as

$$R_N(D) = \inf_p \left\{ \frac{1}{N} I_N(p) \ : \ E\left[\frac{1}{N} d_N(\mathbf{X}, \mathbf{Y})\right] \le D \right\} \qquad (6.22)$$

$$R(D) = \inf_N R_N(D) = \lim_{N \to \infty} R_N(D) \qquad (6.23)$$

and are called the N-tuple rate-distortion function and rate-distortion function, respectively.

We now restrict the distortion measure to be based on a single-letter distortion $d(x, y)$ between corresponding vector components according to

$$d_N(\mathbf{x}, \mathbf{y}) = \sum_{i=1}^{N} d(x_i, y_i)$$

We now state the following theorems which connect the above functions to coding.

THEOREM 6.1 **(Distortion-rate theorem and converse)** *For stationary and ergodic sources (under certain technical conditions) and a single-letter-based distortion measure, for any $\epsilon > 0$ and N sufficiently large, there exists a code C of rate R such that*

$$E\left[\frac{1}{N} d_N(\mathbf{X}, \mathbf{y}_m(\mathbf{X}))|C\right] < D(R) + \epsilon$$

Furthermore, there exists no code of rate R such that

$$E\left[\frac{1}{N} d_N(\mathbf{X}, \mathbf{y}_m(\mathbf{X}))|C\right] < D(R)$$

The distortion-rate function $D(R)$ is an unbeatable lower bound on average distortion for a code of rate R.

The inverse statement of the above theorem is also common.

THEOREM 6.2 **(Rate-distortion theorem and converse)** *For stationary and ergodic sources (under certain technical conditions) and a single-letter-based distortion measure, given $D \ge 0$ and $\epsilon > 0$, for N sufficiently large there exists a code of rate R such that*

$$R < R(D) + \epsilon.$$

Furthermore, there exists no code of rate

$$R < R(D).$$

The rate-distortion function $R(D)$ is an unbeatable lower bound on rate for a code with average distortion D.

The above two theorems are likewise valid for discrete-alphabet sources. The formulas for per-letter distortion and mutual information in Equations (6.18) and (6.19) and the distortion-rate and rate-distortion functions in Equations (6.20) and (6.23) reduce to their scalar counterparts in the case of i.i.d. sources, as introduced in Chapter 5 on scalar quantization.

Rate-distortion theory tells us that a source, whether or not it has memory, can be more efficiently coded if we encode long blocks of data values, the larger the better. This is the interpretation of Equation (6.23) stating that $\lim_{N \to \infty} R_N(D) = R(D)$. The function $R_N(D)$ is the N-tuple rate-distortion function and gives the least possible rate for independent encoding of blocks of N-tuples from the source with average distortion no greater than D. As N increases, $R_N(D)$ grows closer to $R(D)$ for stationary, block-ergodic sources. In fact, the following bound by Wyner and Ziv [1] for the difference between $R_N(D)$ and $R(D)$,

$$R_N(D) - R(D) \leq h_N(\mathbf{X}) - h_\infty(\mathbf{X}) \tag{6.24}$$

$$= \lim_{n \to \infty} \left[\frac{1}{N} I(X_1, X_2, \ldots, X_N; X_{N+1}, \ldots, X_{N+n}) \right]$$

is informative concerning the decrease of rate with N. This rate difference is bounded by the average mutual information per sample that an N-tuple conveys about the infinite remainder of the sequence. As N grows larger, this information must decrease. The bound holds with equality when the average distortion is below some critical distortion D_c, the value of which depends on the particular source and distortion measure. Optimal coding of N-tuples from a source with memory, just as with scalars, requires searches over long sequences of N-tuples to best match a long data stream and is computationally burdensome. A suboptimal approach, again paralleling scalar coding, is to make an instantaneous decision on the reproduction of a single N-tuple. This approach is called vector quantization and is only optimal in the limit of large N.

6.3.1.3   The Gaussian, MSE rate-distortion function

An analytical solution to the rate-distortion minimization arises only in the case of the Gaussian source with MSE distortion measure. It is useful and enlightening to sketch the derivation. Assume that the source is stationary and Gaussian. Let the vector $\mathbf{X} = (X_1, X_2, \ldots, X_N)$ be a block of N samples from the source and denote its N-dimensional mean vector by $\mu = (\mu_1, \mu_2, \ldots, \mu_N)$ and its $N \times N$ covariance matrix by Φ_N. The elements of the covariance matrix may be expressed as

$$\phi_{i,j} = E[(X_i - \mu_i)(X_j - \mu_j)] \equiv C_X(|i - j|), \quad i, j = 0, 1, \ldots, N - 1,$$

where $C_X(k)$ stands for the covariance function with lag k. The probability density function of the Gaussian random vector $\mathbf{X}$ is therefore

$$p_\mathbf{X}(\mathbf{x}) = \frac{1}{(2\pi)^{N/2} |\Phi_N|^{1/2}} \exp -\frac{1}{2} (\mathbf{x} - \mu)^t \Phi_N^{-1} (\mathbf{x} - \mu), \tag{6.25}$$

for the source realization of $\mathbf{X} = \mathbf{x}$. In fact, this may be the only non-trivial probability density formula known for more than one random variable.

The covariance matrix Φ_N is real, symmetric, and non-negative definite. Therefore, there exists a set of N orthonormal eigenvectors $\mathbf{e}_1, \mathbf{e}_2, \ldots, \mathbf{e}_N$ and a corresponding set of N non-negative, not necessarily distinct, eigenvalues $\lambda_1, \lambda_2, \ldots, \lambda_N$ for Φ_N. A given source sequence $\mathbf{X} = \mathbf{x}$ can be represented with respect to the eigenvector basis by

$$\mathbf{x} = P\widetilde{\mathbf{x}} \;, \quad P = [\mathbf{e}_1|\mathbf{e}_2|\ldots|\mathbf{e}_N], \tag{6.26}$$

where P is a unitary matrix ($P^{-1} = P^t$).

The vector $\widetilde{\mathbf{x}}$ is called the transform of the source vector $\mathbf{x}$. Without loss of generality, assume that the source is zero mean, so that $\mu = \mathbf{0}$.[1] The covariance matrix of the transform vector ensemble $\widetilde{\mathbf{X}}$ is

$$\begin{aligned} \Lambda \;&=\; E[\widetilde{\mathbf{X}}\widetilde{\mathbf{X}}^t] \;\;=\; E[P^{-1}\mathbf{X}\mathbf{X}^t P] \\[2mm] &=\; P^{-1}\Phi_N P \;\;=\; \begin{pmatrix} \lambda_1 & & & \\ & \lambda_2 & & 0 \\ & & \ddots & \\ 0 & & & \lambda_N \end{pmatrix} \end{aligned} \tag{6.27}$$

The components of $\widetilde{\mathbf{X}}$ are uncorrelated with variances equal to the eigenvalues. This transformation of $\mathbf{x}$ to $\widetilde{\mathbf{x}}$ is called the (discrete) Karhunen–Loeve transform (KLT). Moreover, since the source vector ensemble is assumed to be Gaussian, the ensemble $\widetilde{\mathbf{X}}$ is also Gaussian with independent components.

When we consider the problem of source encoding $\mathbf{X}$ with the single-letter squared error distortion measure $d(x, \tilde{x}) = (x - \tilde{x})^2$, it is equivalent to encode $\widetilde{\mathbf{X}}$ with the same distortion measure $d(\mathbf{x}, \widetilde{\mathbf{x}}) =| \mathbf{x} - \widetilde{\mathbf{x}} |^2$, because average mutual information and squared error are preserved in the unitary transformation of $\mathbf{X}$ to $\widetilde{\mathbf{X}}$. The N-tuple rate-distortion (or distortion-rate) function is solved for the transform ensemble $\widetilde{\mathbf{X}}$ and is found to be expressed parametrically by

$$R_N = \frac{1}{N} \sum_{n=1}^{N} \max[0, \frac{1}{2} \log \frac{\lambda_n}{\theta}] \tag{6.28}$$

$$D_N = \frac{1}{N} \sum_{n=1}^{N} \min[\theta, \lambda_n] \tag{6.29}$$

As N approaches infinity, the eigenvalues approach in a pointwise fashion a sampling of the discrete-time source power spectral density $S_X(\omega) = \sum_{n=-\infty}^{\infty} \phi(n)e^{-j\omega n}$, where $\{\phi(n)\}$ is the covariance sequence, and the formulas above tend toward the limit of the rate-distortion (or distortion-rate) function:

$$R_\theta = \frac{1}{2\pi} \int_{-\pi}^{\pi} \max\left[0, \frac{1}{2} \log \frac{S_X(\omega)}{\theta}\right] d\omega \tag{6.30}$$

$$D_\theta = \frac{1}{2\pi} \int_{-\pi}^{\pi} \min\left[\theta, \; S_X(\omega)\right] d\omega \tag{6.31}$$

Notice in Equations (6.28) and (6.29) that each transform coefficient whose variance λ_n exceeds θ is encoded with distortion

$$d_n = \theta$$

and rate

$$r_n = \frac{1}{2} \log \frac{\lambda_n}{\theta}.$$

This rate is exactly the rate-distortion function or minimal rate for encoding this coefficient with distortion no greater than θ. The regions of rate assignment are depicted in Figure 6.5 for the analogous continuous spectrum case. This solution therefore points toward a method of optimal coding. First perform a KLT of a vector of arbitrary length N from the source. Then encode each coefficient whose variance $\lambda_n > \theta$ with rate r_n. Each coefficient with variance $\lambda_n \leq \theta$ receives zero rate, so remains uncoded and is set equal to its known mean in the decoder. This solution inspires and justifies the method of transform coding, which is developed in substantial detail in the next three chapters.

For now, we wish to show the rate assignment for the special case when all the variances λ_n exceed the distortion parameter θ. In this case, Equations (6.28) and (6.29) reveal that

$$D_N = \theta$$

and

$$r_n = \frac{1}{2} \log \frac{\lambda_n}{D_N}, n = 1, 2, \ldots, N.$$

Using $R = \frac{1}{N} \sum_{n=1}^{N} r_n$ and some algebraic manipulations, we find that

$$D_N = \left(\prod_{n=1}^{N} \lambda_n \right)^{1/N} 2^{-2R} \tag{6.32}$$

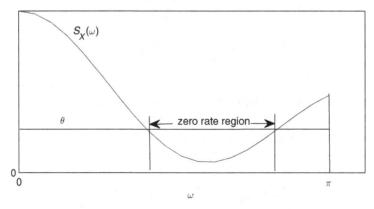

Figure 6.5   Continuous spectral density and parameter level θ dictating the optimal rate assignment versus frequency.

$$r_n = R + \frac{1}{2} \log \frac{\lambda_n}{(\prod_{n=1}^{N} \lambda_n)^{1/N}}, n = 1, 2, \ldots, N, \qquad (6.33)$$

where r_n is the optimal rate to encode the n-th KLT coefficient having variance λ_n.

6.3.2 Vector (block) quantization (VQ)

Consider N values emitted from a data source, $\mathbf{x} = (x_1, x_2, \ldots, x_N)$ and a codebook for K reproduction N-tuples $\mathbf{y}_1, \mathbf{y}_2, \ldots, \mathbf{y}_K$ and a distortion measure $d(\mathbf{x}, \mathbf{y}_m)$ between the source vector $\mathbf{x}$ and any member of the codebook $\mathbf{y}_m, m = 1, 2, \ldots, K$. The encoder searches for $\mathbf{y}_m$ such that $d(\mathbf{x}, \mathbf{y}_m) \leq d(\mathbf{x}, \mathbf{y}_\ell)$ for all $\ell \neq m$ and emits the binary index of m. The rate of the code is then $R = \frac{1}{N} \log_2 K$ bits per source sample. The objective is to choose the codebook of rate R that minimizes the average distortion $E[d(\mathbf{x}, \mathbf{y}_m)]$. This procedure is the same as that depicted in Figure 6.3 and such an encoder is called a vector or block quantizer, or simply VQ. This quantizer induces a rule Q that maps $\mathbf{x}$ to $\mathbf{y}_m$. In lieu of a search, given the codebook $Y = \{\mathbf{y}_1, \mathbf{y}_2, \ldots, \mathbf{y}_K\}$ and the distortion, we can determine a set of decision regions $\mathcal{R}_m$ such that for $\mathbf{x}$ in $\mathcal{R}_m$, $Q(\mathbf{x}) = \mathbf{y}_m$ i.e., $\mathbf{x} \in \mathcal{R}_m$ implies that $d(\mathbf{x}, \mathbf{y}_m) \leq d(\mathbf{x}, \mathbf{y}_\ell)$ for all $\ell \neq m$ or $\mathcal{R}_m = Q^{-1}(\mathbf{y}_m) = \{\mathbf{x} \in \mathcal{R}^N : Q(\mathbf{x}) = \mathbf{y}_m\}$. The regions $\mathcal{R}_m$ partition Euclidean N-space:

$$\bigcup_{m=1}^{K} \mathcal{R}_m = \mathcal{R}^N, \qquad \mathcal{R}_m \cap \mathcal{R}_\ell = \emptyset, \quad \ell \neq m$$

for an arbitrary tie-breaking rule. The encoder identifies $\mathcal{R}_m$ for the given $\mathbf{x}$ and sends the index m in binary. At the decoder m is mapped uniquely to $\mathbf{y}_m$ for the reproduction. Figure 6.6 illustrates in two dimensions the decision regions and their reproduction points.

It is instructive to give an example to illustrate the computation and storage involved in a typical vector quantizer.

Example 6.4 Suppose that we wish to transmit $R = 1$ bit per dimension (or sample) and use vectors of dimension $N = 8$. Then we need a codebook consisting of $M = 2^{RN} = 2^{1*8} = 256$ vectors. We need to search through 256 vectors to find the one with minimum distortion with respect to the current source vector of the same dimension.

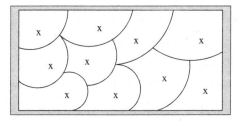

Figure 6.6 Decision regions and their reproduction points marked by 'X' for a hypothetical two-dimensional vector quantizer.

The index of the codebook vector found in this search can be represented by $\log_2 256 = 8$ bits, which realizes 1 bit per dimension, if transmitted raw without entropy coding. Note that increasing the rate to $R = 2$ bits per dimension squares the storage and search space to $2^{16} = 65\,536$ reproduction vectors in the codebook. Doubling the dimension for a fixed rate has the same effect. Therefore, one must be careful about not using too high a rate or dimension in vector quantization in order to hold down the computation and storage cost.

Optimal vector quantizer design involves specific choices of codebook vectors and a specific rule for associating a source vector to a reproduction vector in the codebook. Such a rule defines the decision (Voronoi) regions, which are the regions of N-space that are represented by the codebook vector contained within. The following two conditions are necessary for a quantizer to be optimum. Gray *et al.* [2] have shown that these conditions are sufficient for a local optimum.

1. Given a codebook $\mathbf{y}_1, \mathbf{y}_2, \ldots, \mathbf{y}_K$, the decision regions must be such that

$$\mathcal{R}_m = \{\mathbf{x} : d(\mathbf{x}, \mathbf{y}_m) \leq d(\mathbf{x}, \mathbf{y}_\ell) \quad \text{all } \ell \neq m\} \qquad m = 1, 2, \ldots, K$$

   These regions are called *Voronoi* regions.
2. Given any set of K decision regions $\mathcal{R}_1, \mathcal{R}_2, \ldots, \mathcal{R}_K$ the reproduction points $\mathbf{y}_m, m = 1, 2, \ldots, K$ are such that

$$E[d(\mathbf{x}, \mathbf{y}_m)/\mathbf{x} \in \mathcal{R}_m] = \min_{\mathbf{v}} E[d(\mathbf{x}, \mathbf{v})/\mathbf{x} \in \mathcal{R}_m]$$

Clearly, the decoder knows only that $\mathbf{x}$ is in $\mathcal{R}_m$ and must pick the point $\mathbf{v}$ in $\mathcal{R}_m$ that minimizes the expected distortion. This is easily seen since the quantizer distortion can be expressed as:

$$E[d(\mathbf{x}, Q(\mathbf{x}))/\{\mathcal{R}_m\}] = \sum_{m=1}^{K} \underbrace{E[d(\mathbf{x}, \mathbf{v})/\mathbf{x} \in \mathcal{R}_m]}_{\text{minimize for each } m \text{ over } \mathbf{v}} \times \Pr\{\mathbf{x} \in \mathcal{R}_m\}$$

For squared error $d(\mathbf{x}, \mathbf{v}) = \|\mathbf{x} - \mathbf{v}\|^2$ and $\mathbf{y}_m = E[\mathbf{x}/\mathbf{x} \in \mathcal{R}_m]$.

Proof   Assume that some $\mathbf{v} \neq \mathbf{y}_m$ gives lower error:

$$\begin{aligned}
E[\|\mathbf{x} - \mathbf{v}\|^2/\mathbf{x} \in \mathcal{R}_m] &= E[\|(\mathbf{x} - \mathbf{y}_m) + (\mathbf{y}_m - \mathbf{v})\|^2/\mathbf{x} \in \mathcal{R}_m] \\
&= E[\|\mathbf{x} - \mathbf{y}_m\|^2/\mathbf{x} \in \mathcal{R}_m] - 2(\mathbf{y}_m - \mathbf{v}) \cdot E[(\mathbf{x} - \mathbf{y}_m)/\mathbf{x} \in \mathcal{R}_m] \\
&\quad + E[\|\mathbf{y}_m - \mathbf{v}\|^2/\mathbf{x} \in \mathcal{R}_m] \qquad\qquad (6.34) \\
&\geq E[\|\mathbf{x} - \mathbf{y}_m\|^2] \quad \text{with equality} \quad \mathbf{y}_m = \mathbf{v}
\end{aligned}$$

The cross-product term in Equation (6.34) equals zero by assumption that $\mathbf{y}_m$ is the conditional mean. The concluding statement is a contradiction and so proves that $\mathbf{y}_m$ minimizes the MSE.

For squared error $\mathbf{y}_m$ is the "centroid" of $\mathcal{R}_m$. Condition 2 is called the *centroid condition* for any distortion measure.

Also, we can show for any partition $\{\mathcal{R}_m\}$ and optimal output points $\mathbf{y}_m = E[\mathbf{x}/\mathbf{x} \in \mathcal{R}_m]$ for squared-error distortion, the MSE

$$e^2 = E[\|\mathbf{x} - Q(\mathbf{x})\|^2] = E[\|\mathbf{x}\|^2] - E[\|\mathbf{y}_m\|^2]$$

which means that the average squared value or power in the reproduction ensemble is lower than that in the source. This result contradicts the often used model of the quantization noise vector being independent and additive to the source vector.

Proof

$$e^2 = E[\|\mathbf{x} - Q(\mathbf{x})\|^2] = E[(\mathbf{x} - Q(\mathbf{x}))^T \mathbf{x}] - E[(\mathbf{x} - Q(\mathbf{x}))^T Q(\mathbf{x})]$$

Since $\mathbf{y}_m = Q(\mathbf{x}) = E[\mathbf{x}/\mathbf{x} \in \mathcal{R}_m]$, the second term equals zero. Expanding the first term,

$$
\begin{aligned}
e^2 &= E[\|\mathbf{x}\|^2] - E[(Q(\mathbf{x}))^T \mathbf{x}] \\
&= E[\|\mathbf{x}\|^2] - \sum_{m=1}^{M} E[\mathbf{x}^T \mathbf{y}_m / \mathbf{x} \in \mathcal{R}_m] \text{Pr}\{\mathbf{x} \in \mathcal{R}_m\} \\
&= E[\|\mathbf{x}\|^2] - \sum_{m=1}^{M} \|\mathbf{y}_m\|^2 \text{Pr}\{\mathbf{x} \in \mathcal{R}_m\} \\
&= E[\|\mathbf{x}\|^2] - E[\|Q(\mathbf{x})\|^2]
\end{aligned}
$$

There is an algorithm by Linde *et al.* [3, 4] (LBG algorithm) based on the two necessary conditions for designing a quantizer codebook and decision regions for a known source probability distribution $q(\mathbf{x})$. It is a generalization to vectors of the Lloyd Method I [5] for scalar quantizers and is hence often called the generalized Lloyd algorithm (GLA). The steps of the GLA are stated in Algorithm 6.1.

This algorithm is a fixed point algorithm for Y and $\{\mathcal{R}_m\}$, that is,

$$Y^{i+1} = f(Y^i) \quad \text{and} \quad \{\mathcal{R}_m^{(i+1)}\} = g(\{\mathcal{R}_m^i\})$$

for some functions f and g. Note that $D_{i-1} \geq D_i$, so each iteration cannot increase the error. Since D^i is non-increasing and non-negative, it must have some limit D_∞. Gray *et al.* [2] show that $\{\mathcal{R}_m^{(\infty)}\}$ and Y^∞ have the proper relationship of centroids of own partition and that there is convergence to a fixed point.

A necessary condition for a quantizer to be optimal is that it be a fixed point quantizer. It is locally optimum if no probability is on the boundary of each decision region. For a continuous probability distribution, this presents no problem. However, for a discrete distribution, there may be a problem.

For an unknown source probability distribution use a sequence of source vectors to train the quantizer and insert them into Steps (1) and (3) using numerical averages instead of expectations or ensemble averages. We obtain partitions and centroids with respect to the sequence of training vectors (called a *training sequence*). Then we use

ALGORITHM 6.1 _____

Generalized Lloyd algorithm (GLA)

(0) Initialization: choose some reproduction alphabet (codebook) $Y^0 = \{\mathbf{y}_1^{(0)}, \mathbf{y}_2^{(0)}, \ldots, \mathbf{y}_K^{(0)}\}$.

(1) Given Y^i find minimum distortion decision regions $\mathcal{R}_1^{(i)}, \mathcal{R}_2^{(i)}, \ldots, \mathcal{R}_K^{(i)}$ such that

$$\mathcal{R}_m^{(i)} = \{\mathbf{x}; d(\mathbf{x}, \mathbf{y}_m^{(i)}) \leq d(\mathbf{x}, \mathbf{y}_\ell^{(i)}), \quad \text{all } \ell \neq m\}$$

Compute resulting average distortion

$$D^{(i)} = \sum_{m=1}^{K} E[d(\mathbf{x}, \mathbf{y}_m^{(i)})/\mathbf{x} \in \mathcal{R}_m^{(i)}]\Pr\{\mathbf{x} \in \mathcal{R}_m^{(i)}\}$$

(2) If $(D^{(i-1)} - D^{(i)})/D^{(i)} < \epsilon$ for some $\epsilon > 0$, stop.

(3) For decision regions $\mathcal{R}_1^{(i)}, \mathcal{R}_2^{(i)}, \ldots, \mathcal{R}_K^{(i)}$ in Step (1), find centroids (optimum reproduction points) $\mathbf{y}_m^{(i+1)}$ such that

$$E[d(\mathbf{x}, \mathbf{y}_m^{(i+1)})/\mathbf{x} \in \mathcal{R}_m^{(i)}] = \min_{\mathbf{v}} E[d(\mathbf{x}, \mathbf{v})/\mathbf{x} \in \mathcal{R}_m^{(i)}]$$

(For squared error $\mathbf{y}_m^{(i+1)} = E[\mathbf{x}/\mathbf{x} \in \mathcal{R}_m^{(i)}]$)

$$Y^{i+1} = \{\mathbf{y}_1^{(i+1)}, \mathbf{y}_2^{(i+1)}, \ldots, \mathbf{y}_K^{(i+1)}\}$$

(4) $i \to i + 1$. Go to (1).

_____

those partitions and centroids for other data judged to have similar statistics (i.e., *outside* the training sequence). Algorithm 6.2 states the steps of this algorithm using the training sequence (TS).

Example 6.5 In order to illustrate the design of a vector quantizer, we shall consider a particular case of a two-dimensional, MMSE quantizer with a codebook containing four reproduction vectors. Let us assume that the source vector $\mathbf{X} = (X_1, X_2)$ is Gaussian with components X_1 and X_2 having mean 0, variance 1, and correlation coefficient 0.9. Instead of using integration to calculate distortions and centroids, we shall use averages of the appropriate quantities from a training sequence. We generate a sequence of 10 000 training vectors of two dimensions drawn from the assumed probability distribution. Of course, once we have the training sequence, we do not need to know the probability distribution.

1. Initialization: draw any four vectors from the training sequence to form the initial codebook.
2. For each codevector, calculate squared Euclidean distance from every training vector.
3. For each codevector, identify training vectors closer to it than to any other codevector. These are the training vectors that map to that codevector, thereby defining the codevector's decision region.

ALGORITHM 6.2 ————————————————————————————————————

GLA or LBG algorithm – training sequence version

(0) Initialization: Set $\epsilon \geq 0$, and an initial codebook $Y^0 = \{\mathbf{y}_1^{(0)}, \ldots, \mathbf{y}_K^{(0)}\}$. Given a TS $\widetilde{\mathbf{x}}_j$, $j = 1, 2, \ldots, n$ of N-tuples.
$D_{-1} = \infty$
$i = 0$.

(1) Given Y^i, find minimum distortion decision (Voronoi) regions of TS: $\mathcal{R}_1^i, \mathcal{R}_2^i, \ldots, \mathcal{R}_K^{(i)}$ such that if $\widetilde{\mathbf{x}}_j \in \mathcal{R}_m^i$ then all $d(\widetilde{\mathbf{x}}_j, \mathbf{y}_m^i) \leq d(\widetilde{\mathbf{x}}_j, \mathbf{y}_\ell^i)$ all $\ell \neq m$.
Calculate distortion per sample:

$$D^i = \frac{1}{Nn} \sum_{m=1}^{K} \sum_{\widetilde{\mathbf{x}}_j \in \mathcal{R}_m^i} d(\widetilde{\mathbf{x}}_j, \mathbf{y}_m^i)$$

(2) If $(D_{i-1} - D_i)/D_i < \epsilon$, stop. If not, go to (3).

(3) Find optimal reproduction alphabet for TS partitions. For $m = 1, 2, \ldots, K$ and for all $\widetilde{\mathbf{x}}_j \in \mathcal{R}_m^i$, find $\mathbf{v} \in \mathcal{R}_m$ such that

$$\frac{1}{\|\mathcal{R}_m^i\|} \sum_{\widetilde{\mathbf{x}}_j \in \mathcal{R}_m^i} d(\widetilde{\mathbf{x}}_j, \mathbf{v}) \leq \frac{1}{\|\mathcal{R}_m^i\|} \sum_{\widetilde{\mathbf{x}}_j \in \mathcal{R}_m^i} d(\widetilde{\mathbf{x}}_j, \mathbf{u}) \quad \text{for all } \mathbf{u} \neq \mathbf{v} \text{ in } \mathcal{R}_m^i,$$

where $\|\mathcal{R}_m^i\|$ denotes the number of $\widetilde{\mathbf{x}}_k$'s in $\mathcal{R}_m^i$, and set $\mathbf{v} = \mathbf{y}_m^{(i+1)}$. (For squared error,
$\mathbf{y}_m^{(i+1)} = \frac{1}{\|\mathcal{R}_m^i\|} \sum_{\widetilde{\mathbf{x}}_j \in \mathcal{R}_m^i} \widetilde{\mathbf{x}}_j$).
Set $i = i + 1$, and go to (1).

————————————————————————————————————

4. Calculate average of training vectors in each decision region. These average vectors are the centroids of the decision regions and therefore are the codevectors of the updated codebook.

5. Calculate MSE with respect to training sequence vectors in the decision regions.

6. Calculate fractional decrease in MSE afforded by the updated codebook. If greater than 0.01 (1%), return to Step 2, otherwise go to next step.

7. Report codevectors of last codebook in Step 4 and MSE in Step 5.

Results for the first four and last three iterations are given in Table 6.1. The MSE decreases successively for each iteration until convergence at the last iteration. The last iteration, number 10, produces the final codebook, which lies close to the 45-degree line in the plane. Most likely, the slight deviation from this line is due to the inexactness of the random number generator and perhaps too small a training set. Assuming that the 45-degree line is correct, the boundaries of the decision regions are the lines that are perpendicular to the 45-degree line and pass through the line at points midway between the codevector points. Figure 6.7 depicts this solution for the codevectors and decision regions.

The output of the quantizer is actually the index of the codevector of least distance to the source vector. This index can be represented naturally with two bits. If every index is sent with its natural two-bit representation, the rate of the codebook is $R = \log_2 4/2 = 1$ bit per dimension.

Table 6.1 Calculation results in design of two-dimensional Gaussian quantizer, $\rho = 0.9$

Iteration	Codebook (Step 4)	MSE (Step 5)
1	(0.892,0.880) (−0.279,0.261) (−0.882,−0.880) (0.286,−0.278)	0.321
2	(1.092,1.084) (−0.204,0.160) (−1.089,−1.086) (0.203,−0.188)	0.232
3	(1.166,1.157) (−0.220,0.131) (−1.168,−1.167) (0.213,−0.165)	0.220
4	(1.194,1.189) (−0.259,0.0786) (−1.199,−1.199) (0.253,−0.115)	0.217
⋮	⋮	⋮
8	(1.396,1.392) (−0.398,−0.390) (−1.384,−1.377) (0.412,0.378)	0.168
9	(1.439,1.435) (−0.404,−0.404) (−1.408,−1.398) (0.432,0.406)	0.166
10	(1.468,1.459) (−0.407,−0.408) (−1.421,−1.411) (0.446,0.424)	0.165

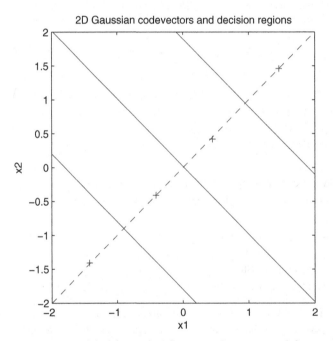

Figure 6.7   Codevectors and decision regions in range $-2 \le x_1, x_2 \le 2$ for two-dimensional Gaussian, $\rho = 0.9$, quantizer in example.

6.3.2.1   Initializing the codebook

The algorithms above converge to a local minimum, but not necessarily a global minimum. Therefore, some care must be exercised in choosing an initial codebook, so that the algorithm does not get trapped in a local minimum. Often, a reasonable start is to use the product quantizer that is a uniform scalar quantizer for each dimension. Another method called the splitting method [4] starts with rate per vector $R = 0$ and increments the rate by one at each stage until the desired rate is reached. It is accomplished by

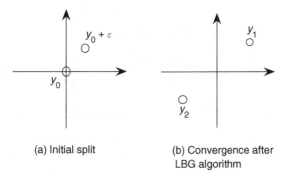

(a) Initial split

(b) Convergence after
LBG algorithm

Figure 6.8   (a) Binary splitting of covectors. (b) Convergence to two best codevectors.

successive splitting into two reproduction vectors obtained at a given rate from the
above algorithm and using them to initialize the algorithm for the next higher rate. We
shall elaborate on this method now and later when we describe tree-structured vector
quantization. The idea is to start with the rate $R' = 0$ codebook, consisting of the cen-
troid of the alphabet or training sequence, usually $\mathbf{y}_o = \mathbf{0}$. Then we split off from $\mathbf{y}_o$
the vector $\mathbf{y}_o + \epsilon$ to obtain two vectors, $\mathbf{y}_o$ and $\mathbf{y}_o + \epsilon$, where ϵ is a small perturbation
vector. These two vectors initialize execution of the GLA or LBG algorithm to con-
verge to the best $R' = 1$ bit codebook with codevectors $\mathbf{y}_1$ and $\mathbf{y}_2$. The initial splitting
and convergence to these two covectors are illustrated in Figure 6.8. Then we split off
again from both vectors to obtain four codevectors $\mathbf{y}_1$, $\mathbf{y}_1 + \epsilon$, $\mathbf{y}_2$ and $\mathbf{y}_2 + \epsilon$. These four
vectors initialize a run of the GLA or LBG algorithm that converges to the best $R' = 2$
bit codebook of four vectors. This process of splitting every codevector into two and
running the VQ design algorithm on the doubled codebook continues until the desired
$M = 2^{R'}$ codevectors are reached. One can see that the basic procedure produces a
higher rate VQ design from a lower one in 1 bit increments. The steps of the algorithm
are delineated in Algorithm 6.3.

6.3.3   Entropy constrained vector quantization

As with the Lloyd method in scalar quantization, the generalized Lloyd method for
vector quantization optimizes a criterion which is inappropriate for optimal coding in
the true rate versus distortion sense. When the indices of the reproduction vectors are
considered as a new source, then a subsequent entropy coding can achieve a rate arbitrarily close to the entropy of this index source, called the index entropy. It is therefore
appropriate to reformulate the objective to minimize the average distortion subject to a
constraint on the index entropy, which is the entropy of the reproduction vector ensemble. In order to accomplish this objective, we first define an "impurity" function between
the source vector $\mathbf{x}$ and its reproduction vector $\mathbf{y} = Q(\mathbf{x})$, as follows

$$j(\mathbf{x}, \mathbf{y}) = d(\mathbf{x}, \mathbf{y}) - \lambda \log P(\mathbf{y}), \tag{6.35}$$

ALGORITHM 6.3

Design of higher rate VQ from a lower rate VQ – splitting algorithm

1. Initialization: let $\mathbf{y}_o = \mathbf{0}$ (or centroid) be the codebook; set $K = 0$.
2. Given codebook of K codevectors, $Y^{(K)} = \{\mathbf{y}_1, \mathbf{y}_2, \dots, \mathbf{y}_K\}$, generate codebook of $2K$ codevectors $Y^{(2K)} = \{\mathbf{y}_1, \mathbf{y}_1\mathbf{y}_1 + \epsilon, \mathbf{y}_2, \mathbf{y}_2 + \epsilon, \dots, \mathbf{y}_K, \mathbf{y}_K + \epsilon\}$. ϵ is a small perturbation vector.
3. Use $Y^{(2K)}$ as initial codebook for LBG or GLA quantization algorithm to generate optimal codebook of $2K$ vectors. Let $K \leftarrow 2K$.
4. Stop if $K = M$, the target codebook size. Otherwise, return to Step 2.

where $\lambda > 0$ is a Lagrange parameter. The objective function to minimize is

$$J = E[j(\mathbf{x}, \mathbf{y})],$$
$$= E[d(\mathbf{x}, \mathbf{y})] - \lambda E[\log P(\mathbf{y})]$$
$$= D + \lambda H(\mathbf{Y}), \tag{6.36}$$

under the constraint that

$$H(\mathbf{Y}) \leq R. \tag{6.37}$$

D and $H(\mathbf{Y})$ are the average distortion per vector and entropy per vector, respectively. The Lagrange parameter λ is evaluated from the entropy constraint in Equation (6.37).

The solution in the vector space is assumed to lie on the boundary $H(\mathbf{Y}) = R$, due to the presupposition of a convex distortion–entropy function. Looking at a graph of a typical convex distortion D versus entropy H of a quantizer, shown in Figure 6.9, we see

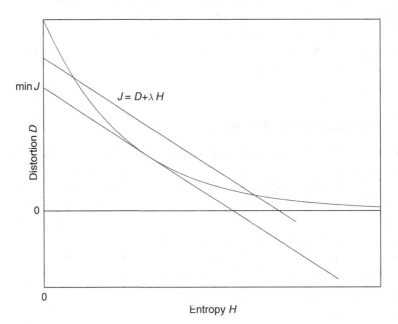

Figure 6.9   Depiction of the objective minimum of an entropy-constrained vector quantizer.

that the minimum of J, for a (H, D) point on this graph, lies on the line $J = D + \lambda H$ that is tangent to the graph. The parameter λ thus turns out to be the negative of the slope of the D versus H characteristic of the quantizer. Every $\lambda > 0$ determines a point on the graph.

For a given set of K reproduction vectors $C = \{\mathbf{y}_1, \mathbf{y}_2, \ldots, \mathbf{y}_K\}$ with respective probabilities $\mathcal{P} = \{P_1, P_2, \ldots, P_K\}$ and a given $\lambda > 0$, the impurity function $j(\mathbf{x}, \mathbf{y})$ partitions the input space into a set of (Voronoi) regions

$$\mathcal{R}_m = \{\mathbf{x} : j(\mathbf{x}, \mathbf{y}_m) \leq j(\mathbf{x}, \mathbf{y}_\ell), \quad \text{all } \ell \neq m\} \quad m = 1, 2, \ldots, K \quad (6.38)$$

The above partition is one of two necessary conditions to achieve a minimum of $J = E[j(\mathbf{x}, \mathbf{y})]$. The other condition is the "centroid" condition that the set of reproduction vectors $C = \{\mathbf{y}_1, \ldots, \mathbf{y}_K\}$ satisfy

$$E[j(\mathbf{x}, \mathbf{y}_m)/\mathbf{x} \in \mathcal{R}_m] = \min_{\mathbf{v}} E[j(\mathbf{x}, \mathbf{v})/\mathbf{x} \in \mathcal{R}_m] \quad (6.39)$$
$$= \min_{\mathbf{v}} \{E[d(\mathbf{x}, \mathbf{v})/\mathbf{x} \in \mathcal{R}_m] - \lambda E[\log P(\mathbf{v})/\mathbf{x} \in \mathcal{R}_m]\},$$

paralleling the previous development with $d(\mathbf{x}, \mathbf{y})$.

Once $\mathbf{x}$ is known to be in a given region, the probability of the reproduction vector is the probability of that region. Therefore, in searching for the $\mathbf{v} \in \mathcal{R}_m$ that minimizes (6.39), only the first term varies while the second term stays constant. We need then to search for the centroid of the decision region $\mathcal{R}_m$ only with respect to the distortion measure $d(\mathbf{x}, \mathbf{v})$. The centroid condition for the measure $j(\mathbf{x}, \mathbf{y})$ is therefore the same as that for $d(\mathbf{x}, \mathbf{y})$ and is expressed by

$$E[j(\mathbf{x}, \mathbf{y}_m)/\mathbf{x} \in \mathcal{R}_m] = \min_{\mathbf{v}} E[d(\mathbf{x}, \mathbf{v})/\mathbf{x} \in \mathcal{R}_m] \quad (6.40)$$

The entropy constraint involves a set of probabilities in the initialization and computation steps. Therefore, we need to modify the GLA in the initialization step (0) and the decision regions definition in step (1) and stopping criterion in step (2) to obtain the entropy-constrained vector quantization (ECVQ) algorithm, originated by Chou *et al.* [6]. The complete algorithm is presented in Algorithm 6.4.

The entropy obtained upon convergence of the algorithm can only be achieved by entropy coding of the indices of the codevectors. The quantity $-\log P_m^{(i)}$ is actually the ideal length in bits of the associated code of the index m. This length averaged over m equals the entropy.

The initialization of the algorithm requires a set of probabilities in addition to the set of reproduction vectors of the unconstrained GLA algorithm. It also requires a guess for the value of λ which achieves the desired rate. Therefore, a good starting point is $\lambda = 0$, which reduces to the unconstrained algorithm where the entropy per vector is allowed to reach its maximum of $\log M$. A set of probabilities is not required in the initialization step for $\lambda = 0.^2$ Once the final sets of codevectors and probabilities are found for $\lambda = 0$, they are used to initialize the algorithm for the next λ which is a small increment from zero. This process is repeated for several increments of λ until all desired points of the distortion-rate curve are found. As λ increases, a greater penalty is exacted for entropy in the objective function, so the corresponding entropy

ALGORITHM 6.4 _____

Entropy constrained vector quantization (ECVQ) algorithm

(0) Choose codebook of reproduction vectors $Y^{(0)} = \{\mathbf{y}_1^{(0)}, \ldots, \mathbf{y}_K^{(0)}\}$ with their respective probabilities $\mathcal{P}^{(0)} = \{P_1^{(0)}, \ldots, P_K^{(0)}\}$. (One can run the GLA algorithm to obtain them.) Set a slope parameter $\lambda > 0$ and $i = 0$.

(1) (a) Given $Y^{(i)}$ and $\mathcal{P}^{(i)}$ calculate the decision regions

$$\mathcal{R}_m^{(i)} = \{\mathbf{x} : d(\mathbf{x}, \mathbf{y}_m^{(i)}) - \lambda \log P_m^{(i)} \le d(\mathbf{x}, \mathbf{y}_\ell^{(i)}) - \lambda \log P_\ell^{(i)} :, \ell \neq m\}$$

for all $m = 1, 2, \ldots, K$.

(b) Update probabilities: calculate $\check{P}_m(i) = Pr\{\mathcal{R}_m^{(i)}\}, m = 1, 2, \ldots, M$.

(c) Calculate resulting average distortion

$$D^{(i)} = \sum_{m=1}^{K} E[d(\mathbf{x}, \mathbf{y}_m^{(i)})/\mathbf{x} \in \mathcal{R}_m^{(i)}]\check{P}_m^{(i)}$$

and entropy

$$H^{(i)} = -\sum_{m=1}^{K} \check{P}_m^{(i)} \log \check{P}_m^{(i)},$$

$$J^{(i)} = D^{(i)} + \lambda H^{(i)}.$$

(2) If $(J^{(i-1)} - J^{(i)})/J^{(i)} < \epsilon$ for some suitable $\epsilon > 0$, stop. Otherwise, set $P_m^{(i)} = \check{P}_m^{(i)}$ and go to next step.

(3) For decision regions $\mathcal{R}_1^{(i)}, \mathcal{R}_2^{(i)}, \ldots, \mathcal{R}_K^{(i)}$ in Step (1), find centroids (optimum reproduction points) $\mathbf{y}_m^{(i+1)}$ such that

$$E[d(\mathbf{x}, \mathbf{y}_m^{(i+1)})/\mathbf{x} \in \mathcal{R}_m^{(i)}] = \min_{\mathbf{v}} E[d(\mathbf{x}, \mathbf{v})/\mathbf{x} \in \mathcal{R}_m^{(i)}]$$

(For squared error $\mathbf{y}_m^{(i+1)} = E[\mathbf{x}/\mathbf{x} \in \mathcal{R}_m^{(i)}]$)

$$Y^{i+1} = \{\mathbf{y}_1^{(i+1)}, \mathbf{y}_2^{(i+1)}, \ldots, \mathbf{y}_K^{(i+1)}\}$$

(4) $i \rightarrow i + 1$. Go to (1).

_____

decreases while distortion increases. The limit of $\lambda = \infty$ gives an entropy or rate of zero and a distortion equal to the variance per vector under the squared error distortion measure. In Chapter 8, we describe the bisection method to converge quickly to a target rate.

There is also a version of this algorithm when probabilities are unknown and a training sequence is available from which to estimate probabilities and calculate distortions and entropies. We can initialize (step (0)) the codevectors and their probabilities by running the LBG algorithm using the given training sequence. The m-th decision region in step (1) is the set of training vectors closest in the $j(\mathbf{x}, \mathbf{y}_m)$ measure to the vectors $\mathbf{y}_m$. One can determine these sets simultaneously for all m by finding the closest codevector to every training vector and grouping the training vectors closest to each $\mathbf{y}_m$ in turn. These groups are the decision regions, whose probabilities can

ALGORITHM 6.5

ECVQ algorithm-training sequence version

1. Run LBG algorithm (Algorithm 6.2) with given training sequence to obtain initial sets of codevectors and their probabilities. Set the slope parameter $\lambda > 0$.
2. For every training vector, find closest codevector $\mathbf{y}_m$ in the measure $j(\mathbf{x}, \mathbf{y}_m) = d(\mathbf{x}, \mathbf{y}_m) - \lambda \log P(\mathbf{y}_m)$. Group together all training vectors with the same closest vector.
3. (a) Estimate group probabilities by number of training vectors in group divided by the total number of training vectors.
   (b) Determine average vector in each group. They are the updated codevectors.
   (c) Using the group probabilities and updated codevectors, calculate entropy H and distortion D with respect to the training sequence. Form $J = D + \lambda H$.
4. Calculate fractional decrease in J from last iteration. If less than some small tolerance $\epsilon > 0$, stop. Otherwise, return to step 2 using current group probabilities and codevectors.

be estimated as the fractions of the training sequence in these regions. We describe the algorithm using a training sequence in Algorithm 6.5, but less formally than previously.

Let us extend Example 6.5 to the case of entropy-constrained vector quantization.

Example 6.6 As before, we assume the source vector $\mathbf{X} = (X_1, X_2)$ is Gaussian with components X_1 and X_2 having mean 0, variance 1, and correlation coefficient 0.9. We generate a sequence of 10 000 training vectors of two dimensions drawn from the assumed probability distribution. We shall utilize the procedure in Algorithm 6.5 to design an entropy-constrained codebook containing four codevectors for this source. This time, we initialize the algorithm with three codevectors and their probabilities. Actually, this is accomplished by running the ECVQ algorithm first with $\lambda = 0$, and initializing the codebook with uniformly distributed codevectors as in the LBG algorithm. The output codevectors and probabilities are used to initialize the next run of the ECVQ algorithm for another $\lambda > 0$. The results for several values of λ are revealed in Table 6.2. The results of the iterations for a given λ are not shown here as they were in Table 6.1.

A plot of the distortion versus rate function (the last two columns in the table) of this quantizer is shown in Figure 6.10.

The codevectors in the codebook for every λ seem to lie again on the 45-degree line in the coordinate plane. As λ increases, distortion increases and entropy decreases. This parameter allows the tradeoff of an increase in distortion for a reduction in rate. One codevector sits near the origin (0,0), while the two other ones move away from the origin and occur with decreasing probability as λ increases. In the limit of large λ, when rate completely dominates the objective function, the minimum distortion and objective function are achieved simultaneously when all source vectors are quantized to (0,0), thereby requiring zero rate.

Table 6.2 Design of two-dimensional Gaussian, $\rho = 0.9$, 3-level ECVQ quantizer for several rates

λ	Codebook	MSE	Entropy (bits)
0.0	(1.195,1.203) (−1.1888,−1.186) (−1.523,−1.515)	0.231	0.767
0.2	(1.214,1.209) (−1.233,−1.233) (−0.012,−0.013)	0.230	0.755
0.4	(1.242,1.241) (−1.304,−1.308) (−0.039,−0.038)	0.238	0.735
0.6	(1.330,1.331) (−1.353,−1.347) (−0.011,−0.015)	0.249	0.701
0.8	(1.368,1.368) (−1.454,−1.454) (−0.041,−0.040)	0.270	0.657
1.0	(1.461,1.460) (−1.576,−1.578) (−0.047,−0.050)	0.313	0.586
1.2	(1.653,1.648) (−1.640,−1.646) (0.002,0.003)	0.378	0.503
1.4	(1.823,1.825) (−1.744,−1.752) (0.021,0.018)	0.459	0.415
1.6	(1.904,1.905) (−1.972,−1.964) (−0.017,−0.016)	0.553	0.324
1.8	(2.153,2.148) (−2.127,−2.124) (0.005,0.005)	0.670	0.220

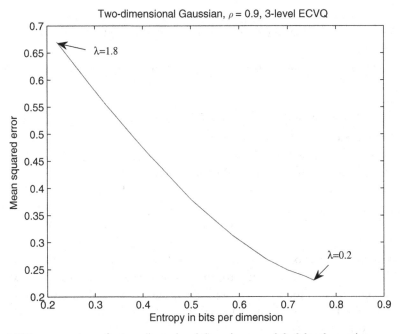

Figure 6.10   MSE versus entropy for two-dimensional Gaussian, $\rho = 0.9$, 3-level quantizer.

When the probability distribution has even symmetry about orthogonal coordinate axes, it is advisable to choose an odd number of quantization points, so that the origin becomes one of the codevectors. The Gaussian distribution has this symmetry, so three quantization points were chosen for the simulation. For an even number of quantization points, one would find that one of them would drop from the codebook as the rate decreases or the parameter λ increases. In programming the algorithm, a step should be inserted to explicitly remove dropped codevectors from the codebook.

We note that these algorithms may be utilized with any vector dimension barring complexity constraints and, in particular, for one dimension. One may therefore use this algorithm to design optimal entropy-constrained scalar quantizers (ECSQ). There may be some benefit for small rates, but not likely for moderate to large rates, where scalar uniform quantizers are so close to optimum.

6.4 Tree-structured vector quantization

One of the difficulties of vector quantization is the search through all members of the codebook necessary to find the best match to the source vector. Such a search is said to be unstructured, because the measure found for the match to one member of the codebook does not tell anything about where to find a better match in continuing the search. A popular way of creating a structured codebook is to guide the search by putting additional test vectors at the intermediate nodes of a binary tree with the codevectors residing at the terminal nodes. The codeword for a source value is the path map of binary symbols through the tree to the codevector at the terminal node. We illustrate this method with a uniform scalar quantizer.

Example 6.7 *Four-level uniform scalar quantization in the interval [0,1].*
The uniform quantizer of four levels in the interval $0 \le x < 1$ has the reproduction (quantization) points $1/8,3/8,5/8,7/8$ with respective decision regions $[0,1/4),[1/4,1/2),[1/2,3/4),[3/4,1)$. Figure 6.11 shows these points and regions. These quantization points would be placed on the terminal nodes of a binary tree. This tree must have two generations of growth from its root. The idea is to build this tree from its root by placing the quantization point for the full interval, its midpoint of 1/2 at the root. Given a source sample x, we test whether $x < 1/2$ or $x \ge 1/2$. We indicate the former circumstance by branching left to the next node and the latter by branching to the right. We label lefthand branches with "0" and righthand ones with "1," as shown in Figure 6.11. The next node on the lefthand ("0") branch indicates $x < 1/2$, so it is labeled with its quantization midpoint of 1/4, while the next node on the righthand ("1") branch indicates $1/2 \le x < 1$, so is labeled with its quantization point of 3/4. The node labeled 1/4 is associated with the region. Succeeding tests at each of these nodes decide whether x is in the lower or upper half of its associated interval. For example, take the node labeled 1/4 associated with [0,1/2). Is x in the lower or upper half of [0,1/2) or, in other words, less than 1/4 or greater than or equal to 1/4? Hence, this node branches to the left and to the right to 1/8 and 3/8, respectively, which are the quantization points in the interval decided upon. That is how the tree in Figure 6.11 was configured. Notice that the decision region gets halved at each stage of testing. In fact, we see that source value is successively approximated to greater accuracy. Notice also that quantization of x takes only two tests, not four as it would for an unstructured 4-level quantizer. The index of the codeword or quantization point to be transmitted is its 2 bit path map, shown in parentheses at the terminal nodes. For example, the path map to 3/8 is 01.

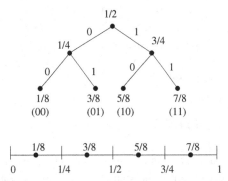

Figure 6.11   Quantization points, decision intervals, and binary tree for uniform quantization in the interval [0,1).

A general tree-structured vector quantizer (TSVQ) is designed in the same way. Each node in the tree is associated with a test codevector and its decision region. This decision region divides optimally into two subregions with their quantization points, which populate the two nodes of the next generation. The design procedure is almost the same as the recursive splitting algorithm for codebook initialization detailed in Algorithm 6.3. There is one crucial difference, however. When a test codevector is split, the GLA algorithm is run only to find the two best reproduction vectors in the same decision region of the test codevector. Therefore, the original decision region is divided into two subregions, each containing its optimal reproduction vector. The design procedure using the squared error criterion (without entropy constraint) and a training sequence is delineated in Algorithm 6.6. The corresponding tree is depicted in Figure 6.12.

Now that the VQ codebook is structured within a tree, we can state the procedure of encoding a source vector using this codebook. We do so in Algorithm 6.7.

The same code tree resides at the decoder, so it follows the path indicated by the transmitted codeword $u_1 u_2 \cdots u_r$ to the codevector at the end of the tree. That codevector is taken to be the reproduction of the source vector $\mathbf{x}$. At each successive stage of the tree, we find a test reproduction vector that is a closer approximation to the source vector. So lowering of the code rate by truncation of the path map codeword always produces a coarser approximation of the source. Such a quantizer is called a *successive approximation quantizer*. In this case, the bitstream (stream of binary symbols) of the code is said to be *embedded* in rate or quality.

The drawbacks of tree versus unstructured (full) search are higher code storage space and slightly degraded performance. For $M = 2^r$ codevectors at the terminal nodes of a tree, storage of the entire tree requires a total $S = \sum_{i=0}^{r} 2^i = 2^{r+1} - 1 = 2M - 1$ elements at the nodes. So the storage cost is roughly $2M$ or double that of an unstructured codebook. Degradation in quality appears to be no more than 1 dB increase in distortion for the same rate. The main advantage of TSVQ is dramatic reduction in encoder computation – only r or $\log_2 M$ binary comparisons versus M to encode a source vector.

ALGORITHM 6.6 _____

Tree-structured VQ design

Generate a training sequence of vectors: TS = $\{\tilde{\mathbf{x}}_1, \tilde{\mathbf{x}}_2, \ldots, \tilde{\mathbf{x}}_n\}$, n large.

(0) Set tree stage $i = 0$. Initialize rate 0 codebook with centroid of TS:

$$\mathbf{y}^{(0)} = \text{cent}(\text{TS}) = \frac{1}{n} \sum_{j=1}^{n} \tilde{\mathbf{x}}_j.$$

Associate decision region (TS) and $\mathbf{y}_1^{(0)}$ with root node of binary tree.

(1) For $k = 1, 2, \ldots, 2^i$, do
  (a) Split node vector $\mathbf{y}_k^{(i)}$ in decision region $\mathcal{R}_k^{(i)}$ to $\mathbf{y}_k^{(i)}$ and $\mathbf{y}_k^{(i)} + \epsilon$, where ϵ is a small perturbation vector.
  (b) Run LBG (GLA) algorithm, initialized with $\mathbf{y}_k^{(i)}$ and $\mathbf{y}_k^{(i)} + \epsilon$, using only TS vectors in $\mathcal{R}_k^{(i)}$. Output is partition of $\mathcal{R}_k^{(i)}$ into two subregions, denoted by $\mathcal{R}_{k1}^{(i)}$ and $\mathcal{R}_{k2}^{(i)}$, and their quantization points $\mathbf{y}_{k1}^{(i)} = \text{cent}(\mathcal{R}_{k1}^{(i)})$ and $\mathbf{y}_{k2}^{(i)} = \text{cent}(\mathcal{R}_{k2}^{(i)})$.
  (c) Associate $(\mathbf{y}_{k1}^{(i)}, \mathcal{R}_{k1}^{(i)})$ with node $2k - 1$ and $(\mathbf{y}_{k2}^{(i)}, \mathcal{R}_{k2}^{(i)})$ with node $2k$ in stage $i + 1$ of tree.
  (d) If $2^{i+1} \geq M$, the target number of codevectors, stop. Otherwise, set $i = i + 1$ and return to step (1).

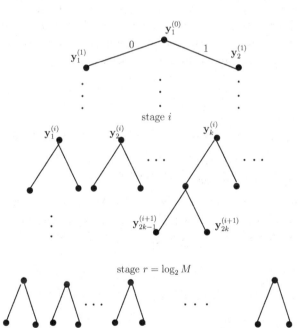

Figure 6.12 Tree-structured VQ design. Bottom nodes contain $M = 2^r$ codevectors of the codebook.

ALGORITHM 6.7 _____

Tree-structured VQ encoding

1. Given a tree-structured codebook of size $M = 2^r$, such as that designed by Algorithm 6.6, and source vector $\mathbf{x}$, determine closer (according to $d(\mathbf{x}, \mathbf{y})$) of the two test vectors at the nodes branching from the root. Advance to that node at stage 1 and record partial path index u_1, where $u_1 = 0$ for lefthand path and $u_1 = 1$ for righthand path.
2. Determine closer to $\mathbf{x}$ of two test vectors at nodes of next stage branching from current node. Advance to that closer node at next stage and record partial path map index $u_2 = 0$ or 1, depending on whether path is lefthand or righthand. Record $u_1 u_2$ as path index to stage 2 of tree.
3. Continue in same way until the two terminal nodes at final stage r, when the reproduction vector and its full path map $u_1 u_2 \cdots u_r$ are found.
4. Send $u_1 u_2 \cdots u_r$ to the decoder.

6.4.1   Variable length TSVQ coding

The TSVQ codes considered thus far consisted of codewords all of the same length. In the design process, this resulted from splitting every node in each stage of the tree until the number of terminal nodes equals the number of desired codevectors. Such a tree is said to be *balanced*. An alternative procedure would be to split only the "best" node among all the end nodes while growing the tree. Leaving aside for the moment the matter of the best node, what would result at any stopping point would be a codebook of varying length codewords. This kind of tree is said to be *unbalanced*. Suppose there are again M codewords in the final codebook and the length in bits of the m-th codeword is $\ell(m)$. Then the average codeword length $\overline{L}$ in the code is

$$\overline{L} = \sum_{m=1}^{M} \ell(m) P_m \text{ bits,} \tag{6.41}$$

where P_m is the probability of the m-th codeword. The rate of the code is therefore

$$R = \frac{\overline{L}}{N} \text{ bits per dimension.}$$

In Figure 6.13 is shown a potential unbalanced TSVQ code tree. The numbers in parentheses at the terminal nodes are the probabilities of the codewords or codevectors associated with those nodes. The average length $\overline{L}$ equals 2.20 bits.

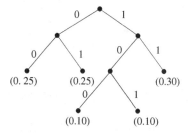

Figure 6.13   Example of unbalanced tree in TSVQ. The numbers in parentheses at the terminal nodes are the probabilities of the codevectors (or codewords) associated with these nodes.

ALGORITHM 6.8

Variable length VQ code tree design

(0) Initialization:
   (a) Generate a training sequence (TS) of N-dimensional vectors.
   (b) Calculate centroid (mean vector of TS for squared error criterion) and put on root of tree.
   (c) Split root vector and run LBG algorithm to obtain optimal codebook of two vectors and their decision regions, which form a partition of the TS.
   (d) Grow two branches from root node and associate with each node at end one of the optimal vectors and its decision region.
(1) For every current end node, select one and split its vector into two and run LBG algorithm using only the TS vectors in its decision region. Calculate λ, the ratio of decrease in distortion to increase in rate, resulting from the split of this node only.
(2) Determine node with maximum λ and grow tree from only that node to form two new nodes (one more end node). Associate the already determined vectors and decision regions with these nodes.
(3) If target rate (average length) or distortion has been reached, stop. Otherwise, return to Step (1).

The best node to split is the one that lowers the average distortion the most for the same increase in average codeword length. Stating this criterion mathematically, let $\Delta\overline{L}$ = average bit length increase and $-\Delta D$ = average distortion decrease. The best node to split is the one that maximizes

$$\lambda = \frac{-\Delta D}{\Delta\overline{L}}. \tag{6.42}$$

Therefore, after the split of the root node, the algorithm calculates λ for all current end nodes and splits only the one having the maximum λ, producing one more end node. It then repeats calculations of λ for all these nodes to seek the one now with the maximum λ, and so continues until the average length or distortion target is reached. This procedure is referred to as *greedy tree growing* and its steps are delineated in Algorithm 6.8.

The resulting code has codewords of variable length. We remind the reader that a code constructed along the branches of a tree with terminal nodes indicating the source symbols (vectors in this case) is prefix-free, so is uniquely decodable. Everything we presented here for binary branching trees can be generalized in the obvious way to α-ary branching trees for some positive integer α.

The flexibility in variable length TSVQ to increase rate with the greatest payoff in distortion reduction leads to its superior performance over fixed-length TSVQ. There is another way to construct variable length trees that results in even better performance that we shall describe now.

6.4.2 Pruned TSVQ

The greedy tree-growing algorithm just described has the limitation that the decision to split and branch at a given node depends only on the payoff from a single stage.

Perhaps if more than one future stage determines this decision, a different node would be chosen. Actually, this scenario was examined by Riskin and Gray [7] and was found to yield some improvment in performance for two stages, but hardly any for greater than two. Another, less unwieldy way to design a variable length TSVQ is first to grow a full tree with fixed-length codewords and to iteratively prune subtrees that yield the smallest decrease in average distortion for the decrease in average rate. Such a procedure is accomplished by the *generalized BFOS algorithm* and is fully explained in Gersho and Gray [8]. Rather than presenting this algorithm here, we shall describe a greedy technique analogous to that used for tree growing.

Suppose that we have designed a full tree TSVQ with length r (binary) codewords. In designing this tree, one can record the decrease in average distortion and increase in rate produced by each split. From these records, one can plot the convex hull of the distortion versus rate (average length) curve. The curve would look similar to that in Figure 6.10. The righthand end of the curve corresponds to the full tree and the lefthand end (rate 0) just the root node. Then starting with the full tree, to obtain smaller rate trees, one selects the best terminal node pair to prune. The criterion is the pair with the smallest ratio of increase in distortion to decrease in rate, that is, the smallest

$$\lambda = \frac{\Delta D}{-\Delta \overline{L}}.$$

The $(\Delta \overline{L}, \Delta D)$ pair for the newly exposed terminal node have been recorded previously when growing the tree. The pruning of the node with smallest λ and the graphical effect on the distortion-rate curve are depicted in Figure 6.14. This process is simply continued until the desired rate or distortion is reached.

6.5   Tree and trellis codes

In tree-structured VQ, representation vectors of the source are placed at every node in the tree. Another use of a tree to encode vectors of n components is to associate components to successive branches of a tree. If one component is placed on every branch, then

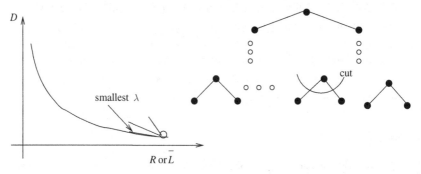

Figure 6.14   Example of first pruning of TSVQ tree and selection of smallest λ in distortion-rate curve.

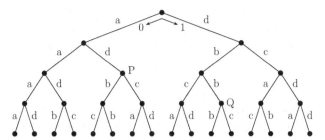

Figure 6.15   Four initial stages of a rate $R = 1$ bit per source letter code tree.

the number of stages (called *depth* or *height*, depending on the orientation) is n. Recall that a balanced tree in TSVQ has $r = \log_2 M$ stages, M being the number of codewords in the codebook. The idea is that a sequence of reproduction letters is read from the branches in following any path through the tree to its terminal node. An example of a four-stage binary tree with a four-letter reproduction alphabet is shown in Figure 6.15. The tree in this figure has one letter on each branch, so the binary path map of four stages, 1 bit per stage, to its terminal node represents a vector of four components. Again the path map serves as the index of the reproduction vector. For example, along the path 1001 is the vector $dbcd$. The number of terminal nodes, 2^n, equals the number of codevectors, so in the example is $2^4 = 16$.

The rate of the tree code with one reproduction letter per branch and two branches per node is $R = 1$ bit per source letter. To obtain different rates, we can change the number of branches per node and the number of source letters per branch. Let the number of letters on each branch be α and the number of branches emanating from each node be 2^b where b is a positive integer. Then the rate is

$$R = b/\alpha,$$

because it requires b bits to specify a branch among 2^b branches, each one containing α letters.

For a given source vector $\mathbf{x}$ of n dimensions, the objective is to determine the n-dimensional vector along a path through the tree that minimizes the distortion with respect to the source vector. The path map of that vector is the codeword ouput. In general, we break the vector $\mathbf{x}$ into segments of length α, so that $\mathbf{x} = (\mathbf{x}_1, \mathbf{x}_2, \ldots, \mathbf{x}_{n_s})$, where $n_s = n/\alpha$ matches the number of tree stages. We put the subvectors $\mathbf{x}_i, i = 1, 2, \ldots, n_s$ of length α in correspondence with reproduction subvectors in the stages of the tree. We denote the reproduction subvector for the k-th branch $(k = 1, \ldots, 2^{bi})$ of the i-th stage of the tree by $\mathbf{y}_{i,k}$. One puts the subvectors of $\mathbf{x}$ in correspondence with their stages in the tree, $\mathbf{x}_1$ to stage 1, $\mathbf{x}_2$ to stage 2, etc. Therefore, one needs to find the sequence of branch indices $(k_1, k_2, \ldots, k_{n_s})$ that minimizes the distortion

$$\sum_{i=1}^{n_s} d(\mathbf{x}_i, \mathbf{y}_{i,k_i}),$$

where $d(\mathbf{x}, \mathbf{y})$ is a distortion measure additive over the stages and is commonly $||\mathbf{x} - \mathbf{y})||^2$. For binary trees $(b = 1)$ and one letter per branch $(\alpha = 1)$, there are $n_s = n$ tree stages. For squared error, the distortion expression then becomes

$$\sum_{i=1}^{n} (x_i - y_{i,k_i})^2.$$

In most cases, one would like to encode long strings of source letters, which means high-dimensional vectors. Since trees grow exponentially with stages, that would mean large trees and would require efficient means to search through the trees to find the minimum distortion codevector. One often gives up on exhaustive searches as too burdensome computationally and seeks restricted search algorithms that achieve nearly the same results. One such algorithm, called the M-algorithm, proceeds through the tree sequentially by stages and keeps only the best M paths in contention.[3] A modification for large n, called the (M, L) algorithm truncates the paths held in memory and keeps only the last L stages, where $L \approx 5M$ usually suffices. A more detailed explanation of this algorithm appears in Section 6.7.

6.5.1 Trellis codes

The exponential growth of a tree code need not happen if there is some structural constraint imposed upon the reproduction letters or subvectors in the branches of the tree. One particularly effective constraint is that the reproductions be generated by a finite-state machine. To illustrate this circumstance, consider the tree code in Figure 6.15. If we look carefully at the reproduction letters that label the branches, we observe that they are functions of the previous two path map symbols and the current symbol. In fact, the correspondences are: $00, 0 \rightarrow a, 00, 1 \rightarrow d, 01, 0 \rightarrow c, 01, 1 \rightarrow b, 10, 0 \rightarrow b$, $10, 1 \rightarrow c, 11, 0 \rightarrow a, 11, 1 \rightarrow d$. By convention, the two bits before the comma are in order of deepest first and the bit after the comma is the current bit. The two most recent past bits may be considered the state of the machine. There are four states: 00, 01, 10, 11. Once a particular state is reached, the branch labels beyond that state always follow the same pattern. For example, look at nodes marked P and Q in the tree. Both are in state 10 and the subtrees rooted at these nodes will be labeled identically through their future stages. So in this case and in the case of any finite-state machine, a tree is a redundant structure upon which to represent the code. One should view the process as a transition from one state to another and an output caused by the current path bit. Symbolically, we rewrite the correspondences as transitions among states with current bit and output on top of the arrow. We also enclose every state's two bits with a box to emphasize that it is a state.

$$\boxed{00} \xrightarrow{0,a} \boxed{00} \qquad \boxed{00} \xrightarrow{1,d} \boxed{01}$$

$$\boxed{01} \xrightarrow{0,c} \boxed{10} \qquad \boxed{01} \xrightarrow{1,b} \boxed{11}$$

$$\boxed{10} \xrightarrow{0,b} \boxed{00} \qquad \boxed{10} \xrightarrow{1,c} \boxed{01}$$

$$\boxed{11} \xrightarrow{0,a} \boxed{10} \qquad \boxed{11} \xrightarrow{1,d} \boxed{11}$$

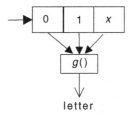

letter

Figure 6.16 Finite-state machine realization of a four-state rate $R = 1$ bit per source letter code in state $\boxed{10}$.

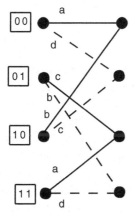

Figure 6.17 One-stage trellis depicting state transitions and outputs of finite-state machine code in Figure 6.16.

A finite-state machine realization of this process is illustrated in Figure 6.16, which shows a binary shift register with three stages (two delay elements) with the input being a path bit and a single output of a reproduction letter a, b, c or d. The register is shown to be in state $\boxed{10}$. When the current bit is input to the register, it causes a shift of one time unit with the third stage bit dropping off the end and an output depending on the 3 bit sequence now in the register. The outputs for this example are given above.

A very useful depiction of this process is a diagram of one stage of a trellis in Figure 6.17. There the nodes correspond to the states and the outputs label the branches. Here, because branch lines cross, instead of branching from a node upward to indicate input "0" and downward to indicate "1," we use the customary convention to mark "0" by a solid branch and "1" by a broken branch. To show the evolution of the code for all time, this single stage is just repeated, because the function of the register contents does not vary with time. The code can therefore be represented with a trellis with the number of stages equal to the numbr of input times or codeword digits (bits). We show the trellis code in Figure 6.18 that corresponds exactly to the tree code in Figure 6.15. It is assumed there that the initial state is $\boxed{00}$. Therefore, only two states are possible for the first stage, $\boxed{0x}$, where x is 0 or 1; and all four states become realizable only at the second stage and beyond. The number of possible paths through r stages is 2^r, the same as in a tree, because there are two possible transitions for each state at every stage. The

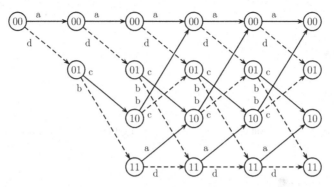

Figure 6.18   Trellis of a four-state rate $R = 1$ bit per source letter code.

generalization of this trellis of code rate $R = 1$ bit per letter to other rates follows in the same way as with a tree code. To put more than one letter on the branches, we postulate α functions of the shift register contents. When only one input bit per unit time enters the register, it creates a code rate of $R = 1/\alpha$ bits per letter. To obtain higher rates, we shift b bits into and along the register per unit time, giving a rate of $R = b/\alpha$ bits per letter. For every b bits in, α reproduction letters are output from the machine. The trellis would have 2^b branches containing α reproduction letters emanating from each node (state). The number of states depends on the number of b-bit storage elements contained in the register. If there are ν b-bit storage elements in the register, the state is the contents of the first $\nu - 1$, so its realizations number $2^{b(\nu-1)}$. ν is called the *constraint length* of the register, because it takes ν input times for an input of b bits to pass through the register. This means that every b-bit input affects the output for ν time units. For r stages of the trellis initially in the all-zero state, there are a total of 2^{br} paths, each one corresponding to a codevector. An example of a finite-state machine generating a $R = 2/3$ code is shown in Figure 6.19. There, $b = 2$ bits are input to a constraint length $\nu = 3$ register having $\alpha = 3$ functions of its contents as outputs. A single stage of the matching trellis (shown in outline form) has $2^{2.2} = 16$ states and $2^2 = 4$ branch transitions to and from each state. On every branch connecting two states reside three reproduction letters.

6.5.2   Encoding and decoding of trellis codes

The trellis representation of a code is considerably more compact than a tree representation. The trellis grows only linearly with the source length rather than exponentially as with a tree. That means that the search for the closest codevector in the trellis to a given source vector is inherently logarithmically less complex than a tree search. Furthermore, the exhaustive search can be accomplished using a recursive dynamic programming algorithm, called the *Viterbi algorithm*. We leave the details of this algorithm and restricted search algorithms to Section 6.7.

The encoding operation of a source vector **x** starts as before by breaking **x** into subvector segments of length α equal to the number of reproduction letters on the

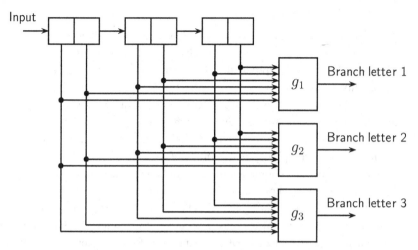

Figure 6.19   Finite-state machine realization of a rate $R = 2/3$ bits per source letter code.

branches of the trellis. The subvectors $\mathbf{x}_1, \mathbf{x}_2, \ldots, \mathbf{x}_{n_s}$ are put sequentially into correspondence to the stages of the trellis indicated by their index. Again, we search through all paths in the trellis to find the codevector distributed along their branches that minimizes the distortion. Assuming that the distortion measure is squared error, the objective is to determine the path $(k_1, k_2, \ldots, k_r)$ that minimizes

$$\sum_{i=1}^{r} ||\mathbf{x}_i - \mathbf{y}_{i,k_i}||^2,$$

where k_i is the index of the branch to stage i. Through the first $v - 1$ stages, these indices range from 1 to 2^{bi}. For every subsequent stage, the range stays constant at 1 to 2^{bv}, because there are 2^b branches from each of $2^{b(v-1)}$ states or nodes. Once the minimum distortion codevector is found, its path map, indicated by the sequence of branch indices, is sent as the codeword. This path map is *not* the particular index sequence $(k_1, k_2, \ldots, k_r)$, which does not account for connected paths, but a sequence of r b-bit words that indicate which branch among 2^b branches to take upon arriving at a state in every stage of the trellis. They are the input bits to the finite-state machine that generates the trellis code. For example, in the $R = 1$ bit per letter trellis in Figure 6.18, suppose that the minimum distortion path contains the sequence of branch letters $dcbd$ from initial state $\boxed{00}$ to state $\boxed{01}$ at stage 4. Following the branches, the path map codeword is 1001.

Decoding a codeword means associating it to its codevector. For a trellis code, the decoder is simplicity itself. We see that a trellis codeword is the sequence of input bits to a finite-state machine whose outputs are the components of its associated codevector. Therefore, only the finite-state machine is needed for decoding. The codeword is received and shifted into the shift register of the machine b bits at a time. The sequence of outputs forms the codevector.

6.5.3   Codevector alphabets

Thus far, there has been no mention of choosing an alphabet for the reproduction letters of the codevectors in a tree or trellis code. Theorems from rate-distortion theory state that the optimal reproduction letters may be drawn at random from the output probability distribution of the test channel that realizes the rate-distortion function. This way of choosing the reproduction alphabet is aptly called *random coding*. The theorems also prove that tree or trellis codes exist which are optimal in that they realize the minimum possible distortion for a given rate limit. Unfortunately, realization of the optimum requires infinite complexity and coding delay ($r \to \infty$, $v \to \infty$). The code construction in the proofs gives some guidance to find good codes and therefore good alphabets. We have to use long input sequences and a moderate number of trellis states. We can take the course of populating the branches of the tree or trellis with variates drawn at random from the optimal test channel output distribution. In the finite-state machine implementation, that means that the mappings of the register contents to outputs vary with input time. The probability is high that a good finite code results, but occasionally the code can be very bad.

The optimal output alphabet is generally continuous for a continuous amplitude source. However, small, finite alphabets can achieve results nearly as good as infinite ones. That was proved theoretically and verified experimentally by Finamore and Pearlman [9]. Analogous to the continuous alphabet case, the optimal output alphabet is drawn from the discrete M-letter output probability distribution of the test channel that realizes the so-called $R_{LM}(D)$ rate-distortion function. $M = 4$ proved to be sufficient for a rate $R = 1$ bit per source letter code and $M = 8$ for a $R = 2$ code. Inferring from these results, Marcellin and Fischer [10] postulated and showed that $M = 2^{R+1}$ letters suffice for a rate R code. Furthermore, they exhibited a different kind of trellis branch association and prescribed scalar quantizer letter values that achieved nearly optimal performance for stationary sources without the drawback of occasionally bad codes. We now explain their method, which is called *trellis coded quantization* or simply TCQ.

6.6   Trellis coded quantization (TCQ)

We shall now describe scalar TCQ, in which the encoder output is one reproduction letter per unit time. The reproduction alphabet of scalar letters is partitioned into a number of sets called *cosets*. The members of a coset consist of maximally separated letters and different cosets are translates of one another. For example, consider the reproduction alphabet of an eight-level uniform quantizer $\mathcal{A} = \{k\Delta :$ $k = -4, -3, -2, -1, 0, 1, 2, 3\}$. If partitioned into four cosets, they would be $D_0 = \{-4\Delta, 0\}$, $D_1 = \{-3\Delta, \Delta\}$, $D_2 = \{-2\Delta, 2\Delta\}$, and $D_3 = \{-\Delta, 3\Delta\}$. The separation of members within each coset of 4Δ is the largest possible for four cosets. A partition of an eight-level uniform quantizer alphabet into four cosets with the same interelement distance of 4Δ is illustrated in Figure 6.20. This idea of a partition of the reproduction alphabet into cosets is borrowed from *trellis coded modulation* (TCM), which was

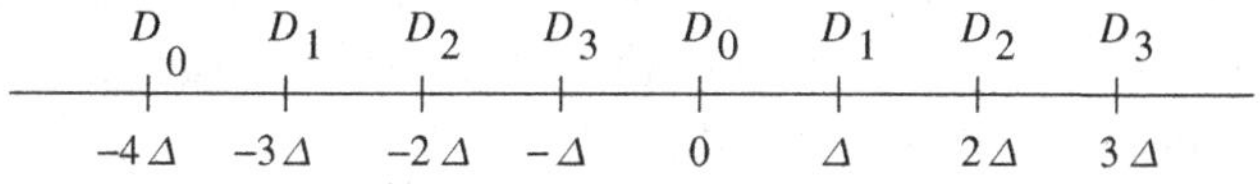

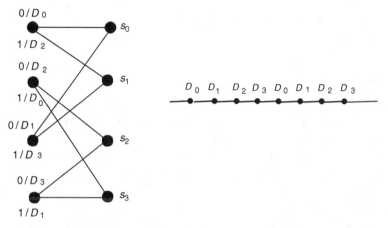

Figure 6.20 Partition of eight-level uniform quantizer levels into four cosets.

Figure 6.21 Trellis and cosets in TCQ.

introduced by Ungerboeck [11]. There the 2^k channel symbols are partitioned into $2^{k-\tilde{k}}$ cosets, each having $2^{\tilde{k}}$ members. (k and $\tilde{k} < k$ are rates in bits per symbol.) In TCQ, we partition the source reproduction alphabet, not the channel symbol alphabet. In the aforementioned example, substituting reproduction letter for channel symbol, $k = 4$ and $\tilde{k} = 2$. In TCQ, the typical partition, and the one used here for simplicity, uses an alphabet of 2^{R+1} reproduction letters and consists of $2^2 = 4$ cosets, each containing 2^{R-1} members. For every encoder input of R bits, a single reproduction letter is produced at the output. The encoder function is actuated as a search through a trellis. What is different about the TCQ trellis is that cosets label the branches, not just individual letters. A single stage of a four-state trellis and the coset members are depicted in Figure 6.21. (We denote the states heretofore by $s_0, s_1, \ldots$, etc.) When a coset labels a branch, it means that any member of that coset will activate the state transition signified by that branch. One often imagines it as parallel transitions and draws it as separate branches connecting the same two states. For example, if we use the cosets of the previous example, the lower ("1") branch from state s_2 contains D_3, so the labels on the parallel branch depictions are $-\varDelta$ and $3\varDelta$. In this trellis, it takes 1 bit to specify whether an upward or downward branch is traversed and $R - 1$ bits to determine which member of the corresponding coset is the reproduction. Referring again to the eight-level uniform quantizer alphabet, the code rate is $R = 2$ bits per sample – 1 bit to specify the coset and 1 bit to specify one of its two members.

The encoding procedure is again to associate successive source letters to the stages of the trellis and search the "parallel transitions" along all the stages of the trellis to find the minimum distortion path. That path is represented as the sequence of upward

("0") and downward ("1") transitions. The reproduction letter creating a transition at any stage is represented by the $R - 1$ bits of its index in the coset designated by the sequence of path bits to the transition at that stage.

With a scalar codebook, the minimum rate of TCQ is $R = 1$ bit per letter. Use of a vector codebook will allow lower rates. That requires design and partitioning of vector reproduction alphabets [12]. One can use lattice codebooks or VQ codebooks designed by the GLA algorithm. We shall not expound further on vector TCQ. Instead, we shall explain how to obtain lower rates using entropy coding in scalar TCQ.

6.6.1   Entropy-coded TCQ

The problem of the 1 bit lower rate limit stems from the need to specify 1 bit to indicate the direction of the branch of the path at each stage of the trellis. The path is equally likely to branch upward or downward, corresponding to 0 or 1 for the path bit, giving 1 bit as the minimum possible entropy and rate. Actually it is not necessary to spend a bit for the branch direction of the path. When in a given state, there are really 2^R possible branches, belonging to two cosets of 2^{R-1} parallel branches each, available to take to a state at the next stage. So if you spend R bits to specify the particular branch, you get the coset to which it belongs and the state to which it connects at the next stage. Then these R bits can be entropy coded to obtain a rate that is probably lower than R.

Let us make this argument more concrete using the trellis and cosets in Figure 6.21. Notice that from each state branches either one of the coset pair $\{D_0, D_2\}$ or one of the pair $\{D_1, D_3\}$. Let us denote these coset pairs as the supersets $\mathcal{F}_0 = D_0 \cup D_2$ and $\mathcal{F}_1 = D_1 \cup D_3$. Therefore, at states s_0 and s_2, we choose an element of $\mathcal{F}_0$ and at states s_1 and s_3, we choose an element of $\mathcal{F}_1$ to proceed to the state at the next stage. Once we have chosen an element in one of the supersets, we determine from the trellis diagram the coset to which it belongs and correspondingly the branch to the state at the next stage.

Since entropy coding has entered the picture, the reproduction alphabet no longer needs to be finite to obtain a finite rate. Consider the infinite set of uniform quantizer levels, $\mathcal{A} = \{i\Delta, i = 0, \pm1, \pm2, \ldots\}$, as the reproduction alphabet. As before, we break this alphabet into four cosets by labeling the levels sequentially $D_0, D_1, D_2, D_3, D_0, D_1, \ldots$, as in Figure 6.21. In other words,

$$D_0 = \{4k\Delta : k = 0, \pm1, \pm2, \ldots\},$$
$$D_1 = \{(4k + 1)\Delta : k = 0, \pm1, \pm2, \ldots\},$$
$$D_2 = \{(4k + 2)\Delta : k = 0, \pm1, \pm2, \ldots\},$$
$$D_3 = \{(4k + 3)\Delta : k = 0, \pm1, \pm2, \ldots\}.$$

The supersets are:

$$\mathcal{F}_0 = D_0 \cup D_2 = \{2k\Delta : k = 0, \pm1, \pm2, \pm3, \ldots\},$$
$$\mathcal{F}_1 = D_1 \cup D_3 = \{(2k + 1)\Delta : k = 0, \pm1, \pm2, \pm3, \ldots\}.$$

Inside a superset, the signed integer k indexes the elements. That is, the k of $2k\Delta$ is the index in $\mathcal{F}_0$ and the k of $(2k + 1)\Delta$ is the index in $\mathcal{F}_1$. Since we know the state, we always know the superset belonging to a given index. So a path through the trellis is specified by a sequence of these indices. The entropy code of this index sequence is then sent to the decoder.

Example 6.8 Suppose that the initial state of the trellis is set to s_0 and the sequence of indices found by the encoder to the sixth stage is −3, −5, 2, 5, 0, −1. So initially we look for −3 in $\mathcal{F}_0$ and find that it corresponds to -6Δ which is in coset D_2, making the next state s_2. The next index -5 belongs then to $\mathcal{F}_1$ and corresponds to -9Δ, which is in D_3. The next state is therefore s_3, designating that the next index 2 is in $\mathcal{F}_1$ and corresponds to 5Δ. The level 5Δ is in D_1, so that the next state is again s_3. Continuing in the same vein, we decode the reproduction levels as $-6\Delta, -9\Delta, 5\Delta, 11\Delta, 0, -\Delta$. The corresponding state sequence is $s_0, s_2, s_3, s_3, s_1, s_2$.

As in regular entropy-coded quantization, the effective rate of the code cannot be determined in advance and depends on the quantization step size Δ. Δ relates well to a quality or distortion target, but not to a rate target. One needs a systematic procedure to run the encoder with different values of Δ with outcomes that converge to the desired rate. Chapter 8 treats these so-called rate control procedures, along with the most common one, the bisection method.

6.6.2 Improving low-rate performance in TCQ

TCQ achieves significant improvement over PCM or DPCM and using 256 states often comes within about 0.2–1.2 dB of the rate-distortion bound for most sources at rates above 1.0 bit per sample. Its performance starts to deteriorate noticeably from that of the rate-distortion bound below about 1.50 to 1.25 bits per sample. The cause is the lack of a level of zero (0) in a superset. As the rate decreases, the 0 reproduction level becomes increasingly probable. So if the encoder happens to be in a state without an outward branch having a 0 level, then it has to choose a higher distortion path than otherwise. A remedy is to add a 0 level to the coset D_1 in the superset $\mathcal{F}_1$. The decoder, being in a known state and receiving a 0, can always distinguish whether this 0 belongs to D_1 or D_0, because the containing supersets cannot occur simultaneously at the same state. The JPEG Part II image compression standard [13] offers a wavelet TCQ codec that uses this coset structure in an eight-state trellis.

6.7 Search algorithms

6.7.1 M-algorithm

A necessary component of tree and trellis coding is efficient means to search for the path of least distortion. An exhaustive search through a tree is impossible to implement,

ALGORITHM 6.9 _____

M-algorithm

The source emits a vector $\mathbf{x} = (x_1, x_2, \ldots, x_n)$ to be encoded using a binary ($b = 1$) tree, such as that in Figure 6.15. Let k_i be the index of the branch at stage i and y_{k_i} denote the reproduction letter belonging to that branch. The reproduction letters y_{k_i} belong to the alphabet $\mathcal{A} = \{a_1, a_2, \ldots, a_A\}$. The distortion criterion is squared error. Choose a value for M, the number of paths to retain.

(1) Starting from the root of the tree, calculate distortion of all paths in the tree up to stage r^*, where r^* is the earliest stage that $M < 2^{r^*}$. The distortion calculation to any stage r is

$$D_r = \sum_{i=1}^{r} (x_i - y_{i,k_i})^2.$$

Notice in this calculation that the i-th component of $\mathbf{x}$ is put into correspondence with stage i in the tree.

(2) For every stage r such that $r^* \leq r < n$,
   (a) determine the M paths having the least distortion D_r and record their binary path maps.
   (b) For each of the M recorded paths, calculate the additional distortion of the two ($2^b = 2$) branches to the next stage. That is, calculate

$$\Delta D_r = (x_{r+1} - y_{k_{r+1}})^2$$

   for every k_{r+1} corresponding to an extension of a stage-r best path. Set $D_{r+1} = D_r + \Delta D_r$.
   (c) Set $r = r + 1$. Return to Step (a) to repeat.
(3) Determine the lowest distortion path among the $2M$ paths at the terminus of the tree. Transmit its path map of n binary symbols.

_____

since the number of paths grows exponentially with depth. Selective search is necessary and almost always results in a path that is nearly the best. A common selective search algorithm in source encoding is the M-algorithm and its truncated version, the (M, L)-algorithm. For the M-algorithm, only the M best (lowest distortion) paths are retained and extended at every stage of the tree. Consider a tree with 2^b branches from each node. When the search reaches a stage r^* such that the chosen $M < 2^{br^*}$, only the M nodes with least distortion paths are selected and extended to $M2^b$ at the next stage and all subsequent stages. Therefore, for a codetree with n_s stages, the number of searches, which is less than $n_s M 2^b$, now grows linearly and no longer grows exponentially with n_s. When the search reaches the end of the tree, the lowest distortion path is selected for transmission. Note that scalar quantization corresponds to $M = 1$, retaining only a single path at each stage. The steps of the M-algorithm are explained in more detail in Algorithm 6.9 for a $R = 1$ bit code on a binary ($b = 1$) tree. The more general cases of rates $R = b/\alpha$ with 2^b branches from each node and α reproduction letters on each branch, and different distortion measures can be easily inferred from this simpler case.

   In practice, M does not have to be too large to obtain nearly as good performance as an exhaustive search. When a tree is populated through a scheme that is not based on an optimal theory, then the performance saturates rather quickly at small values of M

such as 4 or 8. When the tree is populated with random variates drawn from the optimal output test channel distribution of the rate-distortion function, the performance always improves with M, but shows ever smaller improvement above higher values of M such as 128 or 256.

6.7.1.1 (M, L)-algorithm

The lengths of the paths grow longer as the M-algorithm proceeds through the tree. The storage and computational requirements may start to become excessive in some applications. Furthermore, some applications may not tolerate the coding delay of the n_s stages of the full tree. In these cases, a truncated form of the M-algorithm, called the (M, L)-algorithm, is used. There, the largest path held in memory corresponds to L stages of the tree. We envision the computer containing M registers, each holding the path map symbols (binary or 2^b-ary) of a surviving path in the search. Suppose that each storage register can hold no more than L path map symbols. Then a decision must be made for the path map symbol L stages back in the tree and this symbol must be released for transmission. Then when the search for the M best paths at the next stage finds M new path map symbols, the M shift registers holding the symbols of their path maps are shifted once to the left to make room in the rightmost stage for the new path map symbols. See Figure 6.22. From the path symbols dropping out of the left end of the registers, a decision is made for the symbol to be released for transmission. This decision can be made rather arbitrarily, because if L is large enough, the symbols dropping out of the register are highly probable to be all the same. In such a case, one can just pick the symbol dropping out of any one of the registers. A more complex decision scheme is to release for transmission the symbol that occurs the most number of times in this stage. For binary symbols, release "0" or "1," whichever is in the majority.

One of the attractive features of the (M, L)-algorithm is that once the algorithm reaches stage L of the tree, a path map symbol is released for every successive stage of the tree until the full tree length is reached. Then the final transmission of L symbols corresponds to the least distortion among the M paths held in memory. The encoding delay corresponds to the transmission of $L\alpha$ source letters (α letters per stage), whereas for the M-algorithm it corresponds to that for the full source sequence.

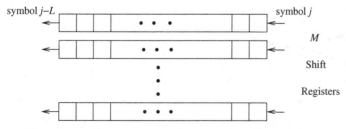

Figure 6.22 M shift registers of length L, each holding a path map symbol sequence. Symbol at time j shifted into every register from the right and left shift of each symbol in every register ejects symbol at time $j - L$ from its left end.

6.7.2   The Viterbi algorithm

A trellis can be efficiently and exhaustively searched through a procedure known as the Viterbi algorithm. This algorithm becomes self-evident once it is realized that only one path to each node (state) needs to be retained at each stage of the trellis. Consider a binary branching trellis and two paths of distortions d_1 and d_2, $d_1 < d_2$, to a node at some stage, as shown in Figure 6.23. Suppose that an extension of these paths has distortion d_3. Then the two extended paths through the node have distortions $d_1 + d_3$ and $d_2 + d_3$ with $d_1 + d_3 < d_2 + d_3$. Therefore, any path containing the d_2 path can never be chosen as the minimum path through the full trellis. Therefore, the d_2 path can be eliminated immediately, once it is known that its distortion is larger than the other path to that node. Extending this realization, we can eliminate all but the minimum distortion paths to every node at this stage. These minimum distortion paths, one for each state, are called *survivors*. Only these survivor paths need to be extended to the next stage, where one determines which of the two paths to each node is the survivor with the smaller distortion. The procedure starts at the first stage and continues through all stages to the end of the trellis, where the binary path map of the least distortion path among the survivors is chosen for transmission. This procedure is a dynamic programming algorithm and was first used in the communications literature by Viterbi to decode convolutional channel codes [14, 15]. The steps of the algorithm are described in Algorithm 6.10. The mode of decision stated in the last step of Algorithm 6.10 is often altered. It is occasionally convenient to force the trellis to end in just one state, the all-zero state, by appending the shift-register input with $v - 1$ zero symbols. Therefore, for each of these zeros, the number of possible states halve until only the all-zero state remains. Therefore, the final decision on the minimum distortion path comes down to choosing one of two paths to a single state. The modifications to the algorithm in this case are readily apparent.

Although the Viterbi algorithm has been stated only for a rate of $R = 1$ bit per source letter, the generalization to other rates follows in the same way as with the M-algorithm for searching of trees. For a rate $R = b/\alpha$ trellis code, the number of states is $K = 2^{b(v-1)}$, the branches emanating from each state number 2^b, and on these branches reside subvectors of the source vector of dimension α. Incremental distortion is calculated between the corresponding subvector from the source and that on any

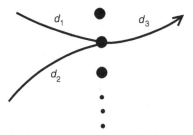

Figure 6.23   Paths through a trellis node.

ALGORITHM 6.10 ───

The Viterbi algorithm

Suppose that we have a binary input, single output letter finite-state machine source coder with v stages and therefore $K = 2^{v-1}$ states, denoted as $s_0, s_1, \ldots, s_{K-1}$. After stage $v - 1$, the matching trellis has K state nodes with two ($2^b = 2$) branches converging to and diverging from each node. The initial state is assumed to be the all-zero state s_0, so at stage $v - 1$ there exists just a single path to every node of the trellis.

1. Calculate the distortion of each path with respect to the source sequence to a state at stage $v - 1$. Denote the minimum distortions of paths to a given state at any stage r as $D_r(s_0), D_r(s_1) \ldots, D_r(s_{K-1})$ and their path maps as $\mathbf{u}_r(s_0), \mathbf{u}_r(s_1), \ldots, \mathbf{u}_r(s_{K-1})$. Here, these path maps are sequences of binary symbols (0's and 1's). Initially $r = v - 1$.
2. For stages $r = v$ to n, do
   for states $s_i, i = 0$ to $K - 1$, at stage r, do
   (a) calculate distortion increments from the two paths converging to the designated state. That is, for state s_i in stage r, calculate $d(x_r, y(\mathbf{u}_{r-1}(s_{k_j}), u_r))$ for $j = 1$ and 2 and $u_r = 0$ or 1. $y(\mathbf{u}_{r-1}(s_{k_j}), u_r)$ is the output letter at stage r for the path map $\mathbf{u}_{r-1}(s_{k_j})$ augmented by the next path symbol u_r that converges to state s_i at stage r. See Figure 6.24 for a pictorial description of this notation pertaining to the four-state trellis in Figure 6.17.
   (b) Determine the minimum of the two path distortions and record its path map to that state. The expressions for the two path distortions are

$$D_r^{(k_1)}(s_i) = D_{r-1}(s_{k_1}) + d(x_r, y(\mathbf{u}_{r-1}(s_{k_1}), u_r)),$$
$$D_r^{(k_2)}(s_i) = D_{r-1}(s_{k_2}) + d(x_r, y(\mathbf{u}_{r-1}(s_{k_2}), u_r)).$$

Therefore,

$$D_r(s_i) = \min_{k_1,k_2} \{D_r^{(k_1)}(s_i), D_r^{(k_2)}(s_i)\}.$$

Let s_{k*} be the state at stage $r - 1$ through which the path having the above minimum distortion passes. The symbol map of the corresponding path to stage r to state s_i is

$$\mathbf{u}_r(s_i) = (\mathbf{u}_{r-1}(s_{k*}), u_r).$$

   (c) Retain $\mathbf{u}_r(s_i)$ and discard other path to state s_i.
3. Determine $i^* = \arg\min_{i=0,1,\ldots,K-1} D_n(s_i)$ and release the associated path map $\mathbf{u}_n(s_{i*})$ as ouput.

───

branch in that stage. The number of data stages is $n_s = n/\alpha$, where n is the source dimension.

The M-algorithm or (M, L)-algorithm can also be applied to searches in the trellis using the Viterbi algorithm. To save memory and calculations, the $M < K$ least distortion paths can be retained at every stage and the path lengths can be truncated to length L ($L > v$), so that one path symbol is released for output at every stage after the paths reach length L, except for the last stage when L symbols are released. A length $L \approx 5M$ seems to allow the procedure to find a nearly optimal path in almost all cases.

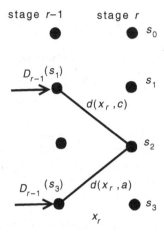

stage $r-1$ stage r

    ●   ● s_0

$D_{r-1}(s_1)$

$\longrightarrow$ ●   ● s_1

$d(x_r, c)$

    ●   ● s_2

$D_{r-1}(s_3)$ $d(x_r, a)$

$\longrightarrow$ ●   ● s_3

x_r

Figure 6.24   Distortions on paths converging to state s_2 at stage r in Viterbi algorithm searching four-state trellis of Figure 6.17. Note that path symbol u_r must be 0 on both paths to converge to state s_2.

6.8   Concluding remarks

In this chapter, the aim was to present the fundamental ideas and methods for coding sources with memory. There are two basic approaches, either to try to remove the memory by prediction or, as we shall see later, by mathematical transformation, or by encoding blocks of data emitted directly from the source. Coding of prediction residuals and vector quantization are among the more common methods, while tree and trellis coding are less common. Tree- and trellis-based methods have received much attention in the research literature, because under certain conditions they can attain optimal performance. During the chapter, we did not prove, but did cite the theory that justifies the possibility of obtaining optimal performance for vector quantization and tree- and trellis-based coding. Currently, the latter are often difficult or impractical to implement for optimal performance, but that situation may change in the future.

Many methods found in practice are enhancements or hybrids of fundamental methods and have been deliberately omitted. For the reader who wishes to delve further into DPCM and adaptive DPCM, the textbook by Jayant and Noll [16] is highly recommended. Similarly, those who wish to know more details on methods of vector quantization should read the authoritative textbook of Gersho and Gray [8]. In Taubman and Marcellin [17] may be found further aspects and results for TCQ, especially in its use in Part 2 of the JPEG2000 image compression standard.

Problems

6.1   (Problem 7.1 [8]) A discrete-time stationary process $\{X_k\}$ has zero mean and correlation values $R_{XX}(0) = 4$, $R_{XX}(1) = 2$, and $R_{XX}(2) = 1$, and $R_{XX}(k) = 0$ for $k > 2$. An optimal first-order predictor is used in a DPCM coder for the input process X_k. The quantizer design assumes approximately uniform quantization of prediction error. The loading factor equals 4 and the number of levels is 128. (Loading factor is the ratio of the overload point (maximum finite threshold) to the standard deviation of the random variable to be quantized.)

Determine the approximate overall SNR of the DPCM system assuming negligible overloading.

6.2 The goal is to use the LBG algorithm to design two MMSE codebooks, of rates $R = 1$ and $R = 2$ bits per sample (dimension) for a two-dimensional Gaussian source of vectors $\mathbf{X} = (X_1, X_2)$, where X_1 and X_2 have zero mean, unit variance and correlation coefficient 0.9. Generate a sequence of 1000 training vectors for use in designing the codebook. (You can generate independent $\mathcal{N}(0, 1)$ variates and use an appropriate transformation.) Choose any reasonable initial codebook. Use a stopping criterion of 1% in relative drop in distortion *and* a maximum of 20 iterations.

Report and plot the location of your final codevectors and specify the final average distortion (averaged over the training set). Submit a copy of your computer source code and program output.

You may use your own source code or the program "lbg.c" in the course website. (http://www.cambridge.org/pearlman). You will have to set codebook and training set sizes in the #define NC and #define NT statements and initialize variables trans1, trans2 in routine create_training_set. To compile with gcc compiler, use -lm option.

6.3 An ECSQ can be designed with the ECVQ algorithm by setting the vector dimension to 1. Using the ECVQ algorithm, we wish to design an ECSQ for a Laplacian probability density and the squared-error criterion. First generate 100 000 or more independent samples from a Laplacian density of variance 2 by computer. Then set the number K of quantization levels and vary the parameter λ until enough (H, D) points are generated to sweep through the possible range of rates $0 \le H \le \log_2 K$.

(a) Plot D versus H for $K = 63, 31, 15, 7$, and 3 on a single graph.

(b) Report for each K and $\lambda = 0.8$ a list of the quantization levels and their probabilities.

6.4 We are required to "compress the data" from an independent, equi-probable binary source. We decide to generate a binary reproduction sequence through

$$\mathbf{v} = \mathbf{u}G.$$

(The sequences $\mathbf{u}$ and and $\mathbf{v}$ are row vectors.)

$$G = \begin{pmatrix} 1 & 0 & 0 & 1 & 1 \\ 0 & 1 & 0 & 1 & 0 \\ 0 & 0 & 1 & 0 & 1 \end{pmatrix}$$

for any given binary sequence. As $\mathbf{u}$ ranges through all its possibilities, we generate all the reproduction sequences $\mathbf{v}$.

(a) Give the rate R of this source code.

(b) Calculate the average distortion per source symbol of this code for the Hamming distortion measure.

(c) (i) Calculate the minimum possible rate which can achieve the average distortion in (b).

(ii) Calculate the minimum distortion possible for the code rate R in (a).

6.5 The diagram shows a decoder of a binary convolution source code. Output "1" is the digit of the first stage; output "2" is the modulo 2 sum of all three stages. Initially the shift register of three stages is filled with 0's and four code digits are shifted sequentially into the register. Then the contents of the register are dumped and replaced with three 0's, whereupon the next four digits are sequentially shifted into the register. The process of dumping, initializing to all 0's, and accepting the next four digits of the input is repeated until the entire binary input sequence is received. Assume that the number of symbols in the binary sequence is a multiple of four.

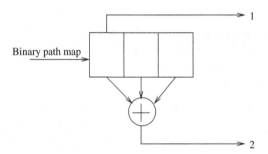

(a) Draw the matching trellis diagram for this decoder. Use the convention of reading the state and contents of the register from right to left (the most recent digit is last). Also use the convention of a solid branch for an input 0 and a dashed branch for an input 1. Label the branches with the corresponding output digits. (You should realize that you need only a trellis of four stages.)

(b) Assume that the encoder performs an exhaustive search of the trellis (Viterbi algorithm) for the best binary reproduction sequence for any given source sequence. The distortion measure is the Hamming distance.

(i) Determine the average Hamming distortion $\overline{d_H}$ (relative frequency of error) per source digit of this code.

(ii) What is the rate R of this code in bits per source symbol?

(c) Compare the performance of the above source code to the best possible performance of a code of this rate or distortion, assuming the binary symbols from the source are equally probable and statistically independent.

Note: In this problem we have a "vector quantization" of a block of binary symbols. The decoder is a shift register instead of a table lookup. The corresponding trellis is easier to search than a table when encoding.

6.6 Delta modulation (DM) is a special type of DPCM with quantization levels, where the predicted value of the next sample is the last sample's reconstructed value. It is realized in the figure below, where the accumulator sums all past values of $\tilde{e}(n)$.

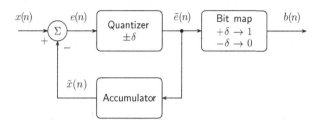

In instantaneous DM, $\tilde{e}(n) = \pm\delta$ is sent immediately, i.e., at time n. However, one can consider delaying the decision by sending instead the sequence, $\tilde{e}(n)$, $n = 0, 1, 2, \ldots, N$ such that $\sum_{n=1}^{N}(x(n) - \tilde{x}(n)^2)$ is minimum. In order to do so, we construct a binary code tree such that all possible $\tilde{x}(n) = \sum_{k=0}^{n} \tilde{e}(k)$.

(a) Construct such a tree for delayed decision encoding to a depth $N = 5$.

(b) Consider the source sequence to be $x(0) = 15$, $x(1) = 23$, $x(2) = 28$, $x(3) = 22$, $x(4) = 8$, $x(5) = -16$. Assume that $\delta = 5$ and $\tilde{x}(0) = 10$. Find the MSE path through the tree and the resultant DM binary code for the sequence.

(c) Compare your code to that of an instantaneous DM code for the source sequence. Also compare the MSEs of these different codes for the same source sequence.

Notes

1. Since the mean is assumed known, it can be subtracted from each sample at the encoder and added back at the decoder.

2. It is good programming practice to choose initially a set of dummy uniform probabilities, so that the entropy calculations stay finite during the initial iteration.

3. M here is not the number of codevectors.

References

1. A. D. Wyner and J. Ziv, "Bounds on the rate-distortion function for stationary sources with memory," *IEEE Trans. Inf. Theory*, vol. 17, no. 5, pp. 508–513, Sept. 1971.
2. R. M. Gray, J. C. Kieffer, and Y. Linde, "Locally optimal block quantizer design," *Inform. Control*, vol. 45, no. 2, pp. 178–198, May 1980.
3. Y. Linde, A. Buzo, and R. M. Gray, "An algorithm for vector quantizer design," *IEEE Trans. Commun.*, vol. 28, no. 1, pp. 84–95, Jan. 1980.
4. R. M. Gray, "Vector quantization," *IEEE ASSP Mag.*, vol. 1, no. 2, pp. 4–29, Apr. 1984.
5. S. P. Lloyd, "Least squares quantization in PCM," *IEEE Trans. Inf. Theory*, vol. 28, no. 2, pp. 129–137, Mar. 1982.
6. P. A. Chou, T. Lookabaugh, and R. M. Gray, "Entropy-constrained vector quantization," *IEEE Trans. Acoust. Speech Signal Process.*, vol. 37, no. 1, pp. 31–42, Jan. 1989.
7. E. A. Riskin and R. M. Gray, "A greedy tree growing algorithm for the design of variable rate vector quantizers," *IEEE Trans. Signal Process.*, vol. 39, no. 11, pp. 2500–2507, Nov. 1991.
8. A. Gersho and R. M. Gray, *Vector Quantization and Signal Compression*. Boston, Dordrecht, London: Kluwer Academic Publishers, 1992.
9. W. A. Finamore and W. A. Pearlman, "Optimal encoding of discrete-time continuous-amplitude, memoryless sources with finite output alphabets," *IEEE Trans. Inf. Theory*, vol. IT-26, no. 2, pp. 144–155, Mar. 1980.
10. M. W. Marcellin and T. R. Fischer, "Trellis coded quantization of memoryless and Gauss-Markov sources," *IEEE Trans. Commun.*, vol. 38, no. 1, pp. 82–93, Jan. 1990.
11. G. Ungerboeck, "Channel coding with multilevel/phase signals," *IEEE Trans. Inform. Theory*, vol. IT-28, no. 1, pp. 55–67, Jan. 1982.
12. T. R. Fischer, M. W. Marcellin, and M. Wang, "Trellis-coded vector quantiztion," *IEEE Trans. Inf. Theory*, vol. 37, no. 6, pp. 1551–1566, Nov. 1991.
13. *Information Technology—JPEG2000 Extensions, Part 2: Core Coding System*, ISO/IEC Int. Standard 15444-2, Geneva, Switzerland, 2001.
14. A. J. Viterbi, "Error bounds for convolutional codes and an asymptotically optimal decoding algorithm," *IEEE Trans. Inform. Theory*, vol. IT-13, no. 4, pp. 260–269, Apr. 1967.
15. J. G. D. Forney, "The Viterbi algorithm," *Proc. IEEE*, vol. 61, no. 3, pp. 268–278, Mar. 1973.
16. N. S. Jayant and P. Noll, *Digital Coding of Waveforms*. Englewood Cliffs, NJ: Prentice Hall, 1984.
17. D. S. Taubman and M. W. Marcellin, *JPEG2000: Image Compression Fundamentals, Standards, and Practice*. Norwell, MA: Kluwer Academic Publishers, 2002.

Further reading

J. B. Anderson and J. B. Bodie, "Tree encoding of speech," *IEEE Trans. Inf. Theory*, vol. 21, no. 4, pp. 379–387, July 1975.

L. Breiman, J. H. Friedman, R. A. Olshen, and C. J. Stone, *Classification and Regression Trees*. Monterey, CA: Wadsworth, 1984.

T. M. Cover and J. A. Thomas, *Elements of Information Theory*. New York, NY: John Wiley & Sons, 1991, 2006.

H. G. Fehn and P. Noll, "Multipath search coding of stationary signals with applications to speech," *IEEE Trans. Commun.*, vol. 30, no. 4, pp. 687–701, Apr. 1982.

P. Kroon and E. F. Deprettere, "A class of analysis-by-synthesis predictive coders for high quality speech coding at rates between 4.8 and 16 Kbits/s," *IEEE J. Sel. Areas Commun.*, vol. 30, no. 4, pp. 687–701, Feb. 1988.

B. Mahesh and W. A. Pearlman, "Multiple-rate structured vector quantization of image pyramids," *J. Visual Commun. Image Represent*, vol. 2, no. 2, pp. 103–113, Jan. 1991.

B. Mazor and W. A. Pearlman, "A trellis code construction and coding theorem for stationary Gaussian sources," *IEEE Trans. Inf. Theory*, vol. 29, no. 6, pp. 924–930, Nov. 1983.

——, "A tree coding theorem for stationary Gaussian sources and the squared-error distortion measure," *IEEE Trans. Inf. Theory*, vol. 32, no. 2, pp. 156–165, Mar. 1986.

J. W. Modestino, V. Bhaskaran, and J. B. Anderson, "Tree encoding of images in the presence of channel errors," *IEEE Trans. Inf. Theory*, vol. 27, no. 6, pp. 677–697, Nov. 1981.

W. A. Pearlman and P. Jakatdar, "A transform tree code for stationary Gaussian sources," *IEEE Trans. Inf. Theory*, vol. 31, no. 6, pp. 761–768, Nov. 1985.

E. A. Riskin, "Optimal bit allocation via the generalized BFOS algorithm," *IEEE Trans. Inf. Theory*, vol. 37, no. 2, pp. 400–402, Mar. 1991.

7 Mathematical transformations

7.1 Introduction

Instead of trying to encode the outputs of a source directly, it is often advantageous to transform the source mathematically into a set of equivalent values which are more easily or efficiently coded. A block diagram of a system employing a transform of the source is depicted in Figure 7.1.

Referring to this figure, the source sequence $\mathbf{x}$ is reversibly transformed to another sequence $\mathbf{y}$, which then can be represented, either with or without loss, by the discrete sequence $\tilde{\mathbf{y}}$. This representation $\tilde{\mathbf{y}}$ is usually a quantization of the transform sequence $\mathbf{y}$, i.e., it is defined as a discrete set of points that can best represent $\mathbf{y}$. Even though the values of $\tilde{\mathbf{y}}$ are discrete, in general they can assume any real value, so it is convenient first to convert it to a sequence of integer indices $\mathbf{m}$, which can be losslessly encoded to the binary sequence $\mathbf{b}$. Assuming perfect reception and decoding of $\mathbf{b}$ to $\mathbf{m}$ at the receiver, the sequence $\tilde{\mathbf{y}}$ is inverse transformed to the output reconstruction sequence $\hat{\mathbf{x}}$. Any differences of $\hat{\mathbf{x}}$ from the source sequence $\mathbf{x}$ result from the particular representation of $\mathbf{y}$ by $\tilde{\mathbf{y}}$ in the encoder.

When the source values are discrete in amplitude, then it is possible to have reversible transforms so that $\mathbf{y}$ also has discrete values. In this case $\tilde{\mathbf{y}}$ can be identical to $\mathbf{y}$, and lossless encoding of $\mathbf{y}$ is then possible. Most often, the source is modeled as continuous in amplitude and $\tilde{\mathbf{y}}$ is a quantization of $\mathbf{y}$, taking on discrete vector or vector component values. The most common measure of distortion between $\mathbf{x}$ and its reconstruction $\hat{\mathbf{x}}$ is the MSE, defined by

$$D_N(\mathbf{x}, \hat{\mathbf{x}}) = \frac{1}{N} \sum_{j=0}^{N-1} (x_j - \hat{x}_j)^2 = \frac{1}{N} ||\mathbf{x} - \hat{\mathbf{x}}||^2,$$

where N is the length of the sequences.

It is very important to control the amount of the above output distortion. Since distortion results from the quantization of the transform sequence, there must be a known relationship between the quantization distortion and the output distortion. When MSE is the distortion measure, these distortions are equal if the transform is a unitary linear transform. A (non-singular) real matrix $\mathbf{P}$ is said to be *unitary* if its transpose equals its inverse, i.e., $P^t = P^{-1}$. For a complex matrix, conjugate transpose $(*t)$

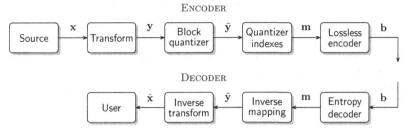

Figure 7.1 Transform coding system.

replaces transpose (t).[1] We shall elaborate on this matter after we introduce some notation.

A general linear transform of a vector $\mathbf{x}$ to a vector $\mathbf{y}$ is representation of its components in a different basis. Consider the basis of $\mathbf{x}$ to be the normal basis with basis vectors $\mathbf{n}_0, \mathbf{n}_1, \ldots, \mathbf{n}_{N-1}$, such that

$$\mathbf{n}_j = (\delta_{j,0}, \delta_{j,1}, \ldots, \delta_{j,N-1})^t, \quad j = 0, 1, 2, \ldots, N-1, \tag{7.1}$$

where

$$\delta_{i,j} = \begin{cases} 1 & \text{if } i = j \\ 0 & \text{otherwise} \end{cases} \tag{7.2}$$

In words, the basis vectors $\mathbf{n}_j$ have 1 in the j-th component and 0 in all the other components. (We have adopted the convention here that vectors are column vectors.) Note that these vectors are orthogonal (dot or inner product equals 0) and have length 1. Such a basis is said to be orthonormal. Then $\mathbf{x}$ can be written as

$$\mathbf{x} = \sum_{j=0}^{N-1} x_j \mathbf{n}_j.$$

Suppose that we wish to express $\mathbf{x}$ in terms of another orthonormal basis $\psi_0, \psi_1, \ldots, \psi_{N-1}$ according to

$$\mathbf{x} = \sum_{j=0}^{N-1} y_j \psi_j.$$

Equating these two representations and forming matrices whose columns are the basis vectors or vector components, we obtain

$$[\mathbf{n}_0|\mathbf{n}_1|\cdots|\mathbf{n}_{N-1}] \begin{pmatrix} x_0 \\ x_1 \\ \vdots \\ x_{N-1} \end{pmatrix} = [\psi_0|\psi_1|\cdots|\psi_{N-1}] \begin{pmatrix} y_0 \\ y_1 \\ \vdots \\ x_{N-1} \end{pmatrix}$$

$$I\mathbf{x} = \Psi\mathbf{y}$$

$$\mathbf{y} = \Psi^{-1}\mathbf{x} \tag{7.3}$$

where Ψ is the matrix having the new basis vectors as columns. (Enclosure with brackets ([ ]) denotes matrix.) The transform in Figure 7.1 is the inverse of the matrix Ψ. The inverse transform that produces the reconstruction is Ψ itself.

The relationship between the statistical properties of the transformed and original source is an important consideration in coding applications. We shall review the relationships most relevant to coding, which are the ones between means and covariances. First, we assume the source $\mathbf{x}$ to be real and wide-sense stationary, so that the sample mean and the autocorrelation function are independent of time or space shifts. Then, for all n, the sample mean and autocorrelation function may be expressed as

$$E[x(n)] = \mu_x \tag{7.4}$$

and

$$R_{xx}(k) = E[x(n)x(n+k)], \tag{7.5}$$

respectively. The autocorrelation function is even ($R_{xx}(k) = R_{xx}(-k)$) and has peak magnitude at $k = 0$. The covariance function is defined as

$$C_{xx}(k) \equiv (E[x(n)] - \mu_x)(E[x(n+k)] - \mu_x) \tag{7.6}$$

The mean square value of $x(n)$ for all n, called the power or energy, is $E[(x(n))^2]$ and is equal to $R_{xx}(0)$, the autocorrelation function at $k = 0$. The variance of $x(n)$ for all n, denoted as σ_x^2, is defined as

$$\sigma_x^2 \equiv (E[x(n)] - \mu_x)^2 \tag{7.7}$$

The variance is seen to be equal to the covariance at lag $k = 0$. That is, $\sigma_x^2 = C_{xx}(0)$. Notice that when the mean is zero, the autocorrelation and covariance functions are equal. Most often, the source process is assumed to have zero mean, because it involves no loss of generality. Being a known quantity, it or its transformed form can always be inserted when convenient. In what follows, we shall assume a zero mean source.

The autocorrelation matrix of a source vector $\mathbf{x} = (x(0), x(1), \ldots, x(N-1))^t$.[2] The autocorrelation matrix contains as elements all N lags of the autocorrelation function of the vector $\mathbf{x}$, which are

$$R_{xx}(k) = E[x(n)x(n+k)], \quad k = 0, 1, 2, \ldots, N - 1.$$

This matrix is defined as

$$\mathcal{R}_x \equiv E[\mathbf{x}\mathbf{x}^t] = \begin{bmatrix} R_{xx}(0) & R_{xx}(1) & R_{xx}(2) & \cdots & R_{xx}(N-1) \\ R_{xx}(1) & R_{xx}(0) & R_{xx}(1) & \cdots & R_{xx}(N-2) \\ R_{xx}(2) & R_{xx}(1) & R_{xx}(0) & \cdots & R_{xx}(N-3) \\ \vdots & \vdots & \vdots & \vdots & \vdots \\ R_{xx}(N-1) & R_{xx}(N-2) & \cdots & R_{xx}(1) & R_{xx}(0) \end{bmatrix}. \tag{7.8}$$

A more compact way to express the elements of this matrix is

$$\mathcal{R}_x = [R_{xx}(|k - l|)]_{k,l=0}^{N-1}.$$

Under a linear transformation to representation in a new basis according to $\mathbf{y} = P\mathbf{x}$, the autocorrelation matrix of $\mathbf{y}$ is calculated to be

$$
\begin{aligned}
\mathcal{R}_y &= E[\mathbf{y}\mathbf{y}^t] \\
&= PE[\mathbf{x}\mathbf{x}^t]P^t \\
&= P\mathcal{R}_x P^t.
\end{aligned}
\tag{7.9}
$$

When P is a unitary matrix, $P^t = P^{-1}$, and the transformation from $\mathcal{R}_x$ to $\mathcal{R}_y$ is a similarity transformation.[3] The diagonal elements of $\mathcal{R}_y$ are the variances of the components of the transform vector $\mathbf{y}$. Expressed mathematically

$$
\sigma_y^2(k) = \mathcal{R}_y(k,k), \ k = 0, 1, \ldots, N - 1.
$$

We prove now that the MSE quantization error equals the MSE reconstruction error, if the transform matrix is unitary. Any matrix whose columns are comprised of orthonormal basis vectors is unitary. From Figure 7.1, $\hat{\mathbf{x}} = \Psi\tilde{\mathbf{y}}$ and $\mathbf{x} = \Psi\mathbf{y}$. Then the reconstruction sum squared error is

$$
\|\mathbf{x} - \hat{\mathbf{x}}\|^2 = \|\Psi\mathbf{y} - \Psi\tilde{\mathbf{y}}\|^2
\tag{7.10}
$$

$$
= \|\Psi(\mathbf{y} - \tilde{\mathbf{y}})\|^2 = (\Psi(\mathbf{y} - \tilde{\mathbf{y}}))^t \Psi(\mathbf{y} - \tilde{\mathbf{y}})
$$

$$
\|\mathbf{x} - \hat{\mathbf{x}}\|^2 = (\mathbf{y} - \tilde{\mathbf{y}})^t \Psi^t \Psi(\mathbf{y} - \tilde{\mathbf{y}})
$$

Since the matrix Ψ is unitary, $\Psi^t\Psi = I$, the identity matrix. Therefore, we conclude that

$$
\|\mathbf{x} - \hat{\mathbf{x}}\|^2 = (\mathbf{y} - \tilde{\mathbf{y}})^t I (\mathbf{y} - \tilde{\mathbf{y}}) = \|\mathbf{y} - \tilde{\mathbf{y}}\|^2.
\tag{7.11}
$$

The above equation states that the sum squared reconstruction error equals the sum squared quantization error. Therefore, the MSE errors of the reconstruction and quantization are the same when the transform is unitary. Another notable property under the same conditions, arrived at through a similar argument, is that the squared lengths (powers) of the source and transform vectors are equal, i.e.,

$$
\|\mathbf{x}\|^2 = \|\mathbf{y}\|^2
\tag{7.12}
$$

7.1.1 Transform coding gain

The motivation for a mathematical transformation of the source is that coding coefficients of the transform results in lower distortion than coding the source values directly. The fractional reduction in distortion is expressed as a *coding gain*. The detailed derivations of the coding gain formulas are presented in the next chapter, Chapter 8, on rate control, but we shall cite the relevant formulas here to help explain differences in the characteristics of the different transforms in this chapter.

First of all, the formulas are exactly true for optimal coding of Gaussian sources and the squared error distortion criterion. We presented the solutions for this case in Section 6.3.1.3 in Chapter 6. Linear transformation of a Gaussian signal process remains Gaussian, but for other sources the transform coefficients have an altered distribution, are not independent, and are only approximately uncorrelated even for a well-chosen transform. We denote their variances by $\sigma_n^2, n = 0, 1, 2, \ldots, N-1$. Nonetheless, we encode the transform coefficients independently and assume that the MSE versus rate function is approximately that of the Gaussian scalar source. For large enough overall target rate per sample R, the coefficients are encoded with different rates according to the prescription that follows the Gaussian solution:

$$r_n = R + \frac{1}{2} \log \frac{\sigma_n^2}{\left(\prod_{n=0}^{N-1} \sigma_n^2\right)^{1/N}}. \tag{7.13}$$

(See Equation (6.33).) This allocation of rate assumes that R is large enough that all coefficients are encoded with non-zero rate. It results in the MMSE distortion of

$$D_{\text{TC}} = \left(\prod_{n=0}^{N-1} \sigma_n^2\right)^{1/N} \cdot 2^{-2R} \tag{7.14}$$

for the given target rate R. (See Equation (6.32).)

Direct optimal coding of the source samples of variance σ_x^2 with the same rate R results in MSE distortion

$$D_{\text{PCM}} = \sigma_x^2 2^{-2R}. \tag{7.15}$$

The ratio of these distortions is defined to be the coding gain. The gain of transform coding over direct PCM coding is therefore

$$G_{\text{TC/PCM}} \equiv \frac{D_{\text{PCM}}}{D_{\text{TC}}} = \frac{\sigma_x^2}{\left(\prod_{n=0}^{N-1} \sigma_n^2\right)^{1/N}} \tag{7.16}$$

For orthonormal transforms,

$$\sigma_x^2 = \frac{1}{N} \sum_{n=0}^{N-1} \sigma_n^2, \tag{7.17}$$

so the coding gain can be expressed as

$$G_{\text{TC/PCM}} = \frac{\frac{1}{N} \sum_{n=0}^{N-1} \sigma_n^2}{\left(\prod_{n=0}^{N-1} \sigma_n^2\right)^{1/N}}. \tag{7.18}$$

The arithmetic mean/geometric mean inequality states that

$$\left(\prod_{n=0}^{N-1} \sigma_n^2\right)^{1/N} \leq \frac{1}{N} \sum_{n=0}^{N-1} \sigma_n^2 \tag{7.19}$$

with equality if and only if all the transform variances σ_n^2 are equal. This equality case means that the source process is white, i.e., samples are mutually uncorrelated or the signal spectrum is flat. Therefore, the coding gain is greater than one, except when the source is white, in which case the transform gives no coding advantage. The objective of transformation then is to produce the smallest possible geometric mean of the transform variances. This happens when the transform packs most of the signal energy into relatively few coefficients.

Although the coding gain formula is strictly true only for larger rates and Gaussian sources, it really is a good measure of the efficacy of a particular transform in a broader context. As long as transforms are compared in optimal coding of the same source at the same rate, the comparison is valid and should hold true for other sources. That is why the coding gain in Equation 7.16 is universally accepted as a measure of a particular transform's coding effectiveness.

7.2 The optimal Karhunen–Loeve transform

Now we present the optimal transform, the Karhunen–Loeve transform (KLT), and its development. Let the source vector $\mathbf{X} = (X_0, X_1, \ldots, X_{N-1})^t$ have zero mean and N-dimensional covariance matrix $\Phi_N = E[\mathbf{XX}^t]$.[4] The elements of this matrix, $[\Phi_N]_{i,j}$ may be expressed as

$$[\Phi_N]_{i,j} = E[X_i X_j] = R_X(|i - j|), \quad i, j = 0, 1, \ldots, N - 1,$$

where $R_X(k)$ is the autocorrelation function with lag k.[5] Note in particular that, for the wide-sense stationary source $\mathbf{X}$, the variance $\sigma_X^2 = R_X(0)$. $X_i, i = 0, 1, 2, \ldots, N - 1$ are the individual components of the vector X. This (Toeplitz) matrix is real, symmetric, and non-negative definite. There exists a set of N orthonormal eigenvectors $\mathbf{e}_1, \mathbf{e}_2, \ldots, \mathbf{e}_N$ and a corresponding set of N non-negative, not necessarily distinct, eigenvalues $\lambda_1, \lambda_2, \ldots, \lambda_N$ for Φ_N. Expressed mathematically, the properties of the eigenvalues and eigenvectors are

$$\Phi_N \mathbf{e_j} = \lambda_j \mathbf{e_j}, \, j = 0, 1, \ldots, N - 1, \tag{7.20}$$

where $||\mathbf{e_j}||^2 = 1$ for all j. A given source sequence $\mathbf{X} = \mathbf{x}$ can be represented with respect to the eigenvector basis by

$$\mathbf{x} = \Psi \mathbf{y}, \quad \Psi = [\mathbf{e}_1 | \mathbf{e}_2 | \ldots | \mathbf{e}_N] \tag{7.21}$$

where Ψ is a unitary matrix, $(\Psi^{-1} = \Psi^t)$.

The vector $\mathbf{y}$ is called the transform of the source vector $\mathbf{x}$. The covariance matrix of the transform vector ensemble $\mathbf{Y}$ is

$$
\begin{aligned}
\Lambda &= E[\mathbf{YY}^t] = E[\Psi^{-1}\mathbf{XX}^t\Psi] \\[2mm]
&= \Psi^{-1}\Phi_N\Psi = \begin{bmatrix} \lambda_1 & 0 & \cdots & 0 \\ 0 & \lambda_2 & \cdots & 0 \\ \vdots & \vdots & \ddots & \vdots \\ 0 & 0 & \cdots & \lambda_N \end{bmatrix}
\end{aligned} \tag{7.22}
$$

The components of $\mathbf{Y}$ are uncorrelated with variances equal to the eigenvalues. This transformation of $\mathbf{x}$ to $\mathbf{y}$ is called the (discrete) KLT, Hotelling, or *principal components* transform. Moreover, when the source vector ensemble is Gaussian, the ensemble $\mathbf{Y}$ is also Gaussian with statistically independent components.

7.2.1   Optimal transform coding gain

The variances of the uncorrelated KLT coefficients are the eigenvalues $\lambda_1, \lambda_2, \ldots, \lambda_N$ of the covariance matrix of the source. By substitution of these variances into (7.16), the KLT coding gain is expresssed as

$$G^{(o)}_{\mathrm{TC/PCM}} = \frac{\sigma_X^2}{\left(\prod_{n=0}^{N-1} \lambda_n\right)^{1/N}}. \tag{7.23}$$

The superscript (o) in $G^{(o)}_{\mathrm{TC/PCM}}$ signifies that this gain is the optimal transform coding gain. An important property of this transform is that it distributes the variance into the fewest low-indexed components. The KLT has the smallest geometric mean of the component variances among all linear transforms of a given source and hence the highest coding gain of any linear transform of the source vector $\mathbf{x}$.

7.3   Suboptimal transforms

The discrete KLT requires knowledge of the covariance function of the source and a solution for the eigenvectors of the $N \times N$ covariance matrix. In general, especially for large N, the solution and the transform are computationally burdensome procedures with no fast algorithms for their execution. Instead, one uses almost always a source-independent transform with a fast execution algorithm. The hope is that the transform, although suboptimal, will give nearly as good a performance as the optimal KLT.

7.3.1   The discrete Fourier transform

One transform that is asymptotically optimal is the discrete Fourier transform (DFT):

$$y(n) = \frac{1}{\sqrt{N}} \sum_{k=0}^{N-1} x(k) e^{-j 2\pi k n/N} \tag{7.24}$$

$$x(k) = \frac{1}{\sqrt{N}} \sum_{n=0}^{N-1} y(n) e^{+j 2\pi k n/N}, \quad k = 0, 1, \ldots, N-1$$

or, in vector-matrix form:

$$\mathbf{y} = F_N \mathbf{x}, \qquad [F_N]_{n,k} = \frac{1}{\sqrt{N}} \exp\{-j 2\pi n k/N\} \tag{7.25}$$

(The indices of the vector components are in parentheses following the variable name, as is customary when representing transform components.) The transform matrix F_N is unitary, that is, $(F_N^*)^t = F_N^{-1}$, and is *circulant*, because its rows and columns are circular shifts of each other. The covariance matrix Λ_y of the transform vector $\mathbf{y}$, related to the covariance matrix Λ_x of the source vector $\mathbf{x}$ by $\Lambda_y = F_N \Lambda_x F_N^{-1}$, is diagonal if and only if $\phi_x(|k - \ell|)$ is periodic with period N. Therefore, in general, the components of a DFT are not uncorrelated, as with the KLT. In most practical cases, the components are weakly correlated.

The DFT has a fast computational algorithm (fast Fourier transform) and is asymptotically optimal in the strong sense that the variance of each $y(n)$ converges pointwise to a value of the spectral density which in turn equals one of the eigenvalues of the covariance matrix [1]. In this limit, the components become uncorrelated.

Although for finite N, the elements $y(n)$ are not uncorrelated, we encode (quantize) them independently as if they were independent. Because the correlations between the components are ignored, we obtain inferior performance to that of the KLT. As N grows larger, the performance of DFT quantization approaches that of KLT quantization. Bounds on the difference in performance between optimal DFT coding and optimal KLT coding have been derived by Pearl [2] and Pearlman [3]. One of the seeming disadvantages of the DFT is that it produces complex coefficients, but, because of the induced Hermitian symmetry for a real input sequence ($y^*(n) = y(N - n)$), there are only N real non-redundant quantities to be encoded.

7.3.2 The discrete cosine transform

Another important transform is the (unitary) discrete cosine transform (DCT) given by

$$
y(n) = \begin{cases} \frac{1}{\sqrt{N}} \sum_{k=0}^{N-1} x(k) & , \quad n = 0 \\ \frac{2}{\sqrt{N}} \sum_{k=0}^{N-1} x(k) \cos \frac{n\pi}{2N}(2k + 1) & , \quad n = 1, 2, \ldots, N - 1 \end{cases} \tag{7.26}
$$

The inverse is the same form with k and n interchanged. The forward and inverse DCT are often expressed more compactly as

$$
y(n) = \frac{2}{\sqrt{N}} \alpha(n) \sum_{k=0}^{N-1} x(k) \cos \frac{n\pi}{2N}(2k + 1), \quad n = 0, 1, \ldots, N - 1 \tag{7.27}
$$

$$
x(k) = \frac{2}{\sqrt{N}} \alpha(k) \sum_{n=0}^{N-1} y(n) \cos \frac{k\pi}{2N}(2n + 1), \quad k = 0, 1, \ldots, N - 1 \tag{7.28}
$$

The weighting function $\alpha(n)$ is defined as

$$
\alpha(n) = \begin{cases} \frac{1}{\sqrt{2}} & , \quad n = 0 \\ 1 & , \quad n \neq 0 \end{cases} \tag{7.29}
$$

The transform matrix is

$$\mathbf{C} = \{C_{nk}\}_{n,k=0}^{N-1}, \quad C_{nk} = \begin{cases} \frac{1}{\sqrt{N}} & \begin{array}{l} n = 0 \\ k = 0, 1, \ldots, N-1 \end{array} \\[2ex] \sqrt{\frac{2}{N}} \cos\frac{n\pi}{2N}(2k+1), & \\[1ex] & \begin{array}{l} n = 1, 2, \ldots, N-1 \\ k = 0, 1, 2, \ldots, N-1 \end{array} \end{cases} \tag{7.30}$$

$\mathbf{C}^{-1} = \mathbf{C}^t = \mathbf{C}$.

The transform elements are real.

This transform is asymptotically optimal in a distributional sense, i.e., for a function f, $f(E[|y(n)|^2], n = 0, 1, \ldots, N-1)$, converges to $f(\lambda_o, \lambda_1, \ldots, \lambda_{N-1})$ as $N \to \infty$, if the eigenvalues are bounded.

Yemini and Pearl [4] proved the asymptotic equivalence of DCT and KLT for all finite order, stationary Markov processes. Ahmed et al. [5] gave empirical evidence that DCT performance is close to KLT even for small values of N in the case of Markov-1 signals.

You can evaluate the DCT through the DFT, which has a fast algorithm. First express the DCT as

$$y(n) = \sqrt{\frac{2}{N}} \text{Re} \left\{ e^{-j\frac{n\pi}{2N}} \alpha(n) \sum_{k=0}^{N-1} x(k) e^{-j\frac{2\pi kn}{2N}} \right\}$$

Note that the summation is a $2N$-point DFT with $x(k) = 0$, $N \leq k \leq 2N - 1$. The $2N$-point DFT is a slight disadvantage along with the exponential (or sine-cosine) multipliers. If the sequence is reordered as

$$\left. \begin{array}{l} w(k) = x(2k) \\ w(N - 1 - k) = x(2k+1) \end{array} \right\} \quad k = 0, 1, \ldots, \frac{N}{2} - 1$$

$$y(n) = \sqrt{\frac{2}{N}} \alpha(n) \text{Re} \left\{ e^{j\frac{\pi n}{2N}} \sum_{k=0}^{N-1} w(k) e^{j2\pi nk/N} \right\}$$

the summation is an N-point DFT of the reordered sequence [6]. Similar procedures apply to the inverses. There are also direct and fast algorithms for computing the DCT that are beyond the scope of this discussion.

7.3.3 The Hadamard–Walsh transform

Another transform often used is the Hadamard–Walsh transform. The transform matrix consists just of 1's and -1's. Therefore the transform is formed by a series of additions and subtractions, making it very fast to compute. It has worse performance than the DFT and DCT and is not asymptotically optimal. Its computational advantages often outweigh its inferior performance when considering certain application scenarios. The transform matrix[6] is defined recursively as follows:

$$\mathbf{H}_2 = \frac{1}{\sqrt{2}} \begin{bmatrix} 1 & 1 \\ 1 & -1 \end{bmatrix} \qquad (7.31)$$

$$\mathbf{H}_{2N} = \frac{1}{\sqrt{2}} \begin{bmatrix} \mathbf{H}_N & \mathbf{H}_N \\ \mathbf{H}_N & -\mathbf{H}_N \end{bmatrix}. \qquad (7.32)$$

For example, to produce $\mathbf{H}_4$,

$$\mathbf{H}_4 = \frac{1}{\sqrt{2}} \begin{bmatrix} \mathbf{H}_2 & \mathbf{H}_2 \\ \mathbf{H}_2 & -\mathbf{H}_2 \end{bmatrix}$$

$$= \frac{1}{2} \begin{bmatrix} 1 & 1 & 1 & 1 \\ 1 & -1 & 1 & -1 \\ 1 & 1 & -1 & -1 \\ 1 & -1 & -1 & 1 \end{bmatrix} \qquad (7.33)$$

Note that the rows (columns) are orthogonal, as they differ from each other in one-half the positions. Also note that they are unit length vectors. Therefore, the matrix $\mathbf{H}_4$ is unitary. By extension, the matrix $\mathbf{H}_{2N}$ is unitary,

$$\mathbf{H}_{2N} \mathbf{H}_{2N}^t = \mathbf{I}_{2N}, \qquad (7.34)$$

where $\mathbf{I}_{2N}$ is the $2N \times 2N$ identity matrix.

7.4 Lapped orthogonal transform

The transforms discussed thus far are applied independently and sequentially to fixed-length blocks of samples from the input. When the coefficients of these blocks are quantized, discontinuities occur at the block boundaries that manifest themselves as data discontinuities in the reconstruction. A transform that ameliorates these boundary artifacts is the *lapped orthogonal transform* (LOT). The idea is to divide the input sequence into blocks of N samples each and apply the transform to a sliding window embracing two blocks containing $L = 2N$ samples and output N coefficients. Then the window slides N samples or one block length to take the next transform of $2N$ samples and output another N coefficients. In this way, each block of N samples is involved in two consecutive transform operations (except for the beginning and end blocks of the input sequence).[7] The region of samples common to two consecutive transforms is called the overlap region.

In order to present the details of the LOT, let the infinite input sequence be denoted by the column vector $\mathbf{x}$ and the length N blocks of the infinite input sequence by the column vectors $\dots, \mathbf{x}_{i-1}, \mathbf{x}_i, \mathbf{x}_{i+1}, \dots$ Let $\mathbf{P}$ be the $L \times N$ transform matrix ($L = 2N$). Then the transform of the input sequence takes the form

$$y = \tilde{\mathbf{P}}^t \mathbf{x} = \begin{pmatrix} \ddots & & & & \\ & \mathbf{P}^t & & \mathbf{0} & \\ & & \mathbf{P}^t & & \\ & \mathbf{0} & & \mathbf{P}^t & \\ & & & & \ddots \end{pmatrix} \begin{pmatrix} \vdots \\ \mathbf{x}_{i-1} \\ \mathbf{x}_i \\ \mathbf{x}_{i+1} \\ \vdots \end{pmatrix} \tag{7.35}$$

We wish to attain perfect reconstruction of the input vector $\mathbf{x}$. Therefore, since

$$\tilde{\mathbf{P}} y = \tilde{\mathbf{P}} \tilde{\mathbf{P}}^t \mathbf{x},$$

we must have

$$\tilde{\mathbf{P}} \tilde{\mathbf{P}}^t = \mathbf{I}. \tag{7.36}$$

The matrix $\tilde{\mathbf{P}}$ must be orthogonal. In order to achieve this property, the rows of the submatrix $\mathbf{P}$ must be orthogonal and the portions of the rows in the overlapped portion must be orthogonal. The following equations express these conditions.

$$\mathbf{P}^t \mathbf{P} = \mathbf{I}_N \tag{7.37}$$
$$\mathbf{P}^t \mathbf{W} \mathbf{P} = \mathbf{I}_N \tag{7.38}$$

where $\mathbf{W}$ is the one-block shift $L \times L$ matrix defined by

$$\mathbf{W} = \begin{pmatrix} \mathbf{0} & \mathbf{I}_{L-N} \\ \mathbf{0} & \mathbf{0} \end{pmatrix}. \tag{7.39}$$

The subscript of $\mathbf{I}$ indicates the dimension of the square identity matrix.

An optimal LOT can only be designed using the covariance matrix of the source. Fixed transforms, independent of source statistics, that provide good performance for most sources of interest, are more desirable, because they compute much more quickly. Malvar [7] originated a fast LOT algorithm that does depend on source statistics, but gives good results nonetheless for real sources such as speech and images. We shall present here only one form of the LOT developed by Malvar. As will be shown later, good performance of a transform in coding depends on the source energy being compacted into a relatively few transform coefficients. The DCT has such a characteristic and has a fast algorithm, so it forms the basis of determining a good LOT. Malvar proposed to build a fast LOT starting with the following matrix

$$\mathbf{P}_0 = \frac{1}{2} \begin{pmatrix} \mathbf{D}_e - \mathbf{D}_o & \mathbf{D}_e - \mathbf{D}_o \\ \mathbf{J}(\mathbf{D}_e - \mathbf{D}_o) & -\mathbf{J}(\mathbf{D}_e - \mathbf{D}_o) \end{pmatrix} \tag{7.40}$$

where $\mathbf{D}_e^t$ and $\mathbf{D}_o^t$ are $N/2 \times N$ matrices consisting of the even and odd rows (basis functions) of the DCT matrix. $\mathbf{J}$ is the $N \times N$ counter identity matrix

$$J = \begin{pmatrix} 0 & 0 & \cdots & & 0 & 1 \\ 0 & 0 & \cdots & 0 & 1 & 0 \\ \vdots & & & & & \vdots \\ 0 & 1 & 0 & \cdots & & 0 \\ 1 & 0 & & \cdots & 0 & 0 \end{pmatrix}. \tag{7.41}$$

This matrix reverses the order of the rows of the matrix it precedes. The LOT transform of the two-block vector $\mathbf{x} = \begin{pmatrix} \mathbf{x}_{i-1} \\ \mathbf{x}_i \end{pmatrix}$ of total length $L = 2N$

$$\mathbf{y} = \mathbf{Z}^t \mathbf{P}_o^t \mathbf{x}, \tag{7.42}$$

Ideally the $N \times N$ matrix $\mathbf{Z}$ should be chosen so that the covariance matrix of $\mathbf{y}$ is diagonal. Therefore, we would wish the columns of $\mathbf{Z}$ to be the normalized eigenvectors of the covariance matrix of $\mathbf{u} = \mathbf{P}_o^t \mathbf{x}$. This covariance matrix is

$$\mathbf{R}_{uu} = E[\mathbf{u}\mathbf{u}^t] = \mathbf{P}_o^t \mathbf{R}_{xx} \mathbf{P}_o, \tag{7.43}$$

where $\mathbf{R}_{xx}$ is the covariance matrix of the source $\mathbf{x}$. Malvar called this solution the quasi-optimal LOT. In order to obtain a fast computational algorithm, not so dependent on source statistics, Malvar developed the following approximation to $\mathbf{Z}$.

$$\mathbf{Z} \approx \begin{pmatrix} \mathbf{I} & \mathbf{0} \\ \mathbf{0} & \mathbf{Z}_2 \end{pmatrix}, \tag{7.44}$$

where $\mathbf{I}$ and $\mathbf{Z}_2$ are each $N/2 \times N/2$ and $\mathbf{Z}_2$ is a cascade of plane rotations

$$\mathbf{Z}_2 = \mathbf{T}_1 \mathbf{T}_2 \cdots \mathbf{T}_{N/2-1}. \tag{7.45}$$

The plane rotations take the form

$$\mathbf{T}_i = \begin{pmatrix} \mathbf{I}_{i-1} & \mathbf{0} & \mathbf{0} \\ \mathbf{0} & \mathbf{Y}(\theta_i) & \mathbf{0} \\ \mathbf{0} & \mathbf{0} & \mathbf{I}_{N/2-i-1} \end{pmatrix}. \tag{7.46}$$

The matrix $\mathbf{Y}(\theta_i)$ is a 2×2 rotation by the angle θ_i defined by

$$\mathbf{Y}(\theta_i) \equiv \begin{pmatrix} \cos\theta_i & \sin\theta_i \\ -\sin\theta_i & \cos\theta_i \end{pmatrix}. \tag{7.47}$$

These 2×2 rotations are implemented by butterfly operations, just as in the fast Fourier transform (FFT). This implementation of the LOT is called the fast LOT. For a 16×8 ($L = 16$, $N = 8$) LOT, Malvar [7] reports that the angles $[\theta_1, \theta_2, \theta_3] = [0.13\pi, 0.16\pi, 0.13\pi]$ achieve maximum coding gain for the first-order Markov source with correlation parameter $\rho = 0.95$. The autocorrelation function of this source is shown below in Equation (7.48). The computational complexity of the LOT is roughly twice that of the DCT.

7.4.1 Example of calculation of transform coding gain

The optimal or ideal coding gain is achieved with the KLT, because it decorrelates the source block completely. The fixed transforms only decorrelate partially for a finite block size N. The coding gain formula in Equation (7.16) is used to assess the efficacy of the different transforms. Here we shall present an example of the calculation of coding gain for one transform, the Hadamard–Walsh, and a block of $N = 4$ samples from a first-order, Gauss–Markov source. The same paradigm can be used for any transform and source model.

A Gauss–Markov source with zero mean has the autocorrelation function

$$R_{xx}(n) = \sigma_x^2 \rho^{|n|}, \quad n = 0, \pm 1, \pm 2, \ldots, \tag{7.48}$$

where ρ is the correlation parameter, $0 \le \rho \le 1$. In order to calculate the coding gain $G_{TC/PCM}$, we must determine the variances of the transform coefficients. One straightforward way is to calculate the covariance or autocorrelation matrix of the transformed source block and pick off its diagonal elements. Although this procedure computes more than needed, it is especially easy to do with a program like MATLAB. So we shall follow this approach, using the formula in Equation (7.9). From Equation (7.8), the 4×4 autocorrelation matrix of the source block is

$$\mathcal{R}_x = \sigma_x^2 \begin{bmatrix} 1 & \rho & \rho^2 & \rho^3 \\ \rho & 1 & \rho & \rho^2 \\ \rho^2 & \rho & 1 & \rho \\ \rho^3 & \rho^2 & \rho & 1 \end{bmatrix}. \tag{7.49}$$

The 4×4 Hadamard–Walsh matrix is

$$\mathbf{H}_4 = \frac{1}{2} \begin{bmatrix} 1 & 1 & 1 & 1 \\ 1 & -1 & 1 & -1 \\ 1 & 1 & -1 & -1 \\ 1 & -1 & -1 & 1 \end{bmatrix} \tag{7.50}$$

We identify $\mathbf{H}_4$ with Φ^{-1} in Equation (7.9). Notice that this matrix is symmetric (equals its transpose). The following matrix calculations produce the autocorrelation matrix of the transform

$$\mathcal{R}_y = \frac{1}{2} \begin{bmatrix} 1 & 1 & 1 & 1 \\ 1 & -1 & 1 & -1 \\ 1 & 1 & -1 & -1 \\ 1 & -1 & -1 & 1 \end{bmatrix} \sigma_x^2 \begin{bmatrix} 1 & \rho & \rho^2 & \rho^3 \\ \rho & 1 & \rho & \rho^2 \\ \rho^2 & \rho & 1 & \rho \\ \rho^3 & \rho^2 & \rho & 1 \end{bmatrix} \frac{1}{2} \begin{bmatrix} 1 & 1 & 1 & 1 \\ 1 & -1 & 1 & -1 \\ 1 & 1 & -1 & -1 \\ 1 & -1 & -1 & 1 \end{bmatrix} \tag{7.51}$$

Assume that the correlation parameter of the source is $\rho = 0.9$ and its variance $\sigma_x^2 = 1$. Substituting into the above formula and calculating $\mathcal{R}_y$ via MATLAB give the result

$$R_y = \begin{bmatrix} 3.5245 & 0 & -0.0000 & -0.0855 \\ 0 & 0.0955 & 0.0855 & 0 \\ -0.0000 & 0.0855 & 0.2755 & 0.0000 \\ -0.0855 & 0.0000 & -0.0000 & 0.1045 \end{bmatrix} \tag{7.52}$$

The diagonal entries are the variances of the transform coefficients as follows:

$$\sigma_{y_0}^2 = 3.5245, \ \sigma_{y_1}^2 = 0.0955, \ \sigma_{y_2}^2 = 0.2755, \ \sigma_{y_3}^2 = 0.1045$$

Substituting these variances into the coding gain formula in Equation (7.16) results in the coding gain as follows:

$$G_{\text{TC/PCM}} = \frac{1}{\left(\prod_{k=0}^{3} \sigma_{y_k}^2 \right)^{(1/4)}} = 3.1867.$$

7.5 Transforms via filter banks

Input signals are usually transformed sequentially in blocks of a given length N. In these circumstances, it is convenient to realize the sequence of transformed blocks using a bank of filters followed by downsampling. Here we show how this is done. Consider first a single block of length N, denoted by $x(0), x(1), \ldots, x(N-1)$. The components of the i-th basis vector ψ_i of the transform are denoted as $\psi_i(0), \psi_i(1), \ldots, \psi_i(N-1)$. Assuming an orthogonal transform, the i-th transform coefficient y_i with this notation becomes

$$y_i = \sum_{n=0}^{N-1} \psi_i(n)x(n). \tag{7.53}$$

We wish now to put this equation into the form of a convolution. First, let us define the time reversed basis functions with components $\tilde{\psi}_i(n) = \psi_i(N-1-n)$. Consider now the following convolution:

$$y_i(\ell) = \sum_{n=0}^{N-1} \tilde{\psi}_i(n)x(\ell - n). \tag{7.54}$$

$y_i(\ell)$ is the output at time m of $x(n)$ fed to the (linear, time-invariant) filter with impulse response $\tilde{\psi}_i(n)$. At time $\ell = N-1$, the output is

$$y_i(N-1) = \sum_{n=0}^{N-1} \tilde{\psi}_i(n)x(N-1-n) \tag{7.55}$$

$$= \sum_{n=0}^{N-1} \psi_i(N-1-n)x(N-1-n) \tag{7.56}$$

$$= \sum_{n=0}^{N-1} \psi_i(n)x(n). \tag{7.57}$$

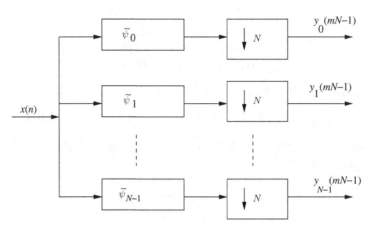

Figure 7.2   Forward N-point block transforms via a bank of N filters.

This is true for all $i = 0, 1, \ldots, N - 1$, so the transform of a block is realized by the outputs at time $N - 1$ of a bank of filters with impulse responses $\tilde{\psi}_i(n)$, as shown in Figure 7.2.

The input $x(n)$ may be viewed as a sequence of blocks of length N. Therefore we can continue to feed the input samples to the filter bank and sample the output stream at times $mN - 1$ for integers $m = 1, 2, \ldots$ until the input samples are exhausted. At every time $mN - 1$, we obtain the transform of the m-th block. This procedure is called downsampling (or decimation) by N and is indicated by the symbol "$\downarrow N$."

The inverse transform can also be computed in a similar way using a filter bank. The inverse transform of a single block takes the form

$$x(n) = \sum_{i=0}^{N-1} \psi_i'(n) y_i, \tag{7.58}$$

where $y_i = y_i(N - 1)$, the outputs of the forward transform filter bank at time $N - 1$. We shall settle the matter of the inverse basis vectors with components $\psi_i'(n)$ later. Defining the function that is one followed by $N - 1$ zeros as

$$\delta_N(n) = \begin{cases} 1, & n = 0 \\ 0, & n = 1, 2, \ldots, N - 1 \end{cases}$$

we can express the terms in the sum in (7.58) as

$$\psi_i'(n) y_i = \sum_{j=0}^{N-1} \psi_i'(j) \delta_N(n - j) y_i. \tag{7.59}$$

This expression is the convolution of $\psi_i'()$ with $y_i \delta_N()$. Therefore, these convolutions, one for each transform coefficient, may be computed using an input of y_i followed by $N - 1$ zeros to a filter with impulse response $\psi_i'(n)$. Summing these outputs reconstructs $x(n)$ for each time $n = 0, 1, 2, \ldots, N - 1$.[8] The filling in of $N - 1$ zeros is called *upsampling* by a factor of N and is indicated by the symbol "$\uparrow N$."

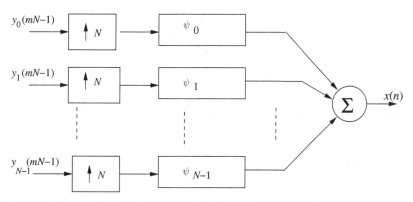

Figure 7.3 Inverse N-point block transforms via a bank of N filters.

Figure 7.3 illustrates the reconstruction of $x(n)$ via input to upsampling operations and a filter bank.

For an orthonormal transform, the basis vectors (forward or inverse) obey the conditions

$$\sum_{n=0}^{N-1} \psi_i(n)\psi_j(n) = \delta(i - j), \tag{7.60}$$

where $\delta(j)$ is the Kronecker delta function, defined as Kronecker delta function

$$\delta(j) = \begin{cases} 1, & j = 0 \\ 0, & j \neq 0 \end{cases} \tag{7.61}$$

This means that the forward and inverse vectors are the same. Therefore,

$$\psi_i'(n) = \psi_i(n), \quad n, i = 0, 1, 2, \ldots, N - 1. \tag{7.62}$$

The impulse response functions in the reconstruction filter bank are just the basis vectors $\psi_i(n)$. For the forward transform, these filter functions were the time-reversed components of the basis vectors, $\tilde{\psi}_i(n) = \psi_i(N - 1 - n)$.

Considering the input to be a sequence of length N blocks as before, at every time $t = mN - 1$, the process of upsampling, bank filtering, and summing is repeated to obtain the reconstruction of each input block.

7.6 Two-dimensional transforms for images

The development so far treated only transformations of one-dimensional sources. Images and two-dimensional data also need compression and two-dimensional transformations. The one-dimensional transformations generalize in a logical way to two dimensions. We essentially add a second coordinate or index to the above formulas. A general two-dimensional transform of a two-dimensional source, $x(k, l), = 1, 2, \ldots, M, l = 1, 2, \ldots, N$ takes the form

$$y(m, n) = \sum_{k=0}^{M-1} \sum_{l=0}^{N-1} x(k, l)\psi(m, n, k, l)$$

$$m = 1, 2, \ldots, M - 1 \quad n = 1, 2, \ldots, N - 1 \tag{7.63}$$

$$x(k, l) = \sum_{m=0}^{M-1} \sum_{n=0}^{N-1} y(m, n)\tilde{\psi}(m, n, k, l)$$

$$k = 1, 2, \ldots, M - 1 \quad l = 1, 2, \ldots, N - 1 \tag{7.64}$$

In order for the transform to be invertible,

$$\sum_{m=0}^{M-1} \sum_{n=0}^{N-1} \psi(m, n, k', l')\tilde{\psi}(m, n, k, l) = \begin{cases} 1, & \text{if } k' = k, l' = l \\ 0, & \text{otherwise} \end{cases} \tag{7.65}$$

The components of the basis vectors take values on a rectangular array of points. There are ways to order them linearly, so that they can be manipulated like the vector components on a one-dimensional point array, as we did before. However, it is not necessary to do so here. Almost everything we learned about the one-dimensional case holds for the two-dimensional case. An important example in point is the constancy of power and MSE between the input and transform array values.

Almost always, separable transforms are used in practice. A transform is separable when the transform kernel is a product of two one-dimensional kernels, namely

$$\psi(m, n, k, l) = \phi_1(m, k)\phi_2(n, l) \quad \forall k, l, m, n$$

To see the simplification arising from this separability, let us rewrite the transform equations with a separable transform.

$$y(m, n) = \sum_{k=0}^{M-1} \sum_{l=0}^{N-1} x(k, l)\phi_1(m, k)\phi_2(n, l)$$

$$= \sum_{k=0}^{M-1} \phi_1(m, k)(\sum_{l=0}^{N-1} x(k, l)\phi_2(n, l)) \tag{7.66}$$

The expression in parentheses in the last equation, which we define as

$$y_2(k, n) \equiv \sum_{l=0}^{N-1} x(k, l)\phi_2(n, l), \tag{7.67}$$

is a transform of row k of the two-dimensional array. If we calculate the transform in place (replacing the source value by its transform value in the corresponding location) for every row, we see that the outer sum in this equation is a transform along the columns according to

$$y(m, n) = \sum_{k=0}^{M-1} y_2(k, n)\phi_1(m, k). \tag{7.68}$$

The result is the coefficient array of the two-dimensional transform in place of the original source array. The inverse transform is also separable and follows the same computational procedure.

Therefore, a separable two-dimensional transform consists of M N-point one-dimensional transforms of rows followed by N M-point transforms of columns. Therefore, $M + N$ (multiply and add) operations compute each transform coefficient. By contrast, $M \times N$ operations are needed to compute each transform coefficient of a non-separable two-dimensional transform.

The common two-dimensional transforms are separable extensions of the one-dimensional ones. In terms of the kernel functions defined above, the two-dimensional DFT is given by

$$\phi_1(m, k) = \frac{1}{\sqrt{M}} \exp\{-j2\pi mk/M\}, \quad k, m = 0, 1, \ldots, M - 1,$$

$$\phi_2(n, l) = \frac{1}{\sqrt{N}} \exp\{-j2\pi nl/N\}, \quad l, n = 0, 1, \ldots, N - 1. \qquad (7.69)$$

The inverse kernels are conjugates of the forward ones with indices interchanged, i.e.,

$$\tilde{\phi}_1(k, m) = \phi_1^*(m, k) \quad \tilde{\phi}_2(l, n) = \phi_2^*(n, l). \qquad (7.70)$$

The kernels for the two-dimensional DCT are given by

$$\phi_1(m, k) = \sqrt{\frac{2}{M}} \alpha(m) \cos \frac{m(2k + 1)\pi}{2M}, \quad k, m = 0, 1, \ldots, M - 1,$$

$$\phi_2(n, l) = \sqrt{\frac{2}{N}} \alpha(n) \cos \frac{n(2l + 1)\pi}{2N}, \quad l, n = 0, 1, \ldots, N - 1. \qquad (7.71)$$

where $\alpha(m)$ or $\alpha(n)$ is the same weighting function defined in (7.29). The inverse kernels are identical to the forward ones, again with interchanged indices.

$$\tilde{\phi}_1(k, m) = \phi_1(m, k) \quad \tilde{\phi}_2(l, n) = \phi_2(n, l) \qquad (7.72)$$

Although these formulas might seem a little daunting, all they mean is that we substitute the appropriate kernel into Equation (7.67) to transform the rows in place and then transform the columns in place using the appropriate kernel substituted into Equation (7.68). Since these steps are one-dimensional transforms, we can use the transform matrices presented previously in Equations (7.21), (7.25), (7.30), and (7.32) to calculate the two-dimensional KLT, DFT, DCT, and Hadamard–Walsh transforms, respectively. Clearly, the same can be done for any separable two-dimensional transform.

7.7   Subband transforms

7.7.1   Introduction

Transforms that decompose the source into non-overlapping and contiguous frequency ranges called *subbands* are called *subband transforms*. A *wavelet transform*, as we shall see, is just a particular kind of subband transform. The source sequence is fed to a bank of bandpass filters which are contiguous and cover the full frequency range. The set of output signals are the subband signals and can be recombined without degradation to produce the original signal. Let us assume first that the contiguous filters, each with bandwidth $W_m, m = 1, 2, \ldots, M$, are ideal with zero attenuation in the pass-band and infinite attenuation in the stop band and that they cover the full frequency range of the input, as depicted in Figure 7.4.

The output of any one filter, whose lower cutoff frequency is an integer multiple of its bandwidth W_m, is subsampled by a factor equal to $V_m = \pi / W_m$ and is now a full-band sequence in the frequency range from $-\pi$ to π referenced to the new, lower sampling frequency. This combination of filtering and subsampling (often illogically called *decimation*) is called M-channel filter bank analysis and is depicted in Figure 7.5. We shall assume that the integer bandwidth to lower frequency relationship (called the *integer-band property*) holds for all filters in the bank so that all outputs are decimated by the appropriate factor. These outputs are called the subband signals or waveforms

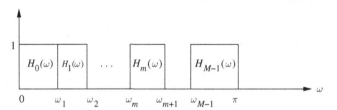

Figure 7.4   Subband filter transfer functions.

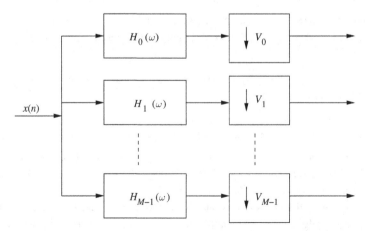

Figure 7.5 M-channel filter bank analysis of source into subbands.

and their aggregate number of samples equals that in the original input waveform. The original input can be reconstructed exactly from the subband signals. The sampling rate of each subband signal is increased to that of the original input by filling in the appropriate number of zero samples (called *upsampling*) and the zero-filled waveform is fed into an ideal filter with gain equal to the subsampling factor covering the original pass-band (called *interpolation*). The sum of these interpolated subband signals equals the original input signal. Figure 7.6 depicts this M-channel filter bank synthesis of the subband signals to reproduce the original signal.

Most often, however, a transform of M subbands is generated by successive application of two-channel filter banks, consisting of half-band filters, to certain subband outputs. The subbands of half-band filters occupy the frequency intervals $[0, \pi/2)$ and $[\pi/2, \pi)$. Figure 7.7 depicts the two-channel analysis and synthesis filter banks.

For example, two stages of applying the two-channel analysis filter bank on the output subband signals creates $M = 4$ subbands of equal width, as shown in Figure 7.8. Synthesis is accomplished by successive stages of application of the two-channel synthesis filter bank. The corresponding synthesis of these $M = 4$ subbands to reconstruct the original signal is shown in Figure 7.9.

A multi-resolution analysis is created by successive application of the two-channel filter bank only to the lowpass subband of the previous stage. Figure 7.10 shows three stages of two-channel filter bank analysis on the lowpass outputs that generates $M = 4$ subbands. Shown also are the frequency occupancies of these subbands, which are of

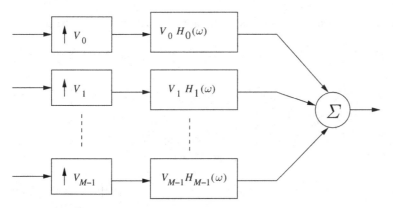

Figure 7.6 M-channel filter bank synthesis of source from subbands.

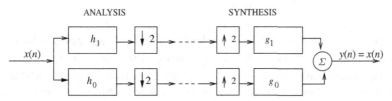

Figure 7.7   Two-channel analysis and synthesis filter banks. h_0 and g_0 denote impulse responses of the lowpass filters; h_1 and g_1 those of the highpass filters.

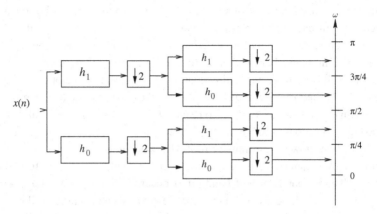

Figure 7.8   Two stages of two-channel analysis for four equal-size subbands.

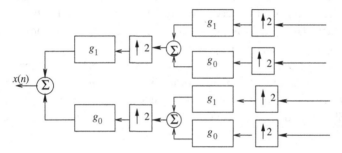

Figure 7.9   Two stages of two-channel synthesis to reconstruct signal from four subbands.

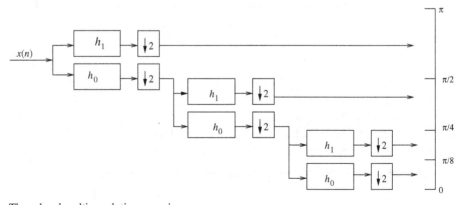

Figure 7.10   Three-level multi-resolution anaysis.

widths $\pi/8$, $\pi/8$, $\pi/4$, and $\pi/2$ in order of low to high frequency. The lowest frequency subband signal is said to be the coarsest or lowest resolution. The subbands are recombined in three stages of two-channel synthesis in the reverse order of the analysis stages, as depicted in Figure 7.11. The two lowest frequency subbands are first combined in the two-channel synthesizer to create the lowpass subband $[0, \pi/4)$ that is the next higher level of resolution. This subband is then combined with the subband $[\pi/4, \pi/2)$ to

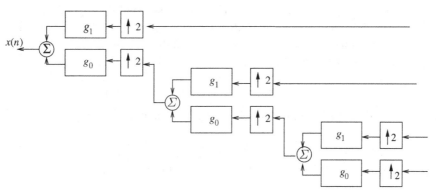

Figure 7.11   Three-level multi-resolution synthesis.

create the next highest level of resolution. Then this $[0, \pi/2)$ subband is combined with the $[\pi/2, \pi)$ high frequency subband to produce the full resolution original signal. So, the subbands that are the output of three stages of two-channel filter bank decomposition are said to contain four levels of resolution. The lowest level is 1/8 the original sequence length, the next level 1/4, the next 1/2 and the final one $1\times$ original length.

7.7.2   Coding gain of subband transformation

As with the matrix transforms visited earlier, transforming a source into subbands provides coding gains. The next chapter derives the actual coding gain in detail, but for our purposes here, we provide a skeletal proof for optimal coding of Gaussian sources and the squared error distortion measure. The objective in coding is to minimize the average distortion (per sample) for a fixed rate per sample.

Let there be N samples from the source and a given code rate of R bits per sample. If b_m bits are given to the m-th subband, then

$$R = \frac{1}{N} \sum_{m=0}^{M-1} b_m \tag{7.73}$$

because the subbands are coded independently.

We assign the same rate r_m to code the coefficients $\mathbf{y}_m$ of the m-th subband, so that $b_m = n_m r_m$. Therefore, the average code rate in bits per sample can be expressed as

$$R = \sum_{m=0}^{M-1} \eta_m r_m. \tag{7.74}$$

where $\eta_m = n_m/N$, the fraction of the number of samples (coefficients) in the m-th subband. Each subband $\mathbf{y}_m$ has samples with the same variance σ_m^2, $m = 0, 1, \ldots, M-1$. We assume that the rate per sample r_m assigned to every subband is non-zero, so that ideal coding of Gaussian coefficients in the m-th subband gives distortion (MSE)

$$d_{Q_m}(r_m) = \sigma_m^2 2^{-2r_m}.$$

Because there are n_m coefficients in the m-th subband, $n_m d_{Q_m}(r_m)$ is the total distortion in coding the m-th subband. After upsampling by $V_m = 1/\eta_m$ and ideal filtering of this subband with amplitude gain V_m, the total distortion contributed by the m-th subband to the reconstruction is

$$d_m(r_m) = V_m n_m d_{Q_m}(r_m) = N\sigma_m^2 2^{-2r_m}, \tag{7.75}$$

and the corresponding distortion per sample in the reconstruction is simply

$$D_m(r_m) = d_{Q_m}(r_m) = \sigma_m^2 2^{-2r_m}. \tag{7.76}$$

The amplitude gain factor of V_m translates to a power gain factor of

$$V_m = \frac{1}{\pi} \int_{\omega_m}^{\omega_{m+1}} V_m^2 \mid H_m(\omega) \mid^2 d\omega,$$

where $V_m = \pi/W_m$ and $W_m = \omega_{m+1} - \omega_{m+1}$. It compensates for the distortion and variance per coefficient decreasing by a factor of V_m, due to the insertion of $V_m - 1$ zeros between consecutive subband samples. Summing over all subbands, the overall average distortion per sample in the reconstruction is

$$D = \sum_{m=0}^{M-1} D_m(r_m) = \sum_{m=0}^{M-1} \sigma_m^2 2^{-2r_m}. \tag{7.77}$$

Necessary conditions to achieve a minimum of D subject to the rate constraint in (7.74) are that

$$\frac{\partial J}{\partial r_m} = 0, \quad m = 0, 1, \ldots, M-1, \tag{7.78}$$

where J is the objective function defined as

$$J = \sum_{m=0}^{M-1} (\sigma_m^2 2^{-2r_m} + \lambda \eta_m r_m), \tag{7.79}$$

and λ is a parameter (Lagrange multiplier) to be set by the rate constraint. Leaving the details of the solution of the simultaneous equations in (7.78) to the next chapter, we find that when R is large enough that the rate per sample r_m for the m-th subband is greater than zero for all m,

$$r_m = R + \frac{1}{2} \log_2 \frac{\sigma_m^2}{\sigma_{\text{WGM}}^2} \tag{7.80}$$

where σ_{WGM}^2 is a weighted geometric mean defined by

$$\sigma_{\text{WGM}}^2 \equiv \prod_{m=0}^{M-1} (V_m \sigma_m^2)^{\eta_m}.$$

The distortion (MSE) per subband coefficient turned out to be

$$d_{Q_m}(r_m) = \sigma_m^2 2^{-2r_m} = \eta_m \sigma_{\text{WGM}}^2 2^{-2R}, \tag{7.81}$$

Therefore, summing over all subbands, the overall average distortion is

$$D = \sum_{m=0}^{M-1} \eta_m \sigma_{\text{WGM}}^2 2^{-2R} = \sigma_{\text{WGM}}^2 2^{-2R}, \tag{7.82}$$

where we use the fact that $\sum_{m=0}^{M-1} \eta_m = 1$.

As before, coding gain is measured relative to PCM coding of the source samples directly. Ideal PCM coding with rate R results in per-sample distortion of

$$D_{\text{PCM}} = \sigma_x^2 2^{-2R} \tag{7.83}$$

The subband coding gain relative to PCM is defined as

$$G_{\text{SB/PCM}} \equiv \frac{D_{\text{PCM}}}{D_{\text{SB}}}, \tag{7.84}$$

Substituting D_{PCM} in (7.83) and D in (7.82) for D_{SB} in (7.84), the ideal coding gain is expressed as

$$G_{\text{SB/PCM}} = \frac{\sigma_x^2}{\prod_{m=0}^{M-1} \left(V_m \sigma_m^2\right)^{\eta_m}} = \frac{\sigma_x^2}{\prod_{m=0}^{M-1} \left(\sigma_m^2 / \eta_m\right)^{\eta_m}}. \tag{7.85}$$

We now prove that the coding gain exceeds one, except in the case of a source with a flat variance spectrum (white source). Following the case of distortion, the mean squared value or power (per sample) of the reconstruction is just the same weighted sum of the subband variances according to

$$\sigma_x^2 = \sum_{m=0}^{M-1} \eta_m V_m \sigma_m^2 = \sum_{m=0}^{M-1} \sigma_m^2.$$

To check this conclusion that the source variance is simply the sum of the subband variances is to observe that a certain frequency band of width $W_m = \pi / V_m$ is selected by its analysis filter and downsampled by V_m. The downsampling following the filter does not change the variance (mean squared value) of any sample. The same goes for every subband in the analysis filter bank. Therefore, the sum of the variances of the subband samples equals the variance of a source sample.

The subband/PCM gain can be written as

$$G_{\text{SB/PCM}} = \frac{\sum_{m=0}^{M-1} \sigma_m^2}{\prod_{m=0}^{M-1} \left(\sigma_m^2 / \eta_m\right)^{\eta_m}}. \tag{7.86}$$

Taking the logarithm of $G_{\text{SB/PCM}}$ and noting again that $\sum_{m=0}^{M-1} \eta_m = 1$, the convexity of the logarithm allows the use of Jensen's inequality

$$\sum_{m=0}^{M-1} \eta_m \log(\sigma_m^2 / \eta_m) \leq \log \left(\sum_{m=0}^{M-1} \eta_m (\sigma_m^2 / \eta_m) \right) = \log \sigma_x^2, \tag{7.87}$$

with equality if and only if σ_m^2/η_m is the same for all subbands. Therefore,

$$G_{\text{SB/PCM}} \geq 1 \qquad (7.88)$$

with equality if and only if σ_m^2/η_m is a constant independent of m. This equality condition is satisfied if the input process is white, just as with any linear transform. Therefore, independent scalar coding of subbands created by an orthonormal filter bank is advantageous over the same kind of scalar coding of the source signal. The gain formula in Equation (7.85) is a useful tool to assess the relative coding efficiencies of different subband decompositions of a source signal.

One may go astray when calculating these gain formulas using practical filters. These formulas rely on ideal bandpass filters with amplitude gain of 1 in the analysis stages and amplitude gain of V_m in the synthesis stages to produce equal source and reconstruction signals (aside from delay). Therefore, the subband variances σ_m^2 are calculated directly from the source power spectrum $S_x(\omega)$ neglecting any filtering. To make this fact concrete, the subband power spectrum in the frequency range $[\omega_{m-1}, \omega_m]$ after downsampling and ideal filtering may be expressed as

$$S_m(\omega) = (1/V_m)S_x(\omega/V_m + \omega_m), \ 0 \leq \omega \leq \pi. \qquad (7.89)$$

The subband variance (for zero mean) is the integral of this spectrum according to

$$\sigma_m^2 = \frac{1}{\pi} \int_0^\pi S_m(\omega)d\omega = \frac{1}{\pi} \int_{\omega_{m-1}}^{\omega_m} S_x(\omega)d\omega. \qquad (7.90)$$

The point here is to calculate the subband variances directly from the source spectrum. Often, the filters are normalized to have unit energy, so that they become orthonormal. In this case, the amplitude factor of $\sqrt{V_m}$ is inserted both in the analysis and synthesis stages, making the filters have unit energy and also producing perfect reconstruction with cascaded gain of V_m as before. The book by Taubman and Marcellin [8] derives gain formulas for orthonormal filters, but uses subband variance after filtering in its formulas. The filtered subband variances, denoted by $\sigma_m'^2$, are related to the unfiltered ones by

$$\sigma_m'^2 = V_m \sigma_m^2, \qquad (7.91)$$

because of the increase in energy by the factor of V_m for the analysis filters. Therefore, in terms of the filtered subband variances, the subband coding gain is

$$G_{\text{SB/PCM}} = \frac{\sigma_x^2}{\prod_{m=0}^{M-1} \left(\sigma_m'^2\right)^{\eta_m}}. \qquad (7.92)$$

Therefore, the seemingly different gain formulas are reconciled. For realizable orthonormal filters with frequency responses $H_m(\omega)$, the filtered subband variances $\sigma_m'^2$ are calculated as

$$\sigma_m'^2 = \frac{1}{\pi} \int_0^\pi |H_m(\omega)|^2 S_x(\omega)d\omega, \quad m = 0, 1, \ldots, M-1. \qquad (7.93)$$

The gain formula in (7.92) with subband variances calculated from realizable filter responses by (7.93) may be used to evaluate the coding efficiencies of different filters, in addition to different arrangements for splitting the source signal into subbands. The following example illustrates a calculation of coding gain.

Example 7.1 Calculation of subband coding gain

Assume that the vector $\mathbf{x}$ of $N = 128$ samples of a stationary source with zero mean and power spectral density

$$S_x(\omega) = A \cos^2(\omega/2), \quad -\pi \leq \omega \leq \pi \tag{7.94}$$

is transformed into three subbands by the filter bank arrangement shown in Figure 7.10 with one less stage. We specify discrete-time Fourier transforms in the frequency region $[-\pi, \pi]$ with the understanding that they are always periodic with period 2π. We wish to calculate the coding gain $G_{\text{SB/PCM}}$. In order to do so, we must determine the variances of the subbands. The variance of individual coefficients at the output of a filter with input spectral density $S_{in}(\omega)$ is governed by the formula

$$\sigma_y'^2 = \frac{1}{\pi} \int_0^\pi S_{in}(\omega) \mid H(\omega) \mid^2 d\omega,$$

where $H(\omega)$ is the (discrete-time) frequency response of the filter. Let us assume that the filters in any stage are ideal half-band with the frequency responses

$$H_0(\omega) = \begin{cases} 1, & |\omega| \leq \pi/2 \\ 0, & \pi/2 < |\omega| < \pi \end{cases} \tag{7.95}$$

$$H_1(\omega) = \begin{cases} 0, & |\omega| \leq \pi/2 \\ 1, & \pi/2 < |\omega| < \pi. \end{cases} \tag{7.96}$$

Discrete-time Fourier transforms are periodic functions of frequency with period 2π, so the filter frequency responses are specified in only one interval of length 2π. After downsampling by a factor of V, the resultant power spectral density $S_{ad}(\omega)$ is related to the power spectral density before downsampling $S_{bd}(\omega)$ by

$$S_{ad}(\omega) = \frac{1}{V} \sum_{k=0}^{V-1} S_{bd}(\frac{\omega - 2\pi k}{V}). \tag{7.97}$$

So for our ideal "brick-wall" filters, passage through one stage of half-band filters and downsampling by a factor of $V = 2$ yield the output spectra

$$S_{y_0}(\omega) = \frac{1}{2} S_x(\omega/2),$$

$$S_{y_1}(\omega) = \frac{1}{2} S_x(\frac{\omega - 2\pi}{2}), \tag{7.98}$$

$$0 \leq \omega < \pi.$$

Because discrete-time power spectra are even 2π-periodic functions, specifying them in the limited range of $0 \leq \omega < \pi$ suffices to specify them for all ω.

Applying the formula in (7.97) once more to the lowpass output of the first stage with spectral density $S_{y_0}(\omega)$ results in the second stage output spectra of

$$S_{y_{00}}(\omega) = \frac{1}{2}S_{y_0}(\omega/2) = \frac{1}{4}S_x(\omega/4) \qquad (7.99)$$

$$S_{y_{01}}(\omega) = \frac{1}{2}S_{y_0}(\frac{\omega - 2\pi}{2}) = \frac{1}{4}S_x(\frac{\omega - 2\pi}{4}) \qquad (7.100)$$

Integrating the spectra in Equations (7.98), (7.99), and (7.100) from $\omega = 0$ to $\omega = \pi$, changing variables, and dividing by π yield the following formulas for the subband variances.

$$\sigma_{y_1}^2 = \frac{1}{\pi} \int_{\omega=\pi/2}^{\pi} S_x(\omega)d\omega.$$

$$\sigma_{y_{01}}^2 = \frac{1}{\pi} \int_{\omega=\pi/4}^{\pi/2} S_x(\omega)d\omega. \qquad (7.101)$$

$$\sigma_{y_{00}}^2 = \frac{1}{\pi} \int_{\omega=0}^{\pi/4} S_x(\omega)d\omega.$$

In plain words, the subband variances are calculated from the portion of the input spectrum in their pass-bands.

Substituting the given input spectrum in Equation (7.94) into the formulas in (7.101) results in the subband variances of

$$\sigma_{y_1}^2 = (A/4\pi)(\pi - 2),$$

$$\sigma_{y_{01}}^2 = (A/8\pi)(\pi + 4 - 4\sin(\pi/4)), \qquad (7.102)$$

$$\sigma_{y_{00}}^2 = (A/8\pi)(\pi + 4\sin(\pi/4)).$$

Upon substituting these variances and corresponding downsampling factors into Equation (7.85), the subband coding gain is determined as

$$G_{\text{SB/PCM}} = \frac{\sigma_x^2}{(2\sigma_{y_1}^2)^{1/2}(4\sigma_{y_{01}}^2)^{1/4}(4\sigma_{y_{00}}^2)^{1/4}} = 1.3053 \qquad (7.103)$$

Clearly, there is not much of a gain in this example. The given source spectrum is a little too flat.

7.7.3   Realizable perfect reconstruction filters

The concept of subband transforms and multi-resolution transforms was presented with ideal filters for the sake of conceptual simplicity. However, ideal filters with zero transition length between pass-band and stop band are not physically realizable. Normally, use of realizable filters will cause aliasing (spectral overlap) due to the finite length in

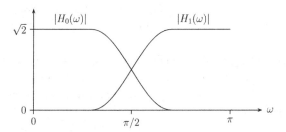

Figure 7.12 Alias-cancelling filters.

this transition region. However, there is a class of filters with frequency response in the transition region that cancels aliasing. A depiction of two such half-band filter responses is shown in Figure 7.12. The mirror symmetry of the magnitudes of the frequency response about $\pi/2$ is called *quadrature mirror* symmetry and the filters are called quadrature mirror or QMF filters. The two-channel case with half-band filters being a basic building block allows us to focus only on this case. In order to fulfill the requirement of perfect reconstruction, we choose linear phase, finite impulse response (FIR) filters. The impulse responses of such filters are either symmetric or anti-symmetric about the midpoint of their support. For an arbitrary FIR filter, $h(n), 0 \leq n \leq L-1$, linear phase dictates that

$$h(n) = \pm h(L-1-n) \tag{7.104}$$

in the time domain and

$$H(\omega) = \pm e^{-j(L-1)\omega} H^*(\omega) \tag{7.105}$$

in the frequency domain. Omitting for now the linear phase condition, the frequency responses of the two analysis filters must obey the following relationships for perfect reconstruction:

$$H_1(\omega) = H_0(\omega + \pi), \tag{7.106}$$

$$G_0(\omega) = H_0(\omega), \quad G_1(\omega) = -H_1(\omega), \tag{7.107}$$

$$H_0^2(\omega) - H_0^2(\omega + \pi) = 2e^{-j(L-1)\omega}, \tag{7.108}$$

where L is even to make the delay odd. The first relationship expresses the quadrature mirror property, which corresponds to $h_1(n) = (-1)^n h_0(n)$ in the time domain. The next relationships in Equation (7.107) assure reconstruction free of aliasing. The last relationship is the power complementarity or allpass property. Actually, any constant amplitude multiplying the phase term on the righthand side of Equation (7.108) can be chosen. The constant of 2 was chosen to make the energy of the filters equal to 1. When we impose the linear phase requirement in (7.105) on the filters, (7.108) becomes

$$|H_0(\omega)|^2 + |H_0(\omega + \pi)|^2 = 2 \tag{7.109}$$

Actually, these properties can be satisfied exactly only for the trivial case of length $L = 2$ FIR filters. However, there are fairly accurate approximations developed for longer odd lengths.

The symmetry and equal lengths of the analysis and synthesis filters are only sufficient, but not necessary for perfect reconstruction. This symmetry makes the low and highpass filters orthogonal.[9] One can achieve perfect reconstruction using biorthogonal filters.[10] One such solution follows.

Let the time-reversal of the impulse response $h(n)$ be denoted $\tilde{h}(n) = h(-n)$. The corresponding discrete-time Fourier transform relationship is $\tilde{H}(\omega) = H^*(\omega)$, with $H(\omega)$ and $\tilde{H}(\omega)$ being the discrete-time Fourier transforms of $h(n)$ and $\tilde{h}(n)$, respectively. The following filter relationships also guarantee perfect reconstruction.

$$H_1(\omega) = e^{-j\omega}\tilde{G}_0^*(\omega + \pi). \tag{7.110}$$

$$\tilde{G}_1(\omega) = e^{-j\omega}H_0^*(\omega + \pi). \tag{7.111}$$

You may check that power complementarity in (7.108) is satisfied. The corresponding impulse response relationships are

$$h_1(n) = (-1)^{1-n}g_0(1-n) \text{ and } \tilde{g}_1(n) = (-1)^{1-n}h_0(1-n) \tag{7.112}$$

If we remove the tildes, or equivalently, replace the time-reversals by their original forms, we revert to the quadrature mirror filters.

7.7.4   Orthogonal wavelet transform

In a two-channel, perfect reconstruction filter bank with orthogonal filters as just above, the only flexibility left is the choice of the prototype lowpass analysis or synthesis filter. The demand of perfect reconstruction defines the other three filters. The orthogonal wavelet filters constitute a certain class of these prototype filters. We shall review the derivation that a wavelet transform and its inverse are realized using a two-channel, perfect reconstruction filter bank composed of these so-called wavelet filters.

The idea is first to consider the discrete-time input signal to the filter bank as coefficients in an orthonormal expansion of a continuous-time signal $x(t)$. Expressed mathematically, let $\theta_n(t), n = \cdots, -1, 0, 1, \cdots$ be an orthonormal set of basis functions, meaning[11]

$$\langle \theta_m(t), \theta_n(t) \rangle = \delta(n - m),$$

where $\delta(n)$ is the Kronecker delta function, defined by

$$\delta(n) = \begin{cases} 1 & \text{if } n = 0 \\ 0 & \text{if } n \neq 0. \end{cases} \tag{7.113}$$

The particular basis of interest here is derived from a so-called scaling function $\phi(t)$ such that

$$\int_{-\infty}^{\infty} \phi(t)dt = 1 \tag{7.114}$$

$$\langle \phi(2^{-k}t), \phi(2^{-k}t - m) \rangle = 2^k \delta(m). \tag{7.115}$$

Note that the inner product relationship above implies unit energy of the scaling function, i.e., $\int_{-\infty}^{\infty} |\phi(t)|^2 dt = 1$. The quantity k is an integer called the scale or dilation level, and m is an integer shift or translation. The meaning of Equation (7.15) is that different shifts of dilations of $\phi(t)$ by factors of 2^k are orthogonal. Furthermore, the basis for $\phi(2^{-(k-1)})(t)$ is the set $\{\phi(2^{-k}t - n)\}$ In particular, for $k = -1$, the set $\{\phi(2t - n)\}$ is an orthogonal basis for $\phi(t)$. Therefore $\phi(t)$ can be expressed as a linear combination of these basis functions, according to

$$\phi(t) = \sum_n q_0(n)\phi(2t - n). \tag{7.116}$$

This expression is called a dilation equation. If we integrate both sides of this equation over all time and use the fact of Equation (7.114), we obtain

$$\sum_n q_0(n) = 2. \tag{7.117}$$

We therefore choose to express the input continuous-time signal in terms of the orthonormal basis $\{\phi(2t - n)\}$ as

$$x(t) = \sum_n x(n)\phi(2t - n). \tag{7.118}$$

Reproducing $x(n)$ for all n is our objective for the cascade of the analysis and synthesis filter banks.

There is a wavelet function $\psi(t)$ associated with the scaling function $\phi(t)$ that is orthonormal to all shifts of itself and orthogonal to all shifts of the scaling function according to

$$\langle \psi(t), \psi(t - m) \rangle = \delta(m) \tag{7.119}$$

$$\langle \psi(t), \phi(t - m) \rangle = 0 \text{ for all } m. \tag{7.120}$$

In addition, it must integrate to zero, i.e., $\int_{-\infty}^{+\infty} \psi(t)dt = 0$. The wavelet function also satisfies a similar dilation equation

$$\psi(t) = \sum_n q_1(n)\phi(2t - n). \tag{7.121}$$

The integral of $\psi(t)$ equals 0 implies that $\sum_n q_1(n) = 0$. This wavelet function together with its integer shifts generates an orthogonal basis for the space orthogonal to that generated by $\phi(t)$ and its integer shifts. These two spaces together constitute the space of the next lower scale $k = -1$. Therefore, since $x(t)$ belongs to this scale, it may be expressed as a linear combination of the shifts of $\phi(t)$ and $\psi(t)$.

$$x(t) = \sum_n x_0(n)\phi(t-n) + \sum_n x_1(n)\psi(t-n), \qquad (7.122)$$

where $x_0(n)$ and $x_1(n)$ are the sample values at time n of the lowpass and highpass subbands, respectively. Now to determine these subband sample values, we reconcile Equations (7.118) and (7.122) for $x(t)$. In Equation (7.122), $x_0(n)$ and $x_1(n)$ must be the orthogonal projections of $x(t)$ onto $\phi(t-n)$ and $\psi(t-n)$, respectively. Taking $x_0(n)$ first,

$$x_0(n) = \langle x(t), \phi(t-n) \rangle \qquad (7.123)$$

$$= \langle \sum_m x(m)\phi(2t-m), \phi(t-n) \rangle \qquad (7.124)$$

$$= \sum_m x(m)\langle \phi(2t-m), \phi(t-n) \rangle \qquad (7.125)$$

$$= \frac{1}{2}\sum_m x(m)q_0(m-2n) \qquad (7.126)$$

The last step follows from substitution of the dilation equation (7.118) and the orthogonality of different shifts of $\phi(2t)$. A similar calculation for $x_1(n) = \langle x(t), \psi(t-n) \rangle$ yields

$$x_1(n) = \frac{1}{2}\sum_m x(m)q_1(m-2n). \qquad (7.127)$$

Recognizing these equations for the subband sequences as correlations, we define

$$h_0(n) = \frac{1}{2}q_0(-n) \text{ and } h_1(n) = \frac{1}{2}q_1(-n). \qquad (7.128)$$

Therefore, the subband sequences may be expressed as convolutions

$$x_0(n) = \sum_m x(m)h_0(2n-m),$$

$$x_1(n) = \sum_m x(m)h_1(2n-m). \qquad (7.129)$$

They may be implemented by feeding the input signal $x(n)$ into the lowpass filter with impulse response $h_0(n)$ and the highpass filter with impulse response $h_1(n)$ and retaining only the even-indexed time samples.

The requirement that $\psi(t)$ and $\phi(t-n)$ must be orthogonal for all n establishes the relationship between $h_0(n)$ and $h_1(n)$. Substituting the dilation Equations (7.116) and (7.121) into $\langle \psi(t), \phi(t-n) \rangle = 0$ yields

$$\frac{1}{2}\sum_\ell q_1(\ell)q_0(\ell-2n) = 0,$$

which implies that the cross-correlation between the $q_1(n)$ and $q_0(n)$ sequences must be zero at even lags. The equivalent Fourier transform relationship is

$$Q_1(\omega)Q_0^*(\omega) + Q_1(\omega+\pi)Q_0^*(\omega+\pi) = 0, \qquad (7.130)$$

where $Q_0(\omega)$ and $Q_1(\omega)$ are the respective discrete-time Fourier transforms of $q_0(n)$ and $q_1(n)$. A solution to this equation is

$$Q_1(\omega) = e^{-j\omega} Q_0^*(\omega + \pi). \qquad (7.131)$$

In the time domain, this condition means that

$$q_1(n) = (-1)^{1-n} q_0(1 - n), \qquad (7.132)$$

or equivalently, in terms of the filter impulse and frequency responses,

$$h_1(n) = (-1)^{n-1} h_0(n - 1), \qquad (7.133)$$

$$H_1(\omega) = e^{-j\omega} H_0(\omega + \pi). \qquad (7.134)$$

The latter equation (7.134) expresses the qmf property.

In order to reconstruct $x(t)$ (and hence $x(n)$) from its subband sequences, we look back to $x(t)$ expressed as in Equation (7.122) and substitute the dilation equations for $\phi(t)$ and $\psi(t)$. Doing so with a few variable replacements and collection of terms yields the following formula:

$$x(t) = \sum_n \left[\sum_\ell x_0(\ell) q_0(n - 2\ell) + \sum_\ell x_1(\ell) q_1(n - 2\ell) \right] \phi(2t - n)$$

The bracketed sums in the above expression must amount to $x(n)$. To prove the required filter bank operations, we separate the two sums within these brackets and define

$$\hat{x}_0(n) = \sum_\ell x_0(\ell) q_0(n - 2\ell), \qquad (7.135)$$

$$\hat{x}_1(n) = \sum_\ell x_1(\ell) q_1(n - 2\ell). \qquad (7.136)$$

Let us define the impulse responses

$$\tilde{h}_0(n) = h_0(-n) \text{ and } \tilde{h}_1(n) = h_1(-n).$$

Using these definitions and substituting the filter impulse responses in (7.128) yield

$$\hat{x}_0(n) = 2 \sum_\ell x_0(\ell) \tilde{h}_0(n - 2\ell), \qquad (7.137)$$

$$\hat{x}_1(n) = 2 \sum_\ell x_1(\ell) \tilde{h}_1(n - 2\ell). \qquad (7.138)$$

We can interpret these equations as inserting zeros between samples of $x_0(n)$ and $x_1(n)$ and filtering the resulting sequences with $\tilde{h}_0(n)$ and $\tilde{h}_1(n)$, respectively. The sum of the outputs reveals $x(n)$, i.e., $x(n) = \tilde{x}_0(n) + \tilde{x}_1(n)$. Therefore, the synthesis filters in Figure 7.7 are

$$g_0(n) = 2\tilde{h}_0(n) \text{ and } g_1(n) = 2\tilde{h}_1(n). \qquad (7.139)$$

Therefore, one prototype lowpass filter $h_0(n)$ determines the other filters to obtain an orthogonal transform. Summarizing the filter determinations in Equations (7.134) and (7.139) reveals that

$$h_1(n) = (-1)^{1-n} h_0(1 - n),$$
$$g_0(n) = 2h_0(-n),$$
$$g_1(n) = 2(-1)^{1+n} h_0(1 + n). \tag{7.140}$$

Example 7.2 The Haar wavelet filters
The simplest orthogonal wavelet transform is the Haar transform. Its prototype lowpass filter is given by[12]

$$h_0(n) = \begin{cases} 1/2, & n = 0, 1 \\ 0, & \text{otherwise} \end{cases} = \frac{1}{2}(\delta(n) + \delta(n - 1)).$$

The corresponding formula in Equation (7.140), reveals the highpass analysis filter as

$$h_1(n) = \begin{cases} -1/2, & n = 0 \\ 1/2, & n = 1 \\ 0, & \text{otherwise} \end{cases} = \frac{1}{2}(-\delta(n) + \delta(n - 1)).$$

From Equations (7.129), the subband signals produced by these filters for the input signal $x(n)$ are expressed by

$$x_0(n) = \frac{1}{2}(x(2n) + x(2n - 1)),$$
$$x_1(n) = \frac{1}{2}(-x(2n) + x(2n - 1)). \tag{7.141}$$

The synthesis lowpass and highpass filters, according to (7.140), are respectively

$$g_0(n) = \begin{cases} 1, & n = -1, 0 \\ 0, & \text{otherwise} \end{cases} = \delta(n + 1) + \delta(n),$$

$$g_1(n) = \begin{cases} -1, & n = 0 \\ 1, & n = -1 \\ 0, & \text{otherwise} \end{cases} = \delta(n + 1) - \delta(n).$$

Notice that the synthesis filters are not causal. FIR filters can always be made causal via suitable delays. Using (7.138), the reconstructed subband signals produced at the ouput of these filters are expressed by

$$\hat{x}_0(2n) = x_0(n), \quad \hat{x}_0(2n + 1) = x_0(n + 1)$$
$$\hat{x}_1(2n) = -x_1(n), \quad \hat{x}_1(2n + 1) = x_1(n + 1) \tag{7.142}$$

We can check that adding the reconstructed subband signals produces the original input signal $x(n)$. Substituting the equations in (7.141) and (7.142), we find that

$$
\begin{aligned}
\hat{x}_0(2n) + \hat{x}_1(2n) &= x_0(n) - x_1(n) &&= x(2n), \\
\hat{x}_0(2n+1) + \hat{x}_1(2n+1) &= x_0(n+1) + x_1(n+1) &&= x(2n+1).
\end{aligned}
$$

Therefore, the original input is reproduced at the filter bank output.

Not previously mentioned is the desirability of having a multiplicity of zeros at $\omega = \pi$ in the lowpass filter transfer function. At least one zero at $\omega = \pi$ is necessary for regularity of the scaling function. Regularity is needed to assure a proper scaling function as the result of iterated application of the filters to produce a pyramid of resolutions, as is done in many wavelet coding methods. More zeros at $\omega = \pi$ give faster decay of the filter response toward $\omega = \pi$ and hence, lessening of artifacts associated with compression. The Haar filter has only a single zero at $\omega = \pi$, so is not entirely satisfactory for coding applications. However, these filters are important building blocks for better filters, as will be seen later.

The Haar filters are the only orthogonal perfect reconstruction filters having linear phase responses. Nonetheless, there have been several such filters developed which have approximate linear phase. We shall not present them here, because current usage has turned mostly to biorthogonal filters, which achieve perfect reconstruction and linear phase responses simultaneously. Several books among the references to this chapter treat the design and implementation of these filters.

7.7.5   Biorthogonal wavelet transform

The orthogonal wavelet transform just described used the same orthogonal basis functions for decomposition (analysis) and reconstruction (synthesis). A biorthogonal wavelet transform results when we have related, but different bases for decomposition and reconstruction. A scaling function $\phi(t)$ and its so-called dual scaling function $\tilde{\phi}(t)$ are needed. The shifts of these functions individually are linearly independent and together form a basis for the space of $x(t)$. But now the basis functions are linearly independent and not orthogonal. These functions also obey dilation relations:

$$
\phi(t) = \sum_n q_0(n)\phi(2t - n) \tag{7.143}
$$

$$
\tilde{\phi}(t) = \sum_n \tilde{q}_0(n)\tilde{\phi}(2t - n). \tag{7.144}
$$

Furthermore, the two scaling functions must satisfy the biorthogonality condition

$$
\langle \phi(t), \tilde{\phi}(t - k) \rangle = \delta(k). \tag{7.145}
$$

We also need two wavelet functions $\psi(t)$ and $\tilde{\psi}(t)$ which are duals of each other. They must meet the following conditions:

$$\langle \psi(t), \tilde{\psi}(t - k) \rangle = \delta(k)$$

$$\langle \psi(t), \tilde{\phi}(t - k) \rangle = 0 \tag{7.146}$$

$$\langle \tilde{\psi}(t), \phi(t - k) \rangle = 0$$

Since $\psi(t)$ and $\tilde{\psi}(t)$ are in the respective spaces spanned by $\{\phi(2t - n)\}$ and $\{\tilde{\phi}(2t - n)\}$, they can be expressed by dilation equations

$$\psi(t) = \sum_n q_1(n) \phi(2t - n) \tag{7.147}$$

$$\tilde{\psi}(t) = \sum_n \tilde{q}_1(n) \tilde{\phi}(2t - n). \tag{7.148}$$

Similar to before, we regard the input sequence $x(n)$ to be coefficients in the expansion of $x(t)$ in the basis $\{\phi(2t - n)\}$:

$$x(t) = \sum_m x(m) \phi(2t - m). \tag{7.149}$$

Although the spaces spanned by $\{\phi(t - n)\}$ and $\{\psi(t - n)\}$ are not orthogonal, we can still write $x(t)$ as

$$x(t) = \sum_n x_0(n) \phi(t - n) + \sum_n x_1(n) \psi(t - n), \tag{7.150}$$

because the second term is the difference between $x(t)$ in the original, finer space expressed in Equation (7.149) and its approximation in the coarser space represented by the first term. The discrete-time subband signals $x_0(n)$ and $x_1(n)$ are now found as projections onto the dual biorthogonal basis as

$$x_0(n) = \langle x(t), \tilde{\phi}(t - n) \rangle \tag{7.151}$$

$$x_1(n) = \langle x(t), \tilde{\psi}(t - n) \rangle. \tag{7.152}$$

Again, substituting $x(t)$ in Equation (7.149) and the dilation equations in (7.148) into the above equations for the subband signals give

$$x_0(n) = \sum_m \sum_\ell x(m) \tilde{q}_0(\ell) \langle \tilde{\phi}(2t - 2n - \ell), \phi(2t - m) \rangle \tag{7.153}$$

$$x_1(n) = \sum_m \sum_\ell x(m) \tilde{q}_1(\ell) \langle \tilde{\phi}(2t - 2n - \ell), \phi(2t - m) \rangle. \tag{7.154}$$

Invoking the biorthogonality condition in (7.145) yields

$$x_0(n) = \frac{1}{2} \sum_m x(m) \tilde{q}_0(m - 2n) \tag{7.155}$$

$$x_1(n) = \frac{1}{2} \sum_m x(m) \tilde{q}_1(m - 2n). \tag{7.156}$$

By defining

$$\tilde{h}_0(n) \equiv \frac{1}{2} \tilde{q}_0(-n) \text{ and } \tilde{h}_1(n) \equiv \frac{1}{2} \tilde{q}_1(-n), \tag{7.157}$$

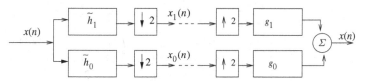

Figure 7.13 Two-channel analysis and synthesis biorthogonal filter banks. $\tilde{h}_0$ and g_0 denote impulse responses of the lowpass filters; $\tilde{h}_1$ and g_1 those of the highpass filters.

we can express $x_0(n)$ and $x_1(n)$ as convolutions

$$x_0(n) = \sum_m x(m)\tilde{h}_0(2n - m) \tag{7.158}$$

$$x_1(n) = \sum_m x(m)\tilde{h}_1(2n - m). \tag{7.159}$$

Therefore, $x_0(n)$ and $x_1(n)$ are the outputs of a two-channel filter bank with FIRs $\tilde{h}_0(n)$ and $\tilde{h}_1(n)$ followed by downsampling by a factor of 2, which is depicted in Figure 7.13.

The reconstruction or synthesis follows the same procedure as in the orthogonal case. It does not involve the dual functions. We start with the expression in (7.150) for $x(t)$, substitute the dilation equations in (7.116) and (7.121), collect terms and change variables to determine the outputs $\hat{x}_0(n)$ corresponding to the input $x_0(n)$ and $\hat{x}_1(n)$ corresponding to the input $x_1(n)$ to reveal that

$$\hat{x}_0(n) = \sum_\ell x_0(\ell)q_0(n - 2\ell) \tag{7.160}$$

$$\hat{x}_1(n) = \sum_\ell x_1(\ell)q_1(n - 2\ell) \tag{7.161}$$

and that

$$x(n) = \hat{x}_0(n) + \hat{x}_1(n). \tag{7.162}$$

Using the filter definitions

$$g_0(n) \equiv q_0(n) \text{ and } g_1(n) \equiv q_1(n), \tag{7.163}$$

we see that the subband reconstructions are produced by upsampling by 2 and filtering with either the impulse response $g_0(n)$ or $g_1(n)$ and then added to give perfect reconstruction of $x(n)$. Figure 7.13 depicts also this synthesis filter bank.

The relationships between the various filters are dictated by the biorthogonal properties of the scaling and wavelet functions and their duals. These relationships are proved by substituting the appropriate dilation equations into a condition in (7.146) or (7.145). These manipulations yield the following correspondences between the property and its constraint on the coefficients of the dilation equations:

Property	Constraint
$\langle \phi(t), \tilde{\phi}(t-n) \rangle = \delta(n)$	$\sum_\ell q_0(\ell)\tilde{q}_0(\ell - 2n) = 2\delta(2n)$
$\langle \psi(t), \tilde{\psi}(t-n) \rangle = \delta(n)$	$\sum_\ell q_1(\ell)\tilde{q}_1(\ell - 2n) = 2\delta(2n)$
$\langle \psi(t), \tilde{\phi}(t-n) \rangle = 0$	$\sum_\ell q_1(\ell)\tilde{q}_0(\ell - 2n) = 0$
$\langle \tilde{\psi}(t), \phi(t-n) \rangle = 0$	$\sum_\ell \tilde{q}_1(\ell)q_0(\ell - 2n) = 0$

Using the filter definitions, the constraints on the coefficient sequences express the biorthogonality of the corresponding filters at even shifts as follows:

$$\sum_\ell g_0(\ell), \tilde{h}_0(2n - \ell) = \delta(2n)$$

$$\sum_\ell g_1(\ell), \tilde{h}_1(2n - \ell) = \delta(2n)$$

$$\sum_\ell q_1(\ell), \tilde{h}_0(2n - \ell) = 0 \tag{7.164}$$

$$\sum_\ell \tilde{h}_1(\ell), g_0(2n - \ell) = 0.$$

The constraint equations are usually expressed in terms of the discrete-time Fourier transforms of the coefficient sequences. We capitalize the letter symbol to denote its Fourier transform, e.g., $q_0(n) \Leftrightarrow Q_0(\omega)$. The discrete Fourier transforms of the constraint equations in the table are listed below.

$$Q_0(\omega)\tilde{Q}_0^*(\omega) + Q_0(\omega + \pi)\tilde{Q}_0^*(\omega + \pi) = 2$$
$$Q_1(\omega)\tilde{Q}_1^*(\omega) + Q_1(\omega + \pi)\tilde{Q}_1^*(\omega + \pi) = 2 \tag{7.165}$$
$$Q_1(\omega)\tilde{Q}_0^*(\omega) + Q_1(\omega + \pi)\tilde{Q}_0^*(\omega + \pi) = 0$$
$$\tilde{Q}_1(\omega)Q_0^*(\omega) + \tilde{Q}_1(\omega + \pi)Q_0^*(\omega + \pi) = 0.$$

The solutions are not unique, but the following relationships satisfy the above constraint equations.

$$\tilde{Q}_1(\omega) = e^{-j\omega}Q_0^*(\omega + \pi)$$
$$Q_1(\omega) = e^{-j\omega}\tilde{Q}_0^*(\omega + \pi). \tag{7.166}$$

Expressing these solutions in terms of the frequency responses of the filters corresponding to these dilation coefficient transforms reveals

$$2\tilde{H}_1(\omega) = e^{-j\omega}G_0^*(\omega + \pi)$$
$$G_1(\omega) = 2e^{-j\omega}\tilde{H}_0^*(\omega + \pi). \tag{7.167}$$

The above frequency response relationships of the filters guarantee cancellation of aliasing. These relationships translate to the following FIR relationships in the time domain:

$$2\tilde{h}_1(n) = (-1)^{1-n}g_0(1 - n)$$
$$g_1(n) = 2(-1)^{1-n}\tilde{h}_0(1 - n) \tag{7.168}$$

Table 7.1 Biorthogonal wavelet filter solutions

Time domain	Frequency domain
Essential relations	
$g_1(n) = 2(-1)^{1-n}\tilde{h}_0(1-n)$ $2\tilde{h}_1(n) = (-1)^{1-n}g_0(1-n)$	$G_1(\omega) = 2e^{-j\omega}\tilde{H}_0^*(\omega+\pi)$ $2\tilde{H}_1(\omega) = e^{-j\omega}G_0^*(\omega+\pi)$
Equal length relations	
$g_0(n) = -(-1)^{1-n}g_1(1-n)$ $\tilde{h}_0(n) = -(-1)^{1-n}\tilde{h}_1(1-n)$	$G_0(\omega) = -e^{-j\omega}G_1^*(\omega+\pi)$ $\tilde{H}_0(\omega) = -e^{-j\omega}H_1^*(\omega+\pi)$

This still leaves open the impulse responses of the filters $g_0(n)$ and $g_1(n)$ that satisfy the constraints. A solution that gives equal lengths of these filters is

$$Q_0(\omega) = -e^{-j\omega}Q_1^*(\omega+\pi) \tag{7.169}$$

or, in terms of the frequency responses,

$$G_0(\omega) = -e^{-j\omega}G_1^*(\omega+\pi). \tag{7.170}$$

The time domain equivalent is

$$g_0(n) = -(-1)^{1-n}g_1(1-n) \tag{7.171}$$

The relationship between $\tilde{h}_0(n)$ and $\tilde{h}_1(n)$, implied by (7.167) and (7.170), is

$$\tilde{h}_0(n) = -(-1)^{1-n}\tilde{h}_1(1-n) \tag{7.172}$$

Table 7.1 lists these time and frequency domain filter solutions.

Solutions exist for unequal lengths of the synthesis or analysis filters. The lengths of the two filters must be both even or both odd. In fact, the most common filters used in wavelet transform image coding are biorthogonal filters of lengths 9 and 7, the so-called biorthogonal CDF(9, 7) filter pair [9]. In the next section, we shall describe only the so-called lifting scheme for construction of filters. For those readers who are interested in a more general and complete treatment, we refer them to one or more of many excellent textbooks, such as [10], [11], and [12].

These two-channel analysis filterbanks can be repeatedly applied to either of the two outputs to divide further into two half-width subbands. The reason is that the property of a single dilation of the scaling and wavelet functions, such as that expressed in (7.116) or (7.121), implies the same property for succeeding stages of dilation. Therefore, subband decompositions, such as those in Figures 7.8 and 7.10, can be created. Analogously, the corresponding synthesis filter banks can be configured in successive stages of two-channel synthesis filter banks.

7.7.6   Useful biorthogonal filters

The so-called *spline 5/3* or simply "5/3" transform is one of two default transforms of the JPEG2000 image compression standard. The other is the *CDF 9/7* or "9/7" transform. The first number refers to the number of taps in the analysis prototype filter $\tilde{h}_0(n)$ and the second to the number in the synthesis prototype filter $g_0(n)$. The 5/3 prototype filter pair are therefore

$$\tilde{h}_0(n) = -\frac{1}{8}\delta(n+2) + \frac{1}{4}\delta(n+1) + \frac{3}{4}\delta(n) + \frac{1}{4}\delta(n-1) - \frac{1}{8}\delta(n-2)$$

$$g_0(n) = \frac{1}{4}\delta(n+1) + \frac{1}{2}\delta(n) + \frac{1}{4}\delta(n-1). \tag{7.173}$$

Note that these filters are symmetric about the center of their support, so are linear phase. They have been delay normalized so that $n = 0$ is their central point. We shall specify them more compactly by just their tap (coefficient) values at times n as follows.

n	0	± 1	± 2
$\tilde{h}_0(n)$	$\frac{3}{4}$	$\frac{1}{4}$	$-\frac{1}{8}$
$g_0(n)$	1	$\frac{1}{4}$	

These filters are especially simple, as they can be implemented with low integer multiplications and bit shift operations.

The CDF 9/7 biorthogonal wavelet transform, on other hand, has irrational coefficients and longer support, so is more complex. Its coefficients are often normalized so that the transform is unitary or orthonormal. This amounts to multiplying the analysis filters and the synthesis filters by $\sqrt{2}$. The coefficients of the prototype filters of the normalized CDF 9/7 biorthogonal wavelet transform to eight decimal places accuracy are:

n	0	± 1	± 2	± 3	± 4
$\sqrt{2}\tilde{h}_0(n)$	0.85269867	0.37740285	-0.11062440	-0.02384946	0.03782845
$\sqrt{2}g_0(n)$	0.78848561	0.41809227	-0.04068941	-0.06453888	

The latter transform is said to be lossy or irreversible, because truncation to finite precision eliminates the capability of perfect reconstruction. The 5/3 filter coefficients are dyadic fractions (integers divided by a power of 2), so can operate with fixed-precision (fixed-point) arithmetic without precision truncation as long as the required precision does not exceed that of the processor, which is normally 64 bits. In such a case, the transform is said to be lossless or reversible. However, reversibility cannot be guaranteed in all cases, because even a single transform step increases the precision by 3 bits and one or more such steps may increase the precision beyond the limit of the processor, depending on the initial precision of the data.

The other two filters, $\tilde{h}_1(n)$ and $g_1(n)$, needed to complete the transform may be calculated using the essential relationships in (7.168), which are repeated in Table 7.1.

7.7.7 The lifting scheme

A popular and efficient method of constructing filters of a biorthogonal wavelet transform is the *lifting scheme* introduced by Sweldens [13]. One starts with a short biorthogonal filter and builds new biorthogonal filters with better frequency localization properties with additional filtering steps, called *lifting* steps. What is produced by these lifting steps are longer filters with more zeros at frequency $\omega = \pi$. We shall describe this scheme briefly.

Referring to Figure 7.14 for illustration, we first see that the downsampled output $x_0^o(n)$ of the lowpass analysis filter is filtered by $p(n)$ and subtracted from the downsampled output $x_1^o(n)$ of the highpass analysis filter. This operation is often called a "prediction" of the highpass signal from the lowpass. This "prediction residual" is then filtered by $u(n)$ and added to the lowpass signal as a kind of update of the lowpass analysis output. This operation is often called an update step. The final lowpass and highpass subband signals are produced by the lifting steps of prediction and update from the initial or "old" subband signals $x_0^o(n)$ and $x_1^o(n)$. Filtering the input signal $x(n)$ with the biorthogonal filters $\tilde{h}_0$ and $\tilde{h}_1$ and downsampling by 2 produced the old subband signals. The mathematical expressions for these steps are:

$$x_1(n) = x_1^o(n) - \sum_k p(k)x_0^o(n-k)$$

$$x_0(n) = x_0^o(n) + \sum_k u(k)x_1(n-k). \tag{7.174}$$

What is especially noteworthy about this conversion from old to new subband signals is that their inversion is readily apparent. Just reverse the order and signs of the prediction and update steps. The update of $x_0^o(n)$ to $x_0(n)$ depends on the new highpass subband signal $x_1(n)$, which is available as an input to the synthesis stage. So we can perform the same update filtering in the synthesis stage and subtract it from the lowpass input $x_0(n)$ to recover the old subband signal $x_0^o(n)$. In turn, the filtering of this old lowpass subband signal, which was subtracted from the old highpass signal $x_1^o(n)$ in the

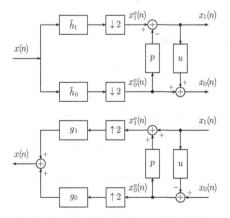

Figure 7.14   Analysis and synthesis stages of the lifting scheme.

analysis stage to yield $x_1(n)$, can now be added back to recover $x_1^o(n)$. This procedure is readily gleaned from the lifting equations (7.174). Given $x_1(n)$ and $x_0(n)$,

$$x_0^o(n) = x_0(n) - \sum_k u(k)x_1(n-k)$$

$$x_1^o(n) = x_1(n) + \sum_k p(k)x_0^o(n-k). \tag{7.175}$$

One notable property is that the prediction and update operations, shown here as linear, can be non-linear and still retain reversibility of the transform. For example, suppose that one or both of the convolutions in Equation (7.174) are truncated downward to integer values. The "old" signals can still be recovered in the same manner as in (7.175), by adding the truncations to the same convolutions. The original input can now be recovered in the usual manner by adding the ouputs of the synthesis filters g_1 and g_0, whose inputs are $x_1(n)$ and $x_0(n)$ upsampled by a factor of 2.

One remarkable fact proved by Sweldens [13] is that the lifting operation achieves a new set of biorthogonal filters, given that the original set is biorthogonal, for arbitrarily chosen prediction and update filters p and u. The filters p and u are chosen either to create more zeros at $\omega = \pi$ (equivalent to the number of vanishing moments) or to shape the corresponding wavelet function $\psi(t)$. The operations of prediction and update can be iterated as many times as necessary to create the desired effect.

7.7.7.1   "Old" filter choices

The choices for the "old" filters should be as simple as possible. The simplest ones that come immediately to mind are the 2-tap Haar filters, which are orthogonal and therefore trivially biorthogonal. We shall later use these filters in another setting. The simplest possible filters are the "do-nothing" and unit delay "do-nothing" filters of $\delta(n)$ and $\delta(n-1)$. After downsampling by two, we get the even-indexed samples for the lowpass channel and the odd-indexed ones for the highpass channel. So the old subband signals are

$$x_1^o(n) = x(2n+1) \text{ and } x_0^o(n) = x(2n).$$

The above is called the "lazy" wavelet transform. With these substitutions into (7.174), the forward wavelet transform is expressed as

$$x_1(n) = x(2n+1) - \sum_k p(k)x(2n-2k)$$

$$x_0(n) = x(2n) + \sum_k u(k)x_1(n-k). \tag{7.176}$$

We have redrawn the analysis stage of Figure 7.14 in Figure 7.15 to show these particular inputs to the lifting steps.

The advantages of lifting are savings of memory and computation. For example, the coefficients of p and u that result in the spline 5/3 filter are

$$p : (1, 1)/2, \quad u : (1, 1)/4.$$

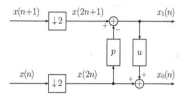

Figure 7.15   Analysis stage of the lifting scheme using the lazy wavelet transform.

Vector notation is used in this instance to indicate the coefficients. Time shifts of these coefficients are usually necessary to obtain the proper registration between the low and highpass subband signals. There are no multiplications and only 3 bit shifts (1 for 1/2 and 2 for 1/4) needed to implement this transform through lifting.

Example 7.3 (Spline 5/3 transform through lifting.) We show by calculation that the Spline 5/3 transform having the prototype filters in Equation (7.173) can be implemented through lifting. Using the prototype filters and the relationships in (7.168), the low and highpass analysis filters to be reproduced by lifting are:

$$\tilde{h}_0(n) = -\frac{1}{8}\delta(n+2) + \frac{1}{4}\delta(n+1) + \frac{3}{4}\delta(n) + \frac{1}{4}\delta(n-1) - \frac{1}{8}\delta(n-2)$$

$$\tilde{h}_1(n) = -\frac{1}{4}\delta(n-1) + \frac{1}{2}\delta(n) - \frac{1}{4}\delta(n-1). \tag{7.177}$$

After filtering and downsampling of the input $x(n)$, we get the following subband signals

$$x_0(n) = -\frac{1}{8}\delta(2n+2) + \frac{1}{4}\delta(2n+1) + \frac{3}{4}\delta(2n) + \frac{1}{4}\delta(2n-1) - \frac{1}{8}\delta(2n-2)$$

$$x_1(n) = -\frac{1}{4}\delta(2n) + \frac{1}{2}\delta(2n+1) - \frac{1}{4}\delta(2n+1). \tag{7.178}$$

The prediction and update filters of the lifting implementation are taken to be

$$p(n) = \frac{1}{2}(\delta(n) + \delta(n+1)) \tag{7.179}$$

$$u(n) = \frac{1}{4}(\delta(n) + \delta(n-1)). \tag{7.180}$$

Substituting these filters into the lifting equations in (7.174), we obtain

$$x_1^{(L)}(n) = x(2n+1) - \frac{1}{2}x(2n) - \frac{1}{2}x(2n+2) \tag{7.181}$$

$$x_0^{(L)}(n) = x(2n) + \frac{1}{4}x_1^{(L)}(n) + \frac{1}{4}x_1^{(L)}(n-1). \tag{7.182}$$

Here, we use the superscript $(\mathcal{L})$ to indicate a signal obtained by lifting. Substituting the equation for $x_1^{(\mathcal{L})}(n)$ into that for $x_0^{(\mathcal{L})}(n)$, we calculate that

$$
\begin{aligned}
x_0^{(\mathcal{L})}(n) &= x(2n) + \frac{1}{4}[x(2n+1) - \frac{1}{2}x(2n) - \frac{1}{2}x(2n+2)] + \frac{1}{4}[x(2n-1) \\
&\quad - \frac{1}{2}x(2n-2) - \frac{1}{2}x(2n)] \\
&= \frac{3}{4}x(2n) + \frac{1}{4}x(2n+1) + \frac{1}{4}x(2n-1) - \frac{1}{8}x(2n+2) \\
&\quad - \frac{1}{8}x(2n-2).
\end{aligned}
\tag{7.183}
$$

We conclude that

$$
x_0^{(\mathcal{L})}(n) = x_0(n) \tag{7.184}
$$
$$
x_1^{(\mathcal{L})}(n) = 2x_1(n). \tag{7.185}
$$

As seen by this example, the lifting scheme yields the same subband signals up to a scale factor. Many times a scale factor is inserted in order to normalize the subband signals, so that their energies sum to the original signal energy. If so, the scaling must be inverted at the input to the synthesis stage to assure that the reconstructed signal exactly equals the source signal.

To obtain the biorthogonal 9/7 filter, a second pair of prediction/update steps are required, namely,

n	-1	0	1
$p_1(n)$	2.2431326	2.2431326	
$u_1(n)$		-0.0749251	-0.0749251
$p_2(n)$	-1.2486247	-1.2486247	
$u_2(n)$		0.6272133	0.6272133

Associated with the filtering steps are two gain factors to multiply the subband signals in order to equalize the source and subband energies. They are $K_0 = 1/K$ and $K_1 = K/2$ for the lowpass and highpass subbands, respectively, where $K = 1.2301741$. Counting these gain multiplications, this transform requires only 6 multiplications (4 identical pairs of values) compared to 16 (9+7) multiplications for direct implementation. Those 16 multiplications may be reduced to 9 if we exploit symmetry.

7.7.8 Transforms with integer output

The transforms encountered up to now, whether or not reversible, do not necessarily produce integer output values. Only integers can be encoded, so real-valued ouputs must be rounded to integers or be quantized prior to coding, thereby incurring loss of accuracy or distortion. Distortion-free or lossless coding is necessary in many circumstances, so given the coding gain inherent in transform coding, a wavelet transform with integer output would offer considerable benefit in coding efficiency besides its natural characteristics of multi-resolution representation and temporal/spatial localization.

Wavelet transforms with integer output were first proposed by Said and Pearlman [14] and then later by Zandi *et al.* [15]. We shall explain the method of Said and Pearlman, because it turns out to be a precedence and special case of lifting.

The starting point is the S-transform, whereby the sequence of integers $c(n)$, $n = 0, 1, \ldots, N - 1$, with N even, can be represented by the two sequences

$$l(n) = \lfloor (c(2n + 1) + c(2n))/2 \rfloor$$
$$h(n) = c(2n + 1) - c(2n), \quad n = 0, 1, \ldots, N/2 - 1, \tag{7.186}$$

where $l(n)$ is the lowpass sequence and $h(n)$ is the highpass. One can see that this transform without the downward truncation is the Haar transform within a scale factor. The truncation does not remove any information, because the sum and difference of two integers are either both odd or both even. When $h(n)$ is even (its least significant bit is 0), doubling $l(n)$ will give $c(2n + 1) + c(2n)$ exactly. When $h(n)$ is odd (its least significant bit is 1), then doubling $l(n)$ will give 1 less than $c(2n + 1) + c(2n)$. The inverse transform is therefore

$$c(2n) = l(n) - \lfloor (h(n) + 1)/2 \rfloor$$
$$c(2n + 1) = h(n) + c(2n). \tag{7.187}$$

$(h(n)/2 = \lfloor (h(n) + 1)/2 \rfloor$, when $h(n)$ is even.)

Observe that the recovery of the original sequence follows that of the lifting scheme. In fact, one can rewrite the forward S-transform as

$$h(n) = c(2n + 1) - c(2n),$$
$$l(n) = c(2n) + \lfloor (h(n) + 1)/2 \rfloor. \tag{7.188}$$

The prediction step is the subtraction of the even sample from the odd sample and the update step is the downward truncation of $(h(n) + 1)/2$. The only difference from the formulas in (7.176) is that these steps involve non-linear operations. As noted earlier, reversibility is assured even when these operations are non-linear.

Said and Pearlman propose to replace $h(n)$ with the difference

$$h_d(n) = h(n) - \lfloor \hat{h}(n) + 1/2 \rfloor,$$

where $\hat{h}(n)$ is a prediction of $h(n)$. This prediction takes the form

$$\hat{h}(n) = \sum_{i=-L_0}^{L_1} \alpha_i \Delta l(n + i) - \sum_{j=1}^{H} \beta_j h(n + j), \tag{7.189}$$

with the definition

$$\Delta l(n) \equiv l(n) - l(n - 1).$$

Table 7.2 Weighting parameters of S+P filters

Predictor	Parameter α_{-1}	α_0	α_1	β_1
A	0	1/4	1/4	0
B	0	2/8	3/8	2/8
C	$-1/16$	4/16	8/16	6/16

The coefficients α_i and β_j are selected either to have small subband entropies or a certain number of zeros at $\omega = \pi$. If $\beta_j = 0$ for all j, past values of $h(n)$ are not involved in the prediction and it reduces to the lifting step of prediction from lowpass values and replacing the highpass sequence with the prediction residual. One such case is the S+P (S+Prediction) A filter with $\alpha_0 = \alpha_1 = 1/4$. This prediction becomes

$$\hat{h}(n) = \frac{1}{4}(l(n-1) - l(n+1)) \tag{7.190}$$

and the updated highpass signal is

$$h_d(n) = h(n) - \lfloor \frac{1}{4}l(n-1) - \frac{1}{4}l(n+1) + \frac{1}{2} \rfloor \tag{7.191}$$

The S+P A, B, and C filters are defined with the weighting parameters of (7.189) listed in Table 7.2. The B and C predictors do not have an interpretation in terms of an added lifting step to include the additional highpass weighting, but nonetheless produce a transform with good entropy reduction properties.

7.7.8.1   Integer versions of known wavelet transforms

Since any biorthogonal wavelet transform can be written using lifting steps, integer versions of any transform can be built by rounding the prediction and update steps to integers. The integerized version of the lifting equations (7.176) becomes

$$x_1(n) = x(2n+1) - \lfloor \sum_k p(k)x(2n-2k) + \frac{1}{2} \rfloor$$
$$x_0(n) = x(2n) + \lfloor \sum_k u(k)x_1(n-k) + \frac{1}{2} \rfloor \tag{7.192}$$

The inversion to recover $x(n)$ follows similarly by reversing the order and the signs.

$$x(2n) = x_0(n) - \lfloor \sum_k u(k)x_1(n-k) + \frac{1}{2} \rfloor$$
$$x(2n+1) = x_1(n) + \lfloor \sum_k p(k)x(2n-2k) + \frac{1}{2} \rfloor \tag{7.193}$$

Table 7.3 Lifting steps to implement transforms [16]

Transform	n	-2	-1	0	1	2
(2,2)	$p(n)$		1/2	1/2		
	$u(n)$			1/4	1/4	
(4,2)	$p(n)$	$-1/16$	9/16	9/16	$-1/16$	
	$u(n)$			1/4	1/4	
(2,4)	$p(n)$		1/2	1/2		
	$u(n)$		$-3/64$	19/64	19/64	$-3/64$
(4,4)	$p(n)$	$-1/16$	9/16	9/16	$-1/16$	
	$u(n)$		$-1/32$	9/32	9/32	$-1/32$

For example, the integerized spline 5/3 transform via lifting is

$$x_1(n) = x(2n+1) - \left\lfloor \frac{x(2n) + x(2n+2) + 1}{2} \right\rfloor \tag{7.194}$$

$$x_0(n) = x(2n) + \left\lfloor \frac{(x_1 n + x_1(n-1))}{4} + \frac{1}{2} \right\rfloor \tag{7.195}$$

This transform is the standard one for JPEG2000, Part 3 (MJP2K, Motion JPEG2000). It is the same as the (2, 2) transform in Calderbank *et al.* [16], where the nomenclature is $(\tilde{N}, N)$. $\tilde{N}$ and N refer to the number of vanishing moments in the analysis and synthesis prototype filters, respectively. We adopt this nomenclature in listing the prediction and update filters for several transforms, including this one, in Table 7.3. These filters can be used either in the regular or integerized lifting equations.

7.8 Concluding remarks

This chapter was written to familiarize the reader with some of the transforms used in signal coding. The intent was not to describe all the transforms used in coding, but to introduce and explain the most common ones in order to give understanding of their place in coding systems. Following chapters will describe the coding systems in which these transforms operate. More comprehensive treatments of transforms may be found in the textbooks by Pratt [17] and Akansu and Haddad [18], for example. One should seek the book by Crochiere and Rabiner [19] for a comprehensive analysis of multi-rate processing. In the case of subband/wavelet transforms specifically, we sought to derive the non-aliasing and perfect reconstruction properties and how the analysis and synthesis filters are connected to wavelets. The attempt was to present the derivations concisely in the simplest possible mathematical language without compromise of rigor. The textbooks by Vetterli and Kovačević [10], Rao and Bopardikar [11], and Strang and Nguyen [12] are recommended for more formal and expansive treatments. We concluded the chapter by describing commonly used wavelet filters and their implementation using the lifting technique.

Problems

7.1 Textbooks on digital signal processing derive the amplitude spectrum of a signal after downsampling (decimation). The operation of downsampling of an input sequence $x(n)$ by an integer factor of V is expressed mathematically by

$$y(n) = x(Vn), n = 0, \pm 1, \pm 2, \ldots$$

For a deterministic sequence $x(n)$ with discrete-time Fourier transform $X(\omega)$ as input to a factor of V downsampler, the output amplitude spectrum is

$$Y(\omega) = \frac{1}{V} \sum_{k=0}^{V-1} X(\frac{\omega - 2\pi k}{V}).$$

However, it is hard to find the corresponding formula for the power spectral density of a stationary random sequence. The objective in this problem is to derive this formula.

Consider now that $x(n)$ is a stationary random sequence and is the input to a factor of V downsampler. Its autocorrelation function and power spectrum are given, respectively, by

$$R_x(k) = E[x(n)x(n+k)]$$

$$S_x(\omega) = \mathcal{F}[R_x(k)] = \sum_{k=-\infty}^{\infty} R_x(k) \cdot e^{-j2\pi k\omega}.$$

(a) Show that the autocorrelation of the downsampled sequence $y(n)$ is

$$R_y(k) = R_x(Vk).$$

(b) Explain why the relationship of the power spectra of the input and output of the downsampler follow the same relationship as the amplitude spectra of a deterministic sequence.

7.2 We wish to derive the relationship of the input and output power spectra of an upsampler, following a similar procedure as the previous problem. The time sequences before and after upsampling by a factor of V are related by

$$y(n) = \begin{cases} x(n/V), & n = kV, k = 0, \pm 1, \pm 2, \ldots \\ 0, & \text{otherwise} \end{cases}$$

or

$$y(Vn) = \begin{cases} x(n), & n = 0, \pm 1, \pm 2, \ldots \\ 0, & \text{otherwise} \end{cases}$$

If their amplitude spectra exist, they would be related by $Y(\omega) = X(V\omega)$. Derive the relationship between their power spectra $S_y(\omega)$ and $S_x(\omega)$.

7.3 Consider the first-order Gaussian Markov source (X_1, X_2) with the covariance matrix

$$\Phi_X = \sigma^2 \begin{pmatrix} 1 & 0.9 \\ 0.9 & 1 \end{pmatrix}$$

where σ^2 is the source variance.
(a) Find the eigenvalues and eigenvectors of the source covariance matrix Φ_X.
(b) Write down the unitary KLT (matrix).
(c) Calculate the theoretically optimal gain over PCM for transform coding sample pairs from this source using the squared-error distortion measure.
(d) Find the optimal rate assignments for the code rates of $R = 2$ and $R = 1/2$ bits per sample.

7.4 Speech and audio data are often transformed in segments of length N from 128 to 512. A common autocorrelation function model of such data is

$$R_{XX}(n) = \sigma_x^2 \rho^{|n|}, \quad n = 0, 1.2, \dots, N - 1, \ 0 < \rho < 1.$$

The data are assumed to have zero mean and variance $\sigma_x^2 = 1$.
(a) Write down an expression for the autocorrelation matrix $\mathcal{R}_X$ for the source for arbitrary N.
(b) Assume that the $N \times N$ forward transform matrix is P, i.e.,

$$\mathbf{y} = P\mathbf{x}$$

where $\mathbf{x}$ and $\mathbf{y}$ are respectively the source and transform vectors of dimension N. Write down the expression for the autocorrelation matrix of the transform $\mathcal{R}_y$ in terms of P and $\mathcal{R}_X$.
(c) Let P be the DCT matrix and let $\rho = 0.95$. By computer (MATLAB recommended), perform the following:
    (i) calculate the transform autocorrelation matrix $\mathcal{R}_y$ for $N = 64$;
    (ii) plot the transform variances versus their indices;
    (iii) calculate the ideal transform coding gain.

7.5 The speech source "s13pc.raw"[13] may be modeled crudely as Markov-1 with unknown variance σ^2 and correlation parameter ρ.
(a) Estimate σ^2 and ρ of this data.
(b) Calculate by computer DCTs of 100 segments of length 64 from this data.
(c) Estimate the variance of every DCT coefficient and plot-estimated variance versus index.
(d) Calculate the estimated ideal transform coding gain. Compare your estimate to the theoretical gain (Problem 7.4) using the same value of ρ.

7.6 Repeat the relevant parts of Problem 7.5 substituting the DFT for the DCT. Why is the DFT coding gain smaller?

7.7 In the H.264/AVC video coding standard, an integer approximation to a DCT matrix is used to transform a 4 × 4 block within a frame prior to coding. This transform matrix is

$$
\mathbf{H} = \begin{bmatrix} 1 & 1 & 1 & 1 \\ 2 & 1 & -1 & -2 \\ 1 & -1 & -1 & 1 \\ 1 & -2 & 2 & -1 \end{bmatrix}
$$

Observe that this matrix is not unitary, but it has the virtue of integer operations being possible in computations with it and its inverse. It is applied separably to four rows and then four columns of a block $\mathcal{B}$. Assume that the two-dimensional, autocorrelation function of the source block is separable, first-order Markov according to

$$
R_{xy}(m, n) = \sigma_B^2 \rho_y^{|m|} \rho_x^{|n|}, \quad m, n = 0, 1, 2, 3
$$

with $\rho_y = 0.95$ and $\rho_x = 0.98$.

(a) Calculate the variances of the transform coefficients in the block.

(b) Calculate $G_{TC/PCM}$, the transform coding gain.

7.8 A source with (discrete-time) power spectrum

$$
S_x(\omega) = Ae^{-10|\omega|/\pi}, \quad -\pi \le \omega < \pi
$$

is the input to a subband coding system. The filter system is the three-stage, recursive low frequency band splitting system shown in Figure 7.10. The filters are ideal half-band low and high frequency filters. Following the method in Example 7.1, calculate the ideal coding gain $G_{SB/PCM}$. When there is one less stage of splitting, what is $G_{SB/PCM}$?

7.9 Repeat the last problem using the Haar wavelet filter pair (Example 7.2) instead of ideal half-band filters.

Notes

1. Many books on mathematics refer to a real matrix with this property as *orthogonal*. We shall use the term *unitary* or *orthonormal* here.

2. The convention is column vectors and the superscript "t" denotes transpose to convert the row vector form to column form.

3. A transform matrix may contain complex elements and produce complex output vectors, such as the DFT to be encountered later. When complex quantities are involved, the conjugate transpose operator $*t$ replaces plain transpose t. Similarly, $\| \mathbf{y} \|^2 = \mathbf{y}^{*t}\mathbf{y}$.

4. We use capital letters to signify a random variable and lower case to denote a particular value of the random variable. In the parlance of information or communications

theory, a random variable together with its probability distribution is often called an *ensemble*.

5. The autocorrelation and covariance functions and matrices are equal in the case of zero mean assumed here.

6. The matrices $\mathbf{H}_{2N}$ are called Hadamard matrices and when the rows are re-ordered according to sequency (analogous to frequency but with ± 1 as the basis) are called Hadamard–Walsh or Walsh–Hadamard matrices.

7. Theoretically, L can be an integer multiple of N other than 2, but the literature does not report any instance of such use.

8. We have ignored the implied time delay of $N - 1$ in reconstructing $x(n)$.

9. Linear phase, FIR filters of even length are necessary to realize orthogonality.

10. The coefficients in an orthogonal basis expansion are found by projection on the same orthogonal basis, whereas coefficients in a biorthogonal basis expansion are found by projection onto another basis, orthogonal to the expansion basis.

11. The notation $\langle f(t), g(t) \rangle$ means the inner product of the functions $f(t)$ and $g(t)$, which in this case is $\int_{-\infty}^{+\infty} f(t)g^*(t)dt$.

12. Kronecker delta function ($\delta()$) notation is given here, because it aids the calculation of the convolutions in the formulas.

13. Available via path datasets/audio/s13pc.raw at http://www.cambridge.org/pearlman

References

1. W. A. Fuller, *Introduction to Statistical Time Series*, 2nd edn. New York, NY: John Wiley & Sons, 1996.
2. J. Pearl, "On coding and filtering stationary signals by discrete Fourier transforms," *IEEE Trans. Inf. Theory*, vol. 19, no. 2, pp. 229–232, Mar. 1973.
3. W. A. Pearlman, "A limit on optimum performance degradation in fixed-rate coding of the discrete Fourier transform," *IEEE Trans. Inf. Theory*, vol. 22, no. 4, pp. 485–488, July 1976.
4. Y. Yemini and J. Pearl, "Asymptotic properties of discrete unitary transforms," *IEEE Trans. Pattern Anal. Mach. Intell.*, vol. 1, no. 4, pp. 366–371, Oct. 1979.
5. N. Ahmed, T. R. Natarajan, and K. R. Rao, "Discrete cosine transform," *IEEE Trans. Comput.*, vol. 23, no. 1, pp. 90–93, Jan. 1974.
6. M. J. Narasimha and A. M. Peterson, "On the computation of the discrete cosine transform," *IEEE Trans. Commun.*, vol. 26, no. 6, pp. 934–936, June 1978.
7. H. S. Malvar, *Signal Processing with Lapped Transforms*. Norwood, MA: Artech House, 1992.
8. D. S. Taubman and M. W. Marcellin, *JPEG2000: Image Compression Fundamentals, Standards, and Practice*. Norwell, MA: Kluwer Academic Publishers, 2002.
9. A. Cohen, I. Daubechies, and J.-C. Feauveau, "Biorthogonal bases of compactly supported wavelets," *Commun. Pure Appl. Math.*, vol. 45, no. 5, pp. 485–560, June 1992.
10. M. Vetterli and J. Kovačević, *Wavelets and Subband Coding*. Englewood Cliffs, NJ: Prentice Hall, 1995.
11. R. M. Rao and A. S. Bopardikar, *Wavelet Transforms: Introduction to Theory and Applications*. Reading, MA: Addison-Wesley Publishing Co., 1998.
12. G. Strang and T. Nguyen, *Wavelets and Filter Banks*. Wellesley, MA: Wellesley-Cambridge Press, 1997.
13. W. Sweldens, "The lifting scheme: a custom-design construction of biorthogonal wavelets," *Appl. Comput. Harmon. Anal.*, vol. 3, no. 2, pp. 186–200, Apr. 1996.
14. A. Said and W. A. Pearlman, "An image multiresolution representation for lossless and lossy compression," *IEEE Trans. Image Process.*, vol. 5, no. 9, pp. 1303–1310, Sept. 1996.
15. A. Zandi, J. Allen, E. L. Schwartz, and M. Boliek, "CREW: compression with reversible embedded wavelets," in *Proceedings of the IEEE Data Compression Conference*, Snowbird, UT, Mar. 1995, pp. 212–221.
16. A. R. Calderbank, I. Daubechies, W. Sweldens, and B.-L. Yeo, "Wavelet transforms that map integers to integers," *Applied and Computational Harmonic Analysis*, vol. 5, no. 3, pp. 332–369, July 1998.
17. W. K. Pratt, *Digital Image Processing*, 2nd edn. New York: J. Wiley & Sons, Inc., 1991.

18. A. N. Akansu and R. A. Haddad, *Multidimensional Signal Decomposition: Transforms, Subbands, and Wavelets,* 2nd edn. San Diego, CA: Academic Press (Elsevier), 2001.
19. R. E. Crochiere and L. R. Rabiner, *Multirate Digital Signal Processing.* Englewood Cliffs, NJ: Prentice-Hall, Inc., 1983.

Further reading

A. N. Akansu and F. E. Wadas, "On lapped orthogonal transform," *IEEE Trans. Signal Process.*, vol. 40, no. 2, pp. 439–442, Feb. 1992.

R. Bellman, *Introduction to Matrix Analysis*, 2nd edn. New York, NY: McGraw-Hill Publishing Co., 1970.

P. S. R. Diniz, E. A. B. da Silva, and S. L. Netto, *Digital Signal Processing.* Cambridge, England: Cambridge University Press, 2002.

N. S. Jayant and P. Noll, *Digital Coding of Waveforms.* Englewood Cliffs, NJ: Prentice Hall, 1984.

W. A. Pearlman, "Performance bounds for subband coding," in *Subband Image Coding*, J. W. Woods, edn. Norwell, MA: Kluwer Academic Publishers, 1991, ch. 1.

W. Sweldens, "The lifting scheme: a construction of second generation wavelets," *SIAM J. Math. Anal.*, vol. 29, no. 2, pp. 511–546, Mar. 1998.

8 Rate control in transform coding systems

Thus far, we have learned principles and described components of coding systems. In this chapter, we shall assemble components based on these principles in order to discover efficient transform coding systems. We start with the most common systems, those employing a transform in the front end prior to coding. The reason that these systems are the most common is that they are the ones that are the most efficient for lossy coding, which is the most ubiquitous kind of coding in consumer use. For example, JPEG (lossy) and JPEG2000 (lossy) for images, MPEG2 and H.264 for video, and MP3 for audio are all lossy transform coding systems.[1]

In Chapter 7, we presented a skeletal block diagram of a transform coding system (Figure 7.1) and presented mathematical expressions and features of transforms commonly used in coding systems. Here, we delve into the details of the coding in that block diagram. As shown in Figure 8.1, the input vector $\mathbf{x}$ of N dimensions is transformed by P^{-1} to the vector $\tilde{\mathbf{x}}$ with components $\tilde{x}_1, \tilde{x}_2, \ldots, \tilde{x}_N$. We consider that the source is composed of a sequence of N-dimensional vectors, so that every transform component $\tilde{x}_n$ in the figure represents a subsource of data. The objective is to encode these component sub-sources in an optimal manner.

8.1 Rate allocation

The motivation behind employing a mathematical transform is to make the component subsources as mutually independent as possible. In practice, that is achieved only in the substantially weaker form of approximate decorrelation. Only in the ideal case of a stationary Gaussian source can the component subsources be made to be mutually statistically independent through the KLT. Nonetheless, the transform does weaken the dependence among the component subsources to a substantial degree, so that they are considered to be independent for the purposes of coding. Therefore, we do encode these transform components independently in practical systems.

In Figure 8.1 every component is depicted as first being quantized to one of a set of quantization levels, whose index is then encoded losslessly. The different components generally have different statistical properties, so that different rates must be assigned in order to meet a target bit budget. The rate of a quantizer will mean the number of bits per sample after the subsequent entropy encoding stage. Assuming perfect reception

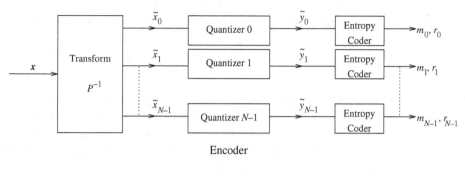

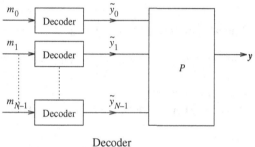

Figure 8.1 A transform coding system.

at the receiver, every coded index is then decoded to an index of a quantization level, which is then mapped (de-quantized) to the associated quantization level. These quantization levels corresponding to transform components are then inverse transformed to an imperfect reconstruction of the source vector.

Consider a single transformed source component $\widetilde{x}_n$ to be quantized to $\widetilde{y}_n$ as depicted in Figure 8.1. Every $\widetilde{x}_n$, $n = 0, 1, \ldots, N-1$ is quantized and encoded independently with rate r_n so that the total rate adds to NR, i.e.,

$$R = \frac{1}{N} \sum_{n=0}^{N-1} r_n \quad \text{(bits/s.s)} \tag{8.1}$$

is the average bit rate per source sample for encoding $\widetilde{\mathbf{x}}$ and hence the input $\mathbf{x}$. The rates add because of the independent quantization and encoding of the components. Each r_n is different because of the different statistics of each $\widetilde{x}_n$.

Let $d_{Q_n}(r_n)$ denote the average distortion of the n-th quantizer Q_n producing rate r_n. We assume the distortion measure is additive, so that the total distortion per sample for independent quantizing and encoding of the components of $\widetilde{\mathbf{x}}$ is

$$D = \frac{1}{N} \sum_{n=0}^{N-1} d_{Q_n}(r_n). \tag{8.2}$$

The problem is now to find the rates $r_n, n = 0, 1, 2, \ldots, N - 1$, assigned to the different quantizer/encoders that minimize D in Equation (8.2) under the rate constraint

$$R = \frac{1}{N} \sum_{n=0}^{N-1} r_n \leq R_T, \tag{8.3}$$

where R_T is the target bit rate. An additional implicit set of constraints is that $r_n \geq 0$ for all n, because rates cannot be negative. Various means have been employed to solve this rate allocation problem and they all start with specifying the distortion-rate functions $d_{Q_n}(r_n)$ either by formulas or tabulations of rate and distortion points. We shall explain the solutions for both kinds of function specification.

8.1.1   Optimal rate allocation for known quantizer characteristics

We start with the initial assumption that the transform components have the same probability distribution, but with different variances. For example, if the source vector $\mathbf{x}$ is Gaussian, then each transform component $\tilde{x}_n$ would be Gaussian with a different variance. When the distortion measure between a component $\tilde{x}_n$ and its quantization $\tilde{y}_n$ is squared error, i.e.,

$$d(\tilde{x}_n, \tilde{y}_n) = (\tilde{x}_n - \tilde{y}_n)^2, \tag{8.4}$$

the distortion-rate functions $d_{Q_n}(r_n)$ can be expressed in terms of the distortion-rate function $\rho(r_n)$ for a unit variance input to a quantizer. Let σ_n^2 be the variance of the component $\tilde{x}_n$. The distortion of the n-th quantizer is then assumed to be

$$d_{Q_n}(r_n) = \sigma_n^2 \rho(r_n) \tag{8.5}$$

and the MSE per sample of the N quantizers is

$$D = \frac{1}{N} \sum_{n=0}^{N-1} \sigma_n^2 \rho(r_n) \tag{8.6}$$

The rates r_n must satisfy the rate budget constraint of R_T bits per sample, so that

$$\frac{1}{N} \sum_{n=0}^{N-1} r_n \leq R_T. \tag{8.7}$$

In order to assure a solution, we assume now that the basic quantizer characteristic $\rho(r)$ is a convex ($\cup$) function of r $\left(\frac{\partial^2 \rho(r)}{\partial r^2} > 0\right)$. Since $\rho(r)$ is also monotonically decreasing with rate r, it is clear that the minimum of D is achieved at the largest possible rate R_T. Ignoring for now the additional constraints of non-negative rates, we seek the minimum of the Lagrangian objective function,

$$J(\lambda) = D + \lambda R$$
$$= \frac{1}{N} \sum_{n=0}^{N-1} (\sigma_n^2 \rho(r_n) + \lambda r_n), \tag{8.8}$$

where λ is the Lagrange multiplier evaluated to satisfy the rate constraint of

$$R_T = \frac{1}{N} \sum_{n=0}^{N-1} r_n. \tag{8.9}$$

An accepted model for the MSE versus rate characteristic of a scalar quantizer of a unit variance random variable is the monotonically non-increasing function

$$\rho(r) = g(r)2^{-ar} \quad , \quad r \geq 0 \tag{8.10}$$

where $g(r)$ is an algebraic function of the rate r, such that $g(0) = 1$ and a is a constant no greater than 2 (see Equation (5.60)). For larger rates r and/or an entropy coded quantizer $a = 2$ is well justified. Monotonicity implies that $\rho(r) \leq \rho(0) = 1$. We make the farther approximation that since $g(r)$ is a much more slowly varying function than the exponential 2^{-ar}, we shall regard it as a constant in the range of interest.

Under these simplifications, the objective function in Equation (8.8) becomes

$$J(\lambda) = \frac{1}{N} \sum_{n=0}^{N-1} (\sigma_n^2 g_n 2^{-ar_n} + \lambda r_n) \tag{8.11}$$

Taking the derivatives with respect to r_n and setting to 0, we obtain the necessary conditions for the minimum as

$$\sigma_n^2 g_n 2^{-ar_n} = \frac{\lambda}{a \ln 2} \equiv \theta \quad n = 0, 1, \dots, N-1 \tag{8.12}$$

Noting that the left side of the above equation equals $d_{Q_n}(r_n)$, these conditions mean that the MSE should be equal among all the quantized transform components, i.e.,

$$d_{Q_n}(r_n) = \theta \quad \forall n. \tag{8.13}$$

Solving Equation (8.12) for the rates r_n, we obtain

$$r_n = \frac{1}{a} \log_2 \frac{\sigma_n^2 g_n}{\theta} \quad n = 0, 1, \dots, N-1. \tag{8.14}$$

The parameter θ is a redefined Lagrange multiplier that is evaluated to satisfy the rate constraint in Equation (8.9).

The solution for the optimal rate allocation in (8.14) is valid when $r_n \geq 0$. However, that may not always be the case for values of θ found for smaller values of the rate target R_T. $r_n \geq 0$ only when $\sigma_n^2 g_n \geq \theta$. Otherwise, r_n will turn out to be negative, clearly an erroneous result. So when $\sigma_n^2 g_n < \theta$, we set r_n to 0 whereby $g_n = 1$ and obtain the distortion $d_{Q_n}(0) = \sigma_n^2$. The solution for the optimal rate allocation is therefore given as

$$r_n = \begin{cases} \frac{1}{a} \log_2 \frac{\sigma_n^2 g_n}{\theta}, & \text{if } \sigma_n^2 g_n > \theta \\ 0, & \text{if } \sigma_n^2 \leq \theta \end{cases}$$
$$n = 0, 1, \dots, N-1. \tag{8.15}$$

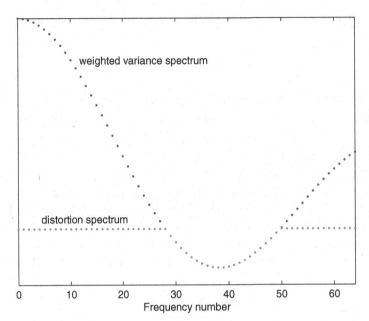

Figure 8.2   Weighted variance and distortion spectrum.

The resulting MSE distortion of each quantized component is

$$d_{Q_n} = \begin{cases} \theta, & \text{if } \sigma_n^2 g_n > \theta \\ \sigma_n^2, & \text{if } \sigma_n^2 \leq \theta \end{cases}$$
$$n = 0, 1, \ldots, N - 1. \tag{8.16}$$

An illustration of the transform distortion (MSE) distribution or spectrum relative to the weighted variance spectrum appears in Figure 8.2. Every component whose weighted variance exceeds the threshold θ has distortion θ. Otherwise, its distortion equals the variance, which is the smallest distortion achieved with zero rate, as the component is quantized (mapped directly) to its known mean value. The value of θ is set by the particular rate target.

One usually sees the above solution with $g_n = 1$ for all n and $a = 2$. Then this rate and distortion allocation is the same as that of the rate-distortion function $R(d)$ of the stationary Gaussian source with squared error distortion measure. So strictly speaking, these formulas with $g_n = 1$ and $a = 2$ are valid only for optimal coding of Gaussian transform components. Therefore, we shall carry on keeping these literal parameters.

Substituting the component rates and distortions in (8.15) and (8.16) in turn into the total rate and distortion equations (8.9) and (8.2), we obtain the resulting rate and distortion for the transform system.

$$R_T = \frac{1}{N} \sum_{n:\sigma_n^2 g_n > \theta} \frac{1}{a} \log \frac{\sigma_n^2 g_n}{\theta} \tag{8.17}$$

$$D = \frac{1}{N} \left[\sum_{n:\sigma_n^2 g_n > \theta} \theta + \sum_{n:\sigma_n^2 \leq \theta} \sigma_n^2 \right]. \tag{8.18}$$

8.1.2 Realizing the optimal rate allocation

8.1.2.1 Bisection procedure

In order to realize the optimal rate allocation for a specific target rate R_T, the parameter θ must be calculated. What is needed first is the set of weighted variances $\{g_n \sigma_n^2\}$. The weights g_n depend on the rates r_n that we are trying to find. We adopt the usual practice of assuming g_n to be a constant $g = 1$ for all n, so that it drops out of consideration. (Equivalently, one can absorb g into the definition of θ for the components with non-zero rate.) Then, by trial and error, you substitute various values of θ into the R_T equation in (8.17) until you obtain the desired R_T. One systematic way is through bisection. Since θ is a distortion parameter, lower values give higher values of R_T and vice versa. Suppose that your desired rate is R_D and you first select a small value of θ, say θ_1, such that the total rate evaluates to $R_1 > R_D$. Then select a second much larger value θ_2 ($> \theta_1$), so that the rate equation evaluates to $R_2 < R_D$. Now use the midpoint value of θ, $\theta_3 = (\theta_1 + \theta_2)/2$, to obtain a new rate value R_3. If $R_1 > R_D > R_3$, then choose the next value of θ as the midpoint between θ_3 and θ_1. Otherwise, if $R_3 > R_D > R_2$, choose the next value as the midpoint between θ_3 and θ_2. Continue in this way until you achieve a value of rate within your tolerance limit of R_D. Figure 8.3 illustrates the first few steps of this bisection procedure $R_1 > R_D > R_3$ and Algorithm 8.1 describes the steps of the procedure in detail.

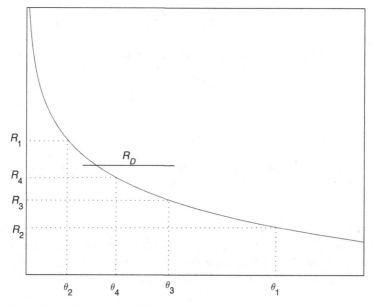

Figure 8.3   Illustration of initial points in bisection procedure.

ALGORITHM 8.1 _____

Bisection procedure to find parameter θ^* for desired rate R_D

1. Let $R(\theta) = \frac{1}{N} \sum_{n:\sigma_n^2 g_n > \theta} \frac{1}{a} \log \frac{\sigma_n^2 g_n}{\theta}$.
2. Set desired rate $R(\theta^*) = R_D$. Set precision parameter $\epsilon > 0$.
3. Choose θ_1 large enough so that $R(\theta_1) < R_D$.
   Choose θ_2 small enough so that $R(\theta_2) > R_D$.
4. Let $\theta_3 = (\theta_2 + \theta_1)/2$.
5. Set $i = 3$.
6. If $R(\theta_i) < R_D, \theta_{i+1} = (\theta_i + \theta_{j*})/2$, $j^* = \arg\min_{j<i} R(\theta_j) > R_D$.
   If $R(\theta_i) > R_D, \theta_{i+1} = (\theta_i + \theta_{j*})/2$, $j^* = \arg\max_{j<i} R(\theta_j) < R_D$.
7. If $|R(\theta_{i+1}) - R_D| < \epsilon$, stop. Let $\theta_{i+1} = \theta^*$.
   Else, set $i = i + 1$ and return to Step 6.

_____

8.1.2.2   Iterative procedure

Another way to calculate the optimal rate allocation is facilitated by ordering the weighted vaiances $g_n \sigma_n^2$ from largest to smallest, and re-indexing them so that

$$g_1 \sigma_1^2 \geq g_2 \sigma_2^2 \geq \cdots \geq g_N \sigma_N^2.$$

As the specified target rate R_T increases, more and more weighted variances exceed the threshold θ. Let us assume that for some rate R, there are N_θ coefficients whose weighted variances exceed θ. Therefore, the rate R can be expressed as

$$R = \frac{1}{N} \sum_{n=0}^{N_\theta - 1} \frac{1}{a} \log \frac{g_n \sigma_n^2}{\theta} \tag{8.19}$$

$$= \frac{1}{Na} \log \frac{\prod_{n=0}^{N_\theta - 1} g_n \sigma_n^2}{\theta^{N_\theta}}. \tag{8.20}$$

Solving for θ and substituting it into the rate formulas

$$r_n = \frac{1}{a} \log \frac{g_n \sigma_n^2}{\theta} \quad n = 0, 1, 2, \ldots, N_\theta - 1,$$

we obtain the rate assignments

$$r_n = \begin{cases} \frac{NR}{N_\theta} + \frac{1}{a} \log \frac{g_n \sigma_n^2}{(\prod_{n=1}^{N_\theta} g_n \sigma_n^2)^{1/N_\theta}}, & 0 \leq n \leq N_\theta - 1 \\ = 0, & N_\theta < n \leq N - 1 \end{cases} \tag{8.21}$$

A simple iterative procedure for determining N_θ is to set initially $N_\theta = N$ and calculate the r_n's according to Equation (8.21). If all the rates r_n are non-negative, then they are the optimal rates. Otherwise, iterate steps of decrementing N_θ by 1 and calculating the rates until all the rates are non-negative. The explicit steps of this procedure follow.

1. Set $N_\theta = N$.
2. Calculate rates r_n using Equation (8.21).
3. If $r_n \geq 0$, for all $n = 0, 1, 2, \ldots, N - 1$, stop. Otherwise, set $N_\theta = N_\theta - 1$ and go to Step 2.

These rates are numbers of bits per transform coefficient, so the question arises whether they should be integers. When you consider the source as a sequence of N-dimensional vectors, the transform coefficients are considered to be independent scalar sources. These scalar sources are quantized and encoded with a variable length entropy code. Each realization of an entropy code has to have an integer length in bits, but the average length should be close to the entropy, which need not be an integer. These rates are the average lengths of the entropy codes, so do not have to be integers.

As we have seen in a previous chapter, uniform threshold quantization is nearly optimal for all but the smallest bit rates. The uniform step size of the quantizer determines the mean squared quantization error. To get a handle on this MSE D_Q, we can use the high rate approximation,

$$D_Q \approx \Delta^2/12,$$

where Δ is the step size. So at least for all but the smallest rates, we can use the same step size Δ to produce approximately the same quantization error $D_Q = \theta$ in every coefficient. The code rates should be close to those calculated by Equation (8.15) for this value of θ.

8.1.3 Fixed level quantization

Although not as efficient, the quantizers following the transform could be fixed in the number of levels with no subsequent entropy coding. Letting K_n be the number of levels for the n-th transform coefficient, the corresponding rate is $r_n = \log_2 K_n$ bits per coefficient. Without entropy coding, these rates must be integers. The model utilized for the distortion versus rate characteristic of entropy-coded quantization is still valid for fixed-level quantization. Therefore, the rate allocation formulas above still apply, but with the additional constraint that values of r_n must be integers. An approximate solution is to round off every final $r_n > 0$ to the nearest integer, but the consequent rate allocation is not guaranteed to be optimal. A more rigorous approach is to use the *marginal returns algorithm*. The idea is to allocate the total bit budget bit-by-bit to the coefficient that exhibits the largest decrease in distortion until the bit budget is exhausted. The steps of the procedure are delineated in Algorithm 8.2.

The search range in this bit allocation algorithm can be reduced by observing that the coefficient having the largest remaining distortion receives the most benefit by adding one more bit. (The derivative of distortion with respect to rate is proportional to the negative of the distortion.) Therefore, it is again advantageous to order and re-index the variance estimates, so that they are decreasing as the index increases. The values of δd_n for the same rate then decrease with increasing n. Therefore, the search can be terminated when one comes to an n such that all higher n's show the same rate allocation

ALGORITHM 8.2 ⎯⎯⎯⎯⎯⎯⎯⎯⎯⎯⎯⎯⎯⎯⎯⎯⎯⎯⎯⎯⎯⎯⎯⎯⎯⎯⎯

Marginal returns algorithm: bit allocation for uncoded, fixed-level quantizers

1. *Initialization.*
   (a) Obtain variance estimates $\hat{\sigma}_n^2$ These variance estimates are the initial distortions.
   (b) Set a bit budget $B = NR_T$ to allocate.
   (c) Set $r_n = 0$ for all $n = 0, 1, 2, \ldots, N - 1$.
2. For $n = 0$ to $n = N - 1$, do
   (a) Increment r_n by 1 bit
   (b) Calculate decrease in distortion in n-th coefficient δd_n
   (c) Decrement r_n by 1 bit
3. Determine $n = n_o$ such that δd_n is a maximum, i.e, $n_o = \arg\max_n \delta d_n$.
4. Increment r_{n_o} by 1 bit.
5. If allocated bits $\sum_{n=0}^{N-1} r_n < B$, the bit budget, return to Step 2. Otherwise, stop.

⎯⎯⎯

thus far. This situation occurs always in laying down the first bit, which must go to the $n = 1$ quantizer.

This algorithm replaces the need for formulas for rate allocation. It is necessary, because formulas are not available for bit allocation under an integer rate constraint. However, the formulas for non-integer rates rounded to integers are informative about the nature of the optimal solution. One should check the results of any algorithm with known formulas as much as possible to assure the integrity and correctness of the calculations.

8.1.4 Optimal bit allocation for arbitrary set of quantizers

We now revisit the minimal solution to the distortion in Equation (8.2) under the rate constraint in Equation (8.3). Here, we focus on the common case where the models of the quantizer distortion-rate functions are arbitrary, but are still assumed to be convex, monotonically decreasing functions of rate. The Lagrangian objective function of Equation (8.8) can be written more generally as

$$J(\lambda) = \frac{1}{N} \sum_{n=0}^{N-1} (d_{Q_n}(r_n) + \lambda r_n) \tag{8.22}$$

where the rates $r_n \geq 0$ for all $n = 0, 2, \ldots, N - 1$.

Again taking the derivatives with respect to every r_n and setting to 0, we obtain

$$d'_{Q_n}(r_n) = -\lambda, \quad r_n \geq 0$$
$$n = 0, 1, 2, \ldots, N - 1 \tag{8.23}$$

where $d'_{Q_n}(r_n)$ indicates the derivative of $d_{Q_n}(r_n)$ with respect to r_n. Since $d_{Q_n}(r_n)$ is monotone decreasing with r_n, the derivative is negative and $\lambda > 0$. (We have eliminated the unconstrained $\lambda = 0$ case.) The equation (8.23) states the condition that the optimal rates r_n occur when the distortion-rate functions of the quantizers have the same

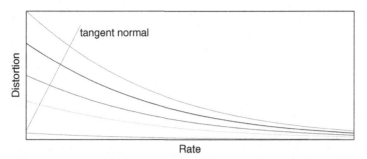

Figure 8.4   Illustration of equal slope condition on quantizers of different components.

slope. A graphical illustration of one such solution appears in Figure 8.4. The value of λ determines the total rate, which increases as λ decreases and vice versa.

In the absence of formulas, one has to search each distortion-rate characteristic for the rate point at which the slope is λ. For a given λ, starting from a large rate point, one would execute an iteration of decreasing rate points and calculations of increasing slope magnitudes until λ is reached. Let us consider one of the quantizer functions $d_{Q}(r)$. Suppose that one has available only a closely spaced, discrete set of points $\{(z_i, d_Q(z_i)\}$. Then empirical slopes are calculated from successive differences and the exact value of λ may be unattainable. In that case, in order not to overshoot the bit budget, one would stop at the smallest slope magnitude that exceeds λ. Expressing this stopping point condition mathematically, we calculate slopes at decreasing rates (increasing i in $z_i, z_{i+1} < z_i$) until we reach

$$\min \left\{ \frac{\delta d_Q(i)}{\delta z_i} \right\} = \min \left\{ \frac{d_Q(z_{i+1}) - d_Q(z_i)}{z_i - z_{i+1}} \right\} \geq \lambda \qquad (8.24)$$

and report z_{i+1}, the smaller of the two rates, as the solution. In Figure 8.5 is an example of seeking the slope of $-\lambda = -0.25$ by calculating slopes starting from rate of 3.75 to 1.25, the first point or largest rate where the slope is less than or equal to -0.25 ($\frac{\delta d_Q(i)}{\delta z_i} \geq 0.25$). Therefore, rate 1.25 is reported as the solution.

Alternatively, for a given λ, a bisection procedure can also be used to locate the rate solution above. It works the same way as described previously by starting with low and high slope magnitudes and continually bisecting the working range until the criterion in Equation (8.24) is met.

This procedure of finding the rate associated with the correct slope must be repeated on every curve $d_{Q_n}(r_n)$, that is, for $n = 1, 2, \ldots, N$. When λ becomes sufficiently large, one or more of the lower curves will not have slopes that large. For the coefficients corresponding to these curves, the rate solution is zero and the bit budget is allocated among the other coefficients. When every curve (or function) $d_{Q_n}(r_n)$ is examined by the above procedure to obtain the rate $r_n(\lambda)$, where the slope magnitude is just greater than or equal to λ, these rates are added to reveal the corresponding total rate in bits,

$$B(\lambda) = \sum_{n=0}^{N-1} r_n(\lambda).$$

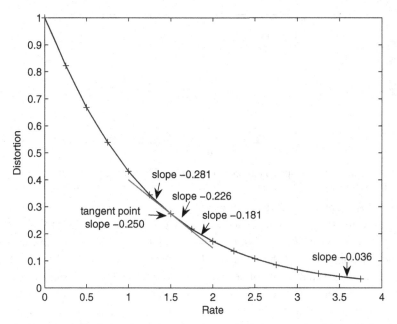

Figure 8.5   Seeking the desired slope by successive slope calculations starting from a high rate.

The steps of this procedure are delineated in Algorithm 8.3.

In many circumstances, one needs $B(\lambda)$ for a range of values of λ. Let us denote these values by λ_j, $j = 1, 2, \ldots, J$, arranged in increasing order

$$\lambda_1 < \lambda_2 < \cdots < \lambda_J.$$

For each curve $d_{Q_n}(r_n)$, we can start from a large rate where the slope magnitude is smaller than λ_1 and move toward smaller rates to calculate slopes to find $r_n(\lambda_1)$ satisfying Equation (8.24) for $\lambda = \lambda_1$. Then we can continue in the same way toward rates smaller than $r_n(\lambda_1)$ to find $r_n(\lambda_2)$, and so forth. Once this has been done for every n, we can store the results in a matrix $[r_n(\lambda_j)]_{n=0, j=1}^{N-1, J}$. Writing out the rows and columns, the matrix is

$$\begin{bmatrix} r_0(\lambda_1) & r_0(\lambda_2) & \cdots & r_0(\lambda_J) \\ r_1(\lambda_1) & r_1(\lambda_2) & \cdots & r_1(\lambda_J) \\ \vdots & \vdots & \vdots & \vdots \\ r_{N-1}(\lambda_1) & r_{N-1}(\lambda_2) & \cdots & r_{N-1}(\lambda_J) \end{bmatrix}.$$

The column sums give the total bit rates $B(\lambda_j) = \sum_{n=0}^{N-1} r_n(\lambda_j)$, corresponding to the slope values λ_j, $j = 1, 2, \ldots, J$.

8.1.5   Building up to optimal rates for arbitrary quantizers

The foregoing procedures tried to approach the optimal rates from above. We can also formulate a procedure to approach the optimal rates from below by allocating bits one

ALGORITHM 8.3

Optimal rate allocation for arbitrary quantizers

> *Notation:*
> - i: index of points increasing with decreasing rate
> - i-th rate in bits to quantizer Q_n: z_n^i ($z_n^{i+1} < z_n^i$)
> - Distortion $d_{Q_n}^i$ for rate z_n^i

1. *Initialization.*
   (a) Obtain sequence of (rate, distortion) points $\{z_n^i, d_{Q_n}^i\}$ for every quantizer Q_n, $n = 0, 1, \ldots, N$.
   (b) Set a rate target B_T.
   (c) Choose $\lambda > 0$.
2. *Main:*
   For $n = 0, 1, 2, \ldots, N - 1$ (for all quantizers)
   (a) set initial i small for large rate
   (b) for steadily increasing i calculate
   $$\frac{\delta d_{Q_n}(i)}{\delta z_n^i} = \frac{d_{Q_n}^{i+1} - d_{Q_n}^i}{z_n^i - z_n^{i+1}}$$
   only until greater than or equal to λ and stop. Let $i + 1 = i_n^*$ be the stopping index at smaller of the rates.
   (c) Report rate solution $r_n(\lambda) = z_n^{i_n^*}$.
3. Number of bits allocated is $B(\lambda) = \sum_{n=0}^{N-1} r_n(\lambda)$. If $B(\lambda)$ not close enough to target B_T, select new λ and return to Step 2. Otherwise, stop.

---

at a time to the quantizer that produces the largest decrease in distortion. This was done previously in the case of fixed-level, uncoded quantizers for every transform coefficient. The same procedure can be used here, but with different initialization conditions. Let us outline the procedure first and justify it later.

Again, let us assume we have a set of (rate, distortion) points for every quantizer. We shall denote them by $\{(z_i^{(n)}, d_Q^{(n)}(z_i^{(n)}))\}$. We assume that successive rate points differ by 1 bit and that the initial rate point is $z_0^{(n)} = 0$ for all n. We form all the distortion reductions that occur by adding 1 bit,

$$\delta d_Q^{(n)}(z_i) = d_Q^{(n)}(z_i) - d_Q^{(n)}(z_{i+1}), i = 0, 1, \ldots; n = 0, 1, 2, \ldots, N - 1.$$

We start with a bit budget B and extract 1 bit and use it for the quantizer with the largest $\delta d_Q^{(n)}(0)$. Then we take the next bit from the budget and give it once again to the quantizer yielding the largest distortion reduction. So in this step, we have to account for the bit given previously and find the maximum among

$$\delta d_Q^{(0)}(0), \delta d_Q^{(1)}(0), \ldots, \delta d_Q^{(n_o)}(1), \ldots, \delta d_Q^{(N-1)}(0),$$

assuming that the previous 1 bit fell to quantizer n_o. So we continue in this way until we reach the bit budget B. Since we allocate every bit individually to the quantizer that reduces the distortion the most, we will have achieved a minimum distortion for the number of bits B. This kind of procedure, while certainly logical and valid, does

ALGORITHM 8.4 _____

Incremental rate allocation to arbitrary set of quantizers

1. *Initialization.*
   (a) Start with array of (rate, slope) points $\{z_i^{(n)}, \delta d_Q^{(n)}(z_i)\}$.
   (b) Set a bit budget B to allocate.
   (c) Set rate allocations $r_n = 0$ for all $n = 0, 1, 2, \ldots, N - 1$.
2. (a) Determine $n = n_o$ where $\delta d_Q^{(n)}(r_n)$ is maximum, i.e.,

   $$n_o = \arg\max_n \delta d_Q^{(n)}(r_n).$$

   (b) Increment r_{n_o} by 1 bit.
3. If allocated bits $\sum_{n=0}^{N-1} r_n < B$, the bit budget, return to Step 2. Otherwise, stop.

_____

not lead us directly to the important equal slope condition for optimal rate allocation in Equation (8.23). That is why it is mentioned only now.[2] The steps of this rate allocation algorithm are presented in Algorithm 8.4.

The search for the largest distortion reduction (slope magnitude) is facilitated by ordering and re-indexing the quantizers by decreasing $\delta d_Q^{(n)}(0)$, so that

$$\delta d_Q^{(0)}(0) > \delta d_Q^{(1)}(0) > \cdots > \delta d_Q^{(N-1)}(0).$$

We also note that for any quantizer, the distortion reductions (slope magnitudes) $\delta d_Q^{(n)}(z_i)$ decrease with rate z_i by virtue of the convex assumption. Hence, the algorithm above never looks for larger slope magnitudes at any higher rate than the current one. Also, by virtue of the additional ordering, the slope magnitudes of the different quantizers at the same rate are decreasing from $n = 0$ to $n = N - 1$, assuming the almost certain circumstance that the rate-distortion curves do not intersect. For example, the first bit always goes to $n = 0$. One can add such conditions to limit the search range in the above algorithm.

8.1.6   Transform coding gain

We shall now assess the benefits of transform coding over PCM, which is independent scalar quantization of the original data source. We assume that the source is stationary and that each component of the source vector $\mathbf{x} = (x_1, x_2, \ldots, x_N)$ is independently (scalar) quantized and encoded with rate R. We assume the MSE distortion-rate model,

$$d_Q(R) = \sigma_x^2 \rho(R), \quad \rho(R) = g(R)2^{-aR},$$

where σ_x^2 is the variance of the source components. This is the same model for quantization of transform coefficients previously encountered in Equations (8.5) and (8.10). The quantity $d_Q(R)$ is therefore the MSE per sample of PCM.

$$D_{\text{PCM}}(R) = \sigma_x^2 \rho(R).$$

Let us compare D_{PCM} to the MSE per sample $D = D_{\text{TQ}}$ of transform quantization with the same rate R. We assume the distortion-rate models of the transform coefficients as in Equations (8.5) and (8.10). We consider the special case when the rate R is large enough that all coefficients receive non-zero rate and $D_{\text{TQ}} = \theta$. Therefore,

$$\sigma_n^2 g_n > \theta \quad \text{for all} \quad n = 0, 1, 2, \ldots, N - 1.$$

Again, we use the shorthand notation $g_n \equiv g(r_n)$. The rate is then

$$R = \frac{1}{N} \sum_{n=0}^{N-1} r_n$$

$$= \frac{1}{N} \sum_{n=0}^{N-1} \frac{1}{a} \log \frac{\sigma_n^2 g_n}{\theta}$$

$$= \frac{1}{aN} \log \frac{\prod_{n=0}^{N-1} \sigma_n^2 g_n}{\theta^N}. \tag{8.25}$$

Solving for θ, we find the transform coding distortion

$$D_{\text{TQ}} = \theta = \left(\prod_{n=0}^{N-1} g_n \sigma_n^2 \right)^{1/N} 2^{-aR}.$$

The transform coding gain over PCM is defined to be the ratio of their MSEs at the same rate R,

$$G_{\text{TQ/PCM}} = \frac{D_{\text{PCM}}}{D_{\text{TQ}}}$$

and equals

$$G_{\text{TQ/PCM}} = \frac{g(R)}{\left[\prod_{n=0}^{N-1} g(r_n) \right]^{1/N}} \cdot \frac{\sigma_x^2}{\left[\prod_{n=0}^{N-1} \sigma_n^2 \right]^{1/N}}. \tag{8.26}$$

Because the transform is unitary,

$$\sigma_x^2 = \frac{1}{N} \sum_{n=0}^{N-1} \sigma_n^2.$$

Substituting into the gain yields

$$G_{\text{TQ/PCM}} = \frac{g(R)}{\left[\prod_{n=0}^{N-1} g(r_n) \right]^{1/N}} \cdot \frac{\frac{1}{N} \sum_{n=0}^{N-1} \sigma_n^2}{\left[\prod_{n=0}^{N-1} \sigma_n^2 \right]^{1/N}}. \tag{8.27}$$

Assuming that the function $g()$ is constant for all rates, we obtain the usual formula for transform coding gain

$$G_{TQ/PCM} = \frac{\frac{1}{N} \sum_{n=0}^{N-1} \sigma_n^2}{\left[\prod_{n=0}^{N-1} \sigma_n^2 \right]^{1/N}}. \tag{8.28}$$

According to the arithmetic-geometric mean inequality [1],

$$\left(\prod_{n=0}^{N-1} \sigma_n^2 \right)^{1/N} \leq \frac{1}{N} \sum_{n=0}^{N-1} \sigma_n^2 , \tag{8.29}$$

with equality if and only if the σ_n^2's are equal (the source process is white or memoryless). The coding gain $G_{TQ/PCM}$ exceeds 1, unless the process is white, in which case there is no advantage to transform coding.

The coding gain $G_{TQ/PCM}$ in Equation (8.28) depends only on the variances of the transform coefficients and, as such, is used often to evaluate the coding effectiveness of a particular transform. The transform with the largest coding gain is the KLT. The variances of the KLT coefficients are the eigenvalues of the covariance matrix of the source, i.e., $\sigma_n^2 = \lambda_n$. Using the KLT, we obtain the ideal coding gain,

$$G_{TQ/PCM}^{(o)} = \frac{\frac{1}{N} \sum_{n=0}^{N-1} \lambda_n}{\left[\prod_{n=0}^{N-1} \lambda_n \right]^{1/N}}. \tag{8.30}$$

If one does not ignore the variation of $g(\ )$ with rate and assumes $g(\ )$ is convex downward, then

$$g(R) \geq \left[\prod_{n=0}^{N-1} g(r_n) \right]^{1/N} ,$$

and $G_{TQ/PCM} \geq 1$ also in this case. Otherwise, the coding gain $G_{TQ/PCM}$ in (8.28) must be tempered by the additional factor in (8.27). The formulas are reasonable approximations to the true coding gain when the rate R is high enough that $\sigma_n^2 > \theta$ for all n.[3] This means that all the components of the transform receive non-zero rate, i.e., they are being quantized and coded and not set directly to zero. The smallest rate for which $\sigma_n^2 > \theta$ for all n is called the critical rate R_c. The θ which produces that rate, $\theta_c = \min_n \sigma_n^2$, is equal to the distortion and is called the critical distortion $\theta_c = D_c$. At smaller rates when one or more coefficients are reconstructed as zero, the actual coding gain is even greater than that predicted by the gain formulas above.

8.2 Subband rate allocation

The preceding development for linear transform coding carries over to the subband transform case with a few changes, due to generally different subband sizes, created by the downsampling, upsampling, and filter operations. We shall consider a general subband decomposition of the source with N samples into M subbands, denoted by vectors $\mathbf{y}_m$, each with n_m coefficients, $m = 0, 1, 2, \ldots, M - 1$. We can envision that each subband was produced by an equivalent analysis filter for its frequency range and downsampling by a factor of $V_m = N/n_m$. To reconstruct the source, each subband is upsampled by a factor of V_m and interpolated by a (conjugate) filter in its corresponding frequency range. Then the sum of these upsampled and interpolated subband processes reconstructs the source. A block diagram of this subband coding system is depicted in Figure 8.6. The frequency responses of the analysis filters are assumed to be ideal with amplitude equal to 1, as depicted in Figure 7.4. The impulse responses of the synthesis filters $g_m(n) = V_m h_m(n)$ guarantee perfect reconstruction in the absence of coding.

We adopt a naive model of the subband processes that is strictly true only when the source is Gaussian and the subband processes are white. That is, we assume that the coefficients within each subband are independent with the same probability distribution. Therefore, each coefficient in a given subband will be quantized and encoded independently and have the same distortion (MSE) versus rate function. Among different subbands, only the variances will differ, so that their distortion-rate functions will be the same unit variance characteristic scaled by the subband variances.

Just as with linear transforms, we wish to allocate the code rate among the subband coefficients, so that the distortion between the source and reconstruction is minimized.

The first step in coding is allocating a given number of bits among the subbands to achieve the minimum distortion. Distortion is again defined to be MSE. Let there be N samples from the source and a given code rate of R bits per sample. If b_m bits are given to the m-th subband, then

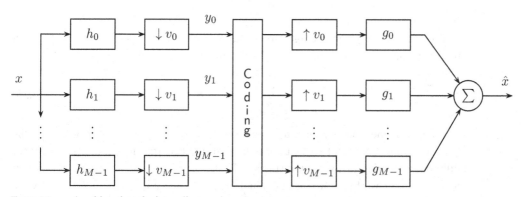

Figure 8.6 A subband analysis, coding, and synthesis system.

$$R \doteq \frac{1}{N} \sum_{m=0}^{M-1} b_m \qquad (8.31)$$

because the subbands are statistically independent and coded independently.

We assign the same rate r_m to code the coefficients of $\mathbf{y}_m$, so that $b_m = n_m r_m$. Therefore, the average code rate in bits per sample can be expressed as

$$R = \sum_{m=0}^{M-1} \eta_m r_m. \qquad (8.32)$$

where $\eta_m = 1/V_m = n_m/N$, the fraction of the number of samples (coefficients) in the subband $\mathbf{y}_m$. Each subband $\mathbf{y}_m$ has samples with the same variance σ_m^2, $m = 1, 2, \ldots, M$. The subband samples (or coefficients) are then quantized to obtain minimum MSE for the given rate r_m. We shall denote this distortion versus rate function by $d_{Q_m}(r_m)$. Often, one wants to include a perceptual weighting of the squared error that varies from subband to subband. In such a case, the MSE quantization becomes multiplied by the weight w_m in its subband. When no perceptual weighting is wanted, set $w_m = 1$ for all m. The perceptually weighted distortion for the sample in the m-th subband is expressed by

$$d_m(r_m) = w_m d_{Q_m}(r_m). \qquad (8.33)$$

Since all n_m samples in the subbands are independently encoded with the same rate r_m, the distortion of the subband is

$$n_m d_m(r_m) = n_m w_m d_{Q_m}(r_m).$$

After upsampling by V_m and filtering with energy gain V_m in each subband, and combining the subbands, the MSE per source sample is

$$D = \frac{1}{N} \sum_{m=0}^{M-1} V_m n_m d_m(r_m)$$

$$= \sum_{m=0}^{M-1} w_m d_{Q_m}(r_m). \qquad (8.34)$$

Therefore, for a given average target rate R, we seek the set of rates $\{r_m\}_{m=1}^{M}$ that minimize the distortion D in (8.34) subject to the average rate constraint

$$\sum_{m=0}^{M-1} \eta_m r_m = R. \qquad (8.35)$$

Note now the analogy of Equations (8.34) and (8.35) to (8.2) and (8.3), the corresponding equations for linear, orthonormal (unitary) transform coding. Except for different weighting factors in the summands, they are the same. These weighting factors could be absorbed into the individual distortion and rate terms to get exactly the same mathematical form. This observation will be useful later in extending the rate allocation algorithms to the subband case. But for now, we will proceed with the same development as before to illuminate the role of these factors.

We seek the set of rates $\{r_m\}$ that minimizes the Lagrangian objective function

$$J(\lambda) = \sum_{m=0}^{M-1} (w_m d_{Q_m}(r_m) + \lambda \eta_m r_m) \tag{8.36}$$

for Lagrange parameter $\lambda > 0$ and rates $r_m \geq 0$ for all $m = 0, 1, 2, \ldots, M - 1$. Differentiating with respect to each r_m, and setting each derivative to 0, we find that

$$V_m w_m d'_{Q_m}(r_m) = -\lambda. \tag{8.37}$$

The weighted slopes of the quantizer distortion-rate characteristics must be equal across the subbands.

Using the unit variance distortion versus rate function $\rho(r)$,

$$d_{Q_m}(r_m) = \sigma_m^2 \rho(r_m)$$

and

$$D = \sum_{m=0}^{M-1} w_m \sigma_m^2 \rho(r_m).$$

We again adopt the following model for $\rho(r)$:

$$\rho(r) = g(r) 2^{-ar},$$

whereby $g(r)$ is a slowly varying function of r and $g(r) \geq g(0) = 1$. In each subband we shall assume that $g(r_m) \approx g_m$ independent of rate. Now substituting this model into (8.37), the solution must obey

$$V_m w_m d_{Q_m}(r_m) = \frac{\lambda}{a \ln 2} \equiv \theta, \tag{8.38}$$

where the quantizer characteristic is given by

$$d_{Q_m}(r_m) = g_m \sigma_m^2 2^{-ar_m}. \tag{8.39}$$

The exponential distortion-rate characteristic gives rise to the condition that the weighted quantizer distortions must be equal across the subbands.

Solving (8.38) for r_m, we obtain

$$r_m = \frac{1}{a} \log \frac{V_m w_m g_m \sigma_m^2}{\theta}, \tag{8.40}$$

for those $r_m > 0$, $m = 0, 1, \ldots, M - 1$.

The parameter θ is set through the rate constraint in Equation (8.32). Let us assume that all $r_m > 0$. Then, from (8.32)

$$R = \frac{1}{a} \sum_m \eta_m \log \frac{V_m w_m g_m \sigma_m^2}{\theta}. \tag{8.41}$$

Heretofore, we shorten the notation to $\sum_m$ or $\prod_m$ to indicate the range of $m = 0, 1, \ldots, M - 1$ in these operators. Solving for θ and recalling that $\sum_m \eta_m = 1$, we find that

$$\theta = \prod_m (V_m w_m g_m)^{\eta_m} 2^{-aR} = \sigma_{WGM}^2 2^{-aR}, \qquad (8.42)$$

where we have defined

$$\sigma_{WGM}^2 \equiv \prod_m (V_m w_m g_m)^{\eta_m}.$$

The subscript indicates that this quantity is a weighted geometric mean of positive numbers.

We return to Equations (8.39) and (8.34) to calculate the distortion per source sample of the subbands and their total. Substituting the solution for θ in (8.42), we see that

$$d_{Q_m}(r_m) = \eta_m \theta \qquad (8.43)$$

$$D = \sum_m \eta_m \theta = \theta = \sigma_{WGM}^2 2^{-aR}. \qquad (8.44)$$

The parameter θ is set through the rate constraint in (8.32). For θ sufficiently large, the solution in Equation (8.40) will yield non-positive r_m. The obtained solution is only valid then for m such that

$$V_m w_m g_m \sigma_m^2 > \theta.$$

Since $g_m \geq 1$, this condition is still met if we set $g_m = 1$. Let us indicate the set of subbands coded at positive rate as

$$J_c = \{m : V_m w_m \sigma_m^2 > \theta\}.$$

The total rate R per sample is allocated only over these subbands, so that

$$R = \sum_{m \in J_c} \eta_m r_m.$$

The general solution for the parameter θ still looks the same,

$$\theta = \sigma_{WGM}^2 \cdot 2^{-aR},$$

except that the definition of σ_{WGM}^2 is now generalized to

$$\sigma_{WGM}^2 = \prod_{m \in J_c} (V_m w_m g_m)^{\eta_m}.$$

When $r_m = 0$ for the m-th subband, every sample therein is set to 0 and has MSE of $g_m \sigma_m^2$. When $r_m < 0$ in this solution, the best we can do is to accept $r_m = 0$ as the solution. For $r_m = 0$, the weighted quantization MSE $w_m g_m \sigma_m^2$ is seen in the reconstruction. For rates $r_m > 0$, $\eta_m \theta = \eta_m w_m d_{Q_m}(r_m)$ is the distortion per sample from subband m seen at the output.

Summarizing, the samples in the subbands $\mathbf{y}_m$, $m = 0, 1, 2, \ldots, M - 1$, should receive the per-sample rates

$$r_m = \begin{cases} \frac{1}{a} \log_2 \frac{V_m w_m g_m \sigma_m^2}{\theta}, & \text{if } V_m w_m \sigma_m^2 > \theta \\ 0, & \text{if } V_m w_m \sigma_m^2 \leq \theta \end{cases}$$
$$m = 0, 1, \ldots, M - 1. \tag{8.45}$$

The resulting perceptually weighted MSE of each quantized component from (8.43) is

$$w_m d_{Q_m} = \begin{cases} \eta_m \theta, & \text{if } V_m w_m \sigma_m^2 > \theta \\ w_m g_m \sigma_m^2, & \text{if } V_m w_m \sigma_m^2 \leq \theta \end{cases}$$
$$m = 0, 1, \ldots, M - 1. \tag{8.46}$$

Normally, the above solutions are seen with $g_m = 1$ for all m and $a = 2$. Furthermore, let there be no perceptual weighting, so that $w_m = 1$ for all m. We now restate the rate and reconstruction distortion above for these special, but important circumstances. With the substitution of the optimal rate allocation to the subbands in Equation (8.45) into (8.32) and (8.34), the target rate and resulting distortion for the reconstructed source become

$$R = \sum_{m:V_m \sigma_m^2 > \theta} \frac{\eta_m}{2} \log \frac{V_m \sigma_m^2}{\theta} \tag{8.47}$$

$$D = \sum_{m:V_m \sigma_m^2 > \theta} \eta_m \theta + \sum_{m:V_m \sigma_m^2 \leq \theta} \sigma_m^2. \tag{8.48}$$

Notice that these equations for overall rate and distortion from a subband transform take the same form as the corresponding equations for linear block transforms. Just substitute number of coefficients N with number of subbands M, θ with θ / V_m, and $1/N$ with η_m inside the sum on m. Please note that the first summation term in D amounts simply to θ.

8.2.1 Practical issues

When we turn off perceptual weighting, the MSE d_{Q_m} in Equation (8.46) is the same value θ in the subbands coded with non-zero rate. For uniform step size quantization at high enough rate or small enough step size Δ, the MSE is

$$d_Q \approx \Delta^2 / 12.$$

Therefore, the step size for every coded subband is

$$\Delta = \sqrt{12\theta}.$$

When we include perceptual weighting for each coded subband, $w_m d_{Q_m}$ must equal the common θ, so the selected step size is $\Delta_m = \sqrt{12\theta / w_m}$ for quantization of the samples in these subbands. Perceptual weighting of the distortion may be realized by multiplying each subband sample in the m-th subband by $\sqrt{w_m}$ and then quantizing the scaled sample with the common step size $\Delta = \sqrt{12\theta}$. In the synthesis filtering stage for the m-th subband, the samples must be rescaled by $1/\sqrt{w_m}$. The scaling of the samples is often done simply by the same scaling of the filter tap values.

As pointed out in the previous chapter, the subband variances are those that are obtained from ideal filtering of the source signal. Therefore, they must be calculated from the source power spectrum in the frequency range of the associated subband. The subband power spectrum in the frequency range $[\omega_m, \omega_{m+1}]$ after downsampling and ideal filtering may be expressed as

$$S_m(\omega) = \frac{1}{V_m} S_x(\omega/V_m + \omega_m), \quad 0 \le \omega \le \pi. \tag{8.49}$$

The subband variance (for zero mean) is the integral of this spectrum according to

$$\sigma_m^2 = \frac{1}{\pi} \int_0^\pi S_m(\omega)d\omega = \frac{1}{\pi} \int_{\omega_m}^{\omega_{m+1}} S_x(\omega)d\omega. \tag{8.50}$$

The energy of the analysis filter $h_m(n)$ is $W_m/\pi = 1/V_m$ and its companion synthesis filter $g_m(n)$ has energy V_m. A pair of ideal filters with the same amplitude gain $\sqrt{V_m}$ will also produce perfect reconstruction and simultaneously be orthonormal. The subband variances after filtering and downsampling will then be $\sigma_m'^2 = V_m\sigma_m^2$. So, when we use an orthonormal filter pair, $\sigma_m'^2$, the filtered subband variance, replaces $V_m\sigma_m^2$ in the formulas developed above for the rate and distortion allocation to subbands.

In practice, ideal filters are not realizable, since they imply infinite length of the impulse response among other impractical attributes. The realizable filters in common use are orthonormal and finite in length (FIR). We can approximate the results of realizable filters by those for ideal orthonormal filters. We therefore replace the weighted variances $V_m\sigma_m^2$ in the formulas for the case of unity amplitude analysis filters by the actual filtered variance, which is the input variance to the subband quantizer. Mathematically, the replacement is

$$V_m\sigma_m^2 \leftarrow \sigma_m'^2 = \frac{1}{\pi} \int_0^\pi |H_m(\omega)|^2 S_x(\omega)d\omega, \quad m = 0, 1, \ldots, M-1. \tag{8.51}$$

For ideal orthonormal filters, the variances at the input to the quantizers $\sigma_m'^2 = V_m\sigma_m^2$. For example, in terms of these variances, the rate and distortion allocations in (8.45) and (8.46) become

$$r_m = \begin{cases} \frac{1}{a}\log_2 \frac{w_m g_m \sigma_m'^2}{\theta}, & \text{if } w_m\sigma_m'^2 > \theta \\ 0, & \text{if } w_m\sigma_m'^2 \le \theta \end{cases}$$
$$m = 0, 1, \ldots, M-1 \tag{8.52}$$

$$w_m d_{Q_m} = \begin{cases} \eta_m\theta, & \text{if } w_m\sigma_m'^2 > \theta \\ w_m g_m \eta_m \sigma_m'^2, & \text{if } w_m\sigma_m'^2 \le \theta \end{cases}$$
$$m = 0, 1, \ldots, M-1. \tag{8.53}$$

8.2.2 Subband coding gain

We wish now to calculate the gain for coding in subbands as opposed to direct PCM coding of the source. Consider now the special case of rate R sufficiently large that $V_m w_m \sigma_m^2 > \theta$ for all m. Using the constraint on subband rates that the $\sum_m \eta_m r_m = R$, we found in (8.42) that

$$
\theta = \sigma_{\text{WGM}}^2 2^{-aR}, \quad \sigma_{\text{WGM}}^2 \equiv \prod_{m=0}^{M-1} \left(V_m w_m g_m \sigma_m^2 \right)^{\eta_m}.
$$

Substituting θ into Equation (8.45), the rate per sample r_m for the m-th subband becomes

$$
r_m = R + \frac{1}{a} \log_2 \frac{V_m w_m g_m \sigma_m^2}{\sigma_{\text{WGM}}^2}, \quad m = 0, 1, 2, \ldots, M-1, \tag{8.54}
$$

and the distortion after reconstruction becomes simply

$$
D_{\text{SB}} = \theta = \sigma_{\text{WGM}}^2 2^{-aR}. \tag{8.55}
$$

When the samples of the (stationary) source with variance σ_x^2 are directly and independently coded using the same technique and target rate R, the average distortion is

$$
D_{\text{PCM}} = g(R) \sigma_x^2 2^{-aR}. \tag{8.56}
$$

The subband coding gain for PCM is defined to be

$$
G_{\text{SB/PCM}} \equiv \frac{D_{\text{PCM}}}{D_{\text{SB}}}, \tag{8.57}
$$

the ratio of the MSEs of direct coding of the source and coding of the subbands. This gain, by substitution of Equations (8.56) and (8.55), is expressed as

$$
G_{\text{SB/PCM}} = \frac{g(R)\sigma_x^2}{\sigma_{\text{WGM}}^2} = \frac{\sigma_x^2}{\prod_{m=0}^{M-1} \left(V_m w_m \sigma_m^2 \right)^{\eta_m}} \times \frac{g(R)}{\prod_{m=0}^{M-1} (g_m)^{\eta_m}}. \tag{8.58}
$$

If the function $g(\ )$ is assumed to be constant, which is the case for ideal scalar coding and a reasonable approximation for practical scalar coding, and recalling that $g_m = g(r_m)$, the second fraction equals one and the coding gain is

$$
G_{\text{SB/PCM}} = \frac{\sigma_x^2}{\prod_{m=0}^{M-1} \left(V_m w_m \sigma_m^2 \right)^{\eta_m}} \tag{8.59}
$$

Although this formula gives an actual coding gain for high rates, it cannot be proved for general sets of weights $\{w_m\}$ that the coding gain $G_{\text{SB/PCM}}$ is truly a gain and exceeds unity. However, for no perceptual weighting, when $w_m = 1$ for all m, the subband-PCM gain can be written as

$$G_{\text{SB/PCM}} = \frac{\sigma_x^2}{\prod_{m=0}^{M-1} (V_m \sigma_m^2)^{\eta_m}} = \frac{\sigma_x^2}{\prod_{m=0}^{M-1} (V_m \sigma_m^2)^{1/V_m}}, \tag{8.60}$$

where we note that

$$\sigma_x^2 = \sum_{m=0}^{M-1} \sigma_m^2,$$

Since $\sum_m \eta_m = 1$, the convexity of the logarithm allows the use of Jensen's inequality

$$\sum_{m=0}^{M-1} \eta_m \log(V_m \sigma_m^2) \leq \log\left(\sum_{m=0}^{M-1} \eta_m \cdot V_m \sigma_m^2 \right) = \log \sigma_x^2, \tag{8.61}$$

with equality if and only if $V_m \sigma_m^2$ is the same for all subbands. Therefore,

$$G_{\text{SB/PCM}} \geq 1 \tag{8.62}$$

with equality if and only if $V_m \sigma_m^2$ is a constant independent of m. This equality condition is satisfied if the input process is white, just as with any linear transform. Therefore, independent scalar coding of subbands is advantageous over the same kind of scalar coding of the source signal. The gain formula in (8.60) is a useful tool to assess the relative coding efficiencies of different subband decompositions of a source signal.

8.2.2.1 Coding gain with practical filters

One may want to evaluate the coding efficiency of subband decomposition using different filters. The coding gain in Equation (8.60) will not allow this evaluation, because it uses the approximation of ideal filters. As mentioned before, the rate allocation formulas can be adjusted for the case of realizable orthonormal filters by substitution of $V_m \sigma_m^2$ with the filtered variances

$$\sigma_m'^2 = \frac{1}{\pi} \int_0^\pi | H_m(\omega) |^2 S_x(\omega) d\omega, \quad m = 0, 1, \ldots, M - 1. \tag{8.63}$$

Therefore, with this replacement, the subband coding gain expression becomes

$$G_{\text{SB/PCM}} = \frac{\sigma_x^2}{\prod_{m=0}^{M-1} (\sigma_m'^2)^{\eta_m}}, \tag{8.64}$$

with $\sigma_m'^2$ in (8.63).

8.3   Algorithms for rate allocation to subbands

As we have shown, the formulas that allocate rate optimally to subbands and transform coefficients take the same form and can be equated using the proper assocation of quantities. These formulas rely on a mathematical model for the distortion versus rate function. The underlying principle of the rate allocation is contained in Equation (8.37), which states that the weighted slopes of the distortion versus rate function of the quantization must be equal across the subbands. The weighting by the downsampling factor of each subband distinguishes it from the corresponding block transform condition in (8.23). Assume without loss of generality that $w_m = 1$ for no perceptual weighting.[4] Then all the rate allocation algorithms developed for block transforms can be applied directly to the subband case when we weight the quantizer distortion in each subband by its downsampling factor V_m.

Because quantization is done in the subbands, the formulas and algorithms are sensitive to the amplitude scaling of the analysis filters. We have assumed them to have constant amplitude of 1 in their frequency ranges. Expressed mathematically,

$$H_m(\omega_m) = \sum_n h_m(n) = 1, \quad m = 0, 1, 2, \ldots, M - 1.$$

The recommended practice is to make the analysis–synthesis pairs orthonormal, so that

$$\frac{1}{\pi} \int_0^\pi \mid H_m(\omega_m) \mid^2 d\omega = \sum_n h_m^2(n) = 1, \quad m = 0, 1, 2, \ldots, M - 1.$$

Assuming that the m-th filter is ideally bandlimited to $W_m = \omega_{m+1} - \omega_m$, multiplying its amplitude by $\sqrt{V_m}$ makes its energy equal to 1. When this is done for all M filters, the factors of V_m are now absorbed into the distortion function of every quantizer, so that the weighting by V_m in the equal slope condition in Equation (8.37) disappears. For the same reason, $V_m \sigma_m^2$ in the formulas for rate allocation gets replaced by $\sigma_m'^2$, where we interpret $\sigma_m'^2$ to be the subband's variance after filtering and downsampling.

One additional step to make the algorithms for block and subband transforms correspond is to replace rate per transform coefficient r_n with rate per subband $b_m = n_m r_m$. In Chapter 11.6, "Subband/wavelet coding systems", we present a rate allocation algorithm to tree blocks that can be applied without change to subbands.

In practice, realizable perfect reconstruction half-band filters are the prototypes. As mentioned above, the high and lowpass prototype filters are normally scaled in amplitude to be orthonormal. This scaling is easy to do if the numbers are rational, but the required irrational $\sqrt{2}$ scaling cannot be done within finite precision. However, when lossless compression capability is required, precision loss is not permissible. We must use integer arithmetic and filters that output integer values. Therefore, arbitrary scaling is generally not possible. We shall discuss ways in Chapter 12.4 to circumvent arbitrary scaling, so that lossless compression can be realized.

8.4   Conclusions

This chapter has presented several formulas and numerical procedures used for allocating rate to transform and subband coefficients in signal compression systems. The formulas required stationary statistics, statistical independence of transform or subband coefficients, and distortion measures based on MSE. The numerical procedures relax some, but not all of these assumptions. Furthermore, the developed formulas are really only approximations, because they neglect the aliasing of neighboring subbands caused by quantization error. Nevertheless, the formulas and numerical procedures presented herein do give reasonably good results in most circumstances. Furthermore, they lay the foundation for understanding the rate allocation problem and the mathematical and numerical tools needed for its solution under more general statistical models and distortion measures.

Problems

8.1 An approximate analysis [2] reveals that the MSE versus rate characteristic for optimum uniform step size quantization of a unit variance Gaussian scalar source is

$$\rho(r) = \begin{cases} 2^{-1.57r}, & r \le 1.186 \\ 1.42 \cdot 2^{-2r}, & r > 1.186 \end{cases}$$

Using the Gauss–Markov source and optimal bit assignment of Problem 7.3 and this $\rho(r)$, calculate the average distortion D for this transform coding scheme for the rates $R = 2$ and $R = 1/2$.

8.2 Consider a stationary random sequence $\{x(n)\}$ with autocorrelation function

$$R_x(k) = \sigma_x^2 a^{|k|}, \ k = 0, \pm 1, \pm 2, \ldots$$

We wish to transform blocks of $N = 8$ samples and quantize the coefficients with minimum MMSE for the given rate R. The selected transform is the DFT.

(a) Letting $a = 0.96$, calculate the variances of the DFT coefficients in the block. (MATLAB is recommended.)

(b) Assuming optimal quantizers for MMSE, determine the optimal allocation of rates per coefficient to yield the following average rate per sample R: 3, 2, 1, 0.5, 0.25.

(c) Now suppose you do not have the optimal rate-distortion characteristic $\rho(r) = 2^{-2r}$, but instead just a series of (rate, distortion) points found from quantization of a unit variance coefficient. Just to test our formulas, let us calculate these points from the optimal $\rho(r)$ by evaluating this function at 80 uniformly spaced points in the range of $r = 0$ to $r = 8$.

Now determine the optimum rate allocations requested in the previous part using Algorithm 8.3 for optimal rate allocation starting with a set of discrete (rate, distortion) points.

8.3 Repeat the previous problem using the non-optimal $\rho(r)$ in Problem 8.1.

8.4 Consider a subband coding system (SBC) for encoding a speech source modeled by the discrete-time stationary third-order autoregressive random process given by

$$x(n) = a_1 x(n-1) + a_2 x(n-2) + a_3 x(n-1) + v(n), \quad n = 0, \pm 1, \pm 2 \ldots$$

with $a_1 = 1.231, a_2 = -0.625, a_3 = 0.119$. The members of the sequence $\{v(n)\}$ are i.i.d. random variables with zero mean and variance σ_v^2. Consider that the ideal analysis filter bank splits the source spectrum, which is bandlimited from 0 to 8000 Hz, into four subbands in the frequency ranges (in Hz):

$$
\begin{array}{lllll}
i. & [0-1000] & [1000-2000] & [2000-3000] & [3000-4000] \\
ii. & [0-500] & [500-1000] & [1000-2000] & [2000-4000]
\end{array}
$$

The sampling rate $f_s = 8000$ Hz (*critical sampling*). (To associate to discrete-time frequency: $\pi = f_s/2$.) Assume that the subbands are optimally quantized for a Gaussian source and MSE.

(a) For each splitting arrangement, determine the optimal rate assignments to the subbands for the rates $R = 2.0$ and $R = 0.5$ bits per sample. Calculate the coding gain in each case.

(b) For each splitting arrangement, calculate $G_{SB/PCM}$.

Notes

1. Parentheses enclosing the word "lossy" follow JPEG and JPEG2000, because they also have lossless modes. The generic term of "transform" includes both the traditional decorrelating transforms such as DCT and the subband transforms such as wavelet.
2. One can see intuitively that the operating points travel down the distortion-rate curves trying to catch up in slope to the slope leader as quickly as possible with the addition of each bit. Hence, the points tend to converge to slope equality.
3. Since $g(r_n) \geq 1$, this rate assures $g(r_n)\sigma_n^2 > \theta$.
4. The perceptual weighting factor can be absorbed into the distortion slope by $\sqrt{w_m}$ scaling.

References

1. R. Bellman, *Introduction to Matrix Analysis*, 2nd edn. New York, NY: McGraw-Hill Publishing Co., 1970.
2. R. C. Wood, "On optimum quantization," *IEEE Trans. Inf. Theory*, vol. IT-15, no. 2, pp. 248–252, Mar. 1969.

Further reading

L. Breiman, J. H. Friedman, R. A. Olshen, and C. J. Stone, *Classification and Regression Trees*. Monterey, CA: Wadsworth, 1984.

B. Mahesh and W. A. Pearlman, "Multiple-rate structured vector quantization of image pyramids," *J. Visual Commun. Image Represent*, vol. 2, no. 2, pp. 103–113, Jan. 1991.

W. A. Pearlman, "Performance bounds for subband coding," in *Subband Image Coding*, ed. J. W. Woods. Norwell, MA: Kluwer Academic Publishers, 1991, ch. 1.

W. A. Pearlman, "Information theory and image coding," in *Handbook of Visual Communications*, eds. H.-M. Hang and J. W. Woods. San Diego, CA: Academic Press, 1995, ch. 2.

R. P. Rao and W. A. Pearlman, "On entropy of pyramid structures," *IEEE Trans. Inf. Theory*, vol. 37, no. 2, pp. 407–413, Mar. 1991.

S. Rao and W. A. Pearlman, "Analysis of linear prediction, coding, and spectral estimation from subbands," *IEEE Trans. Inf. Theory*, vol. 42, no. 4, pp. 1160–1178, July 1996.

E. A. Riskin, "Optimal bit allocation via the generalized BFOS algorithm," *IEEE Trans. Inf. Theory*, vol. 37, no. 2, pp. 400–402, Mar. 1991.

Y. Shoham and A. Gersho, "Efficient bit allocation for an arbitrary set of quantizers," *IEEE Trans. Acoust. Speech Signal Process.*, vol. 36, no. 9, pp. 1445–1453, Sept. 1988.

A. K. Soman and P. P. Vaidyanathan, "Coding gain in paraunitary analysis/synthesis systems," *IEEE Trans. Signal Process.*, vol. 5, no. 41, pp. 1824–1835, May 1993.

D. S. Taubman and M. W. Marcellin, *JPEG2000: Image Compression Fundamentals, Standards, and Practice*. Norwell, MA: Kluwer Academic Publishers, 2002.

J. W. Woods and T. Naveen, "A filter based bit allocation scheme for subband compression of HDTV," *IEEE Trans. Image Process.*, vol. 1, no. 3, pp. 436–440, July 1992.

J. W. Woods and S. D. O'Neil, "Subband coding of images," *IEEE Trans. Acoust. Speech Signal Process.*, vol. 34, no. 5, pp. 1278–1288, Oct. 1986.

9 Transform coding systems

9.1 Introduction

In previous chapters, we described mathematical transformations that produce nearly uncorrelated elements and pack most of the source energy into a small number of these elements. Distributing the code bits properly among these transform elements, which differ statistically, leads to coding gains. Several methods for optimal rate distribution were explained in Chapter 8. These methods relied on knowledge of the distortion versus rate characteristics of the quantizers of the transform elements. Using a common shape model for this characteristic and the squared error distortion criterion meant that only the variance distribution of the transform elements needed to be known. This variance distribution determines the number of bits to represent each transform element at the encoder, enables parsing of the codestream at the decoder and association of decoded quantizer levels to reconstruction values. The decoder receives the variance distribution as overhead information. Many different methods have arisen to minimize this overhead information and to encode the elements with their designated number of bits. In this chapter, we shall describe some of these methods.

9.2 Application of source transformations

A transform coding method is characterized by a mathematical transformation or transform of the samples from the source prior to encoding. We described the most common of these transforms in Chapter 7. The stream of source samples is first divided into subblocks that are normally transformed and encoded independently. The reason for transforming subblocks of the image is that an image is non-stationary and the coding can be adapted to the changing statistics on a blockwise basis. For one-dimensional sources like audio or biological signals, the size of the blocks is typically 1024 or 2048. For two-dimensional sources such as images, the blocks are squares with dimensions typically 8×8 or 16×16. Normally, a separable two-dimensional transform operates on each block independently. The input samples, customarily called *pixels*, typically are recorded as 8–16 bit non-negative integers. The transform coefficients have a much larger range of floating point values, which may be negative or positive. Figure 9.1 illustrates an image divided into square blocks and its block transforms placed in the positions of their associated image blocks. The transform used in this figure is the DCT,

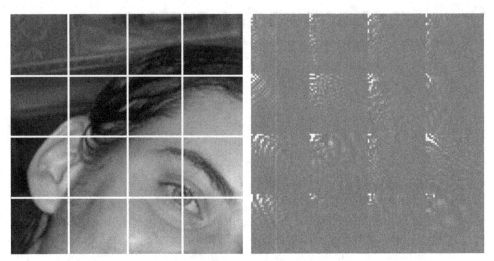

Figure 9.1   A 128 × 128 image divided into 32 × 32 blocks (left) and a rendition of its blockwise DCT (right). For display purposes of the latter, the range of the transform was scaled and translated to the [0, 255] interval with a middle gray value of 128.

the mathematical formula for which is given in Equation (7.71). We shall describe adaptive methods for coding these block transforms of images. These methods can also be applied to one-dimensional block transforms with obvious adjustments.

9.2.1   Model-based image transform coding

9.2.1.1   Spectral modeling

Dividing the image into small subblocks enables adaptation of the encoding to local characteristics. One method of adaptation involves generation of a different statistical model for each image block. The earliest methods for coding transform blocks assumed a model for the autocorrelation function or power spectral density of the image block. The most common model was the separable, two-dimensional, zero mean, Markov-1 autocorrelation function of

$$R_{xy}(m, n) = \sigma_{xy}^2 \rho_y^{|m|} \rho_x^{|n|}, \quad m, n = 0, \pm 1, \pm 2, \ldots, \quad (9.1)$$

where the block dimensions are $M \times M$, σ_{xy}^2 is the block variance, and ρ_x and ρ_y are the horizontal and vertical correlation parameters. The variances of the transform coefficients can be calculated from this autocorrelation function. Therefore, only four parameters, the block variance, the two correlation parameters, and the block mean need to be sent as overhead for each block for the purpose of decoding. These parameters can be estimated by various means prior to transformation and coding. This method was not really satisfactory, due mainly to the restriction of the model, but also due to the relatively large overhead for small blocks. Using 8 bits to send every parameter results in overhead of 0.5 bits per sample for an 8 × 8 block or 0.125 bits per sample for a 16 × 16 block. Some economies can be achieved by differential coding of parameters between adjacent blocks. Use of small blocks is desirable, because it enhances the adaptability of the coding to local signal characteristics.

One way to overcome the limitations of the Markov-1 autocorrelation model is to increase the model order. Of course that results in higher overhead, but also should reduce the number of bits for the same distortion. However, estimation of higher-order two-dimensional models, even separable ones, and the calculation of the transform variances are computationally very intensive and probably inaccurate due to insufficient data for the desirable small block sizes. A way to circumvent this complexity is to linearly order the coefficients through a zigzag scan of the block and estimate a one-dimensional higher-order model. Such a scheme was devised by Pearlman [1] to achieve adaptive cosine transform image coding with constant block distortion.

A zigzag scan of a block transform is depicted in Figure 9.2. This type of scan is motivated by the observation that the energies of transform coefficients tend to decrease from low frequency at the upper left toward high frequency at the lower right. The smoother the decrease in energy, the lower the model order needed to accurately represent an autocorrelation model corresponding to the variances of these coefficients ordered according to this scan. In [1], the linearly ordered coefficients were transformed by the inverse DCT to give an artificial one-dimensional sequence. The forward DCT of this sequence restores the linear scan transform sequence whose variances we wish to estimate. To do so, Burg's algorithm for maximum entropy spectral estimation [2] is used to estimate the parameters of an autoregressive (AR) model that generates the artificial sequence. From this model, the autocorrelation function is derived from which the variances of the linearly ordered transform coefficients are calculated. Such a distribution of transform variances is often called a *variance spectrum*. Because the DCT was chosen for the one-dimensional transform, the inverse DCT, forward DCT, and the variance calculations can be executed with fast computational algorithms. The AR

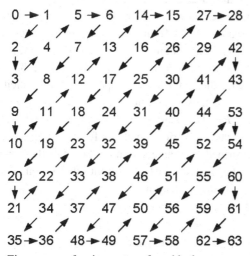

Figure 9.2   Zigzag scan of an image transform block.

model parameters and the mean of every subblock must be written as overhead into the compressed bitstream. For p AR parameters for each $M \times M$ subblock, this overhead rate amounts to $(p + 1)b/M^2$ bits per sample, where b is the number of bits to code each parameter. If the parameters are left uncoded, the overhead would represent a significant portion of the codestream for small blocks and small total code rates.

As in almost all modern transform coding systems, the choice of block transform in [1] was the separable, two-dimensional DCT. The estimation of a different variance spectrum for each block allows the achievement of the same distortion for every block (with some exceptions) by assigning a different rate to each one. The optimal rate R_T and distortion (MSE) D for a block are given in Equations (8.17) and (8.18) for one-dimensional indexing of a block of size $N = M^2$. Rewriting these equations for two dimensions and setting the g and a parameters to their standard values of 1 and 2, respectively, the optimal rate and distortion are:

$$R_T = \frac{1}{M^2} \sum_{m,n:\sigma_{m,n}^2 > \theta} \frac{1}{2} \log \frac{\sigma_{m,n}^2}{\theta} \tag{9.2}$$

$$D = \frac{1}{M^2} \left[\sum_{m,n:\sigma_{m,n}^2 > \theta} \theta + \sum_{m,n:\sigma_{m,n}^2 \leq \theta} \sigma_{m,n}^2 \right] \tag{9.3}$$

$$m, n = 0, 1, 2, \ldots, M - 1,$$

where $\sigma_{m,n}^2$ is the variance of the transform coefficient at coordinates (m, n) within the block. The estimated variances replace the actual variances $\sigma_{m,n}^2$ in these equations. Each summand term specifies the optimally assigned rate or distortion for the transform coefficient located at coordinates (m, n). The aim is to attain the same D for every block. However, a solution may not exist for some blocks, because too few of their variances would be higher than the parameter θ required for the target D. These blocks would be assigned such a small rate that its boundaries would show prominently when situated adjacent to a high variance block. To ameliorate this effect, the rate of every block was not allowed to fall below a certain minimum rate R_{min}, unless, of course, the block was naturally empty of content. In simulations, $R_{min} = 0.25$ bits per sample proved to be a good choice. Once the distortion D is specified, the implicit parameter θ can be determined from (9.3) and substituted into (9.2) to determine the rates to each coefficient and the overall rate R_T. The overall rate per sample includes R_T plus the bits needed to represent the variances. Various methods for encoding the coefficients with the assigned rates are described in Chapters 4 and 5. Some of these methods will be revisited in a later section, but first we wish to present other schemes for spectral modeling and reduction of overhead information.

9.2.1.2   Classified spectral modeling

Representing the variance spectrum of every block with a different AR model may lead to an unacceptable amount of overhead. This overhead can be alleviated by sorting the subblocks into a small number of classes and determining a spectral model to represent each class. If the number of subblocks in each class of an $N \times N$ image is the same and there are K classes, the overhead rate for the AR spectral modeling described above is reduced by a factor of $(N/M)^2/K$, which is the ratio of the number of subblocks to the number of classes. There are many methods in the literature for classification that could be used for this purpose. Chen and Smith [3] devised a simple and effective scheme for classification that relied on clustering subblocks according to their range of AC energy. The AC energy of a subblock is the sum of the squares of the transform coefficients excluding the DC (0,0) term. They calculated the AC energy of every subblock, ordered these energies from largest to smallest and marked the successive points where 0, 1/4, 1/2, 3/4, and all the energies are contained. Class 1 was designated as all subblocks whose AC energies fall between 0 and 1/4; class 2 was designated as the subblocks with energies between 1/4 and 1/2; class 3 was designated as the subblocks with energies between 1/2 and 3/4; and class 4 was designated as the subblocks with the remaining smallest energies. This procedure yields four classes of equal size belonging to different energy ranges. Therefore, each subblock must be labeled with its class number and this label must also be encoded along with its associated subblock. The average of the squares of the same coordinate AC coefficients in a given class form the estimates of the variances needed for the bit assignments. The DC (0,0) coefficient of every subblock is encoded with 8 or 9 bits to ensure absence of block boundary discontinuities.

Zonal sampling

A method akin to the variance classification above is called *zonal sampling*. In this method, classes or zones are established within a transform block according to similarity of their expected variances. The estimated variance of a zone is the average of the squared values of its transform coefficients, again excluding the DC coefficient. Every coefficient in a particular zone receives the same number of bits as dictated by its estimated variance. For every subblock, a number of variances equal to the number of zones, a zone map, and a mean value are sent as overhead. The encoder has the flexibility to generate new variance estimates for every block or accept the variances of a contiguous block and send an indicator of this acceptance to the decoder.

9.2.2 Encoding transform coefficients

The choice of the coding method for the transform coefficients depends on the transmission mode and the affordability of computation. Earlier transform coding systems employed fixed-length coding of 16×16 subblocks. The modern systems, capable of very fast and cheap computation and transmitting in digital packets, use highly adaptive forms of variable-length entropy coding in smaller subblocks.

For either method, the starting point is to obtain the optimal bit assignment $r_{m,n}$ to the coefficient to be coded. Using the variance estimate $\hat{\sigma}_{m,n}^2$ obtained by one of the methods above, the optimal number of code bits is

$$r_{m,n} = \begin{cases} \frac{1}{2} \log \frac{\hat{\sigma}^2_{m,n}}{\theta}, & \hat{\sigma}^2_{m,n} > \theta \\ 0, & \hat{\sigma}^2_{m,n} \le \theta, \end{cases} \qquad (9.4)$$

where the parameter θ is set from the block's rate or distortion target by Equation (9.3). Direct, iterative methods of solution are delineated in Chapter 8 for the cases of integer and non-integer rates $r_{m,n}$ for all (m, n) in the block.

Most transform blocks contain a substantial percentage of coefficients which are either low or zero in value. It often saves rate to establish a threshold and code only the coefficients whose magnitudes exceed the threshold and set the remainder to zero. The magnitude minus the threshold is coded and an additional bit conveys the sign. The threshold and a binary map of the coded (1) and uncoded (0) coefficients are sent as overhead. The binary map is efficiently run-length coded using a zigzag scan through the block. No variances need to be conveyed to the decoder for the coefficients indicated by zeros in the map. The rate is allocated to the threshold-adjusted coefficients. The threshold can be adjusted on a blockwise basis to adapt to local activity and the target rate. This type of coding is called *threshold coding*.

9.2.2.1   Fixed-length coding

In fixed-length coding, a transform coefficient is quantized to achieve MMSE. Therefore, the value of the coefficient to be coded at coordinates (m, n) is normalized through scaling by $\hat{\sigma}_{m,n}$ and quantized to

$$L(m, n) = 2^{\lceil r_{m,n} \rceil}$$

levels using a Lloyd–Max non-uniform step size quantizer. $\lceil r_{m,n} \rceil$ bits are then written to the bitstream to convey the index of the quantizer bin. The decoder receives the index and $\hat{\sigma}_{m,n}$, looks up the quantization point in that bin and scales it by $\hat{\sigma}_{m,n}$ to reveal the reconstructed value.

9.2.2.2   Entropy coding

If transmission of variable length codewords can be accommodated, then uniform step size quantization followed by entropy coding of the quantizer indices results in the lowest rate for a given MSE distortion. Achieving the same distortion for every coefficient whose variance exceeds a rate-dependent threshold θ brings the lowest rate for the block. That is the interpretation of the rate-distortion function given parametrically by the Equations (9.2) and (9.3). An approximate formula for the step size Δ that attains a MSE $\overline{e^2}$ is

$$\Delta = \sqrt{12\overline{e^2}}.$$

Therefore, all above-threshold coefficients should be quantized with the same step size

$$\Delta = \sqrt{12\theta}. \qquad (9.5)$$

The bin index q and reconstruction value y of a quantizer with input x are given by

$$q = \left\lfloor \frac{x}{\Delta} + \xi_e \right\rfloor, \tag{9.6}$$

$$y = (q + \xi_d)\Delta, \tag{9.7}$$

where $\xi_e = 1/2$ and $\xi_d = 0$ for a midtread quantizer and $\xi_e = 0$ and $\xi_d = 1/2$ for a midrise quantizer with midpoint reconstructions.

According to the assumed quantizer rate versus distortion model, entropy coding of the sequence of bin indices will produce codeword lengths (in bits) proportional to the logarithm of the variances of the coefficients, according to Equation (9.4). Methods of entropy coding, such as Huffman and Arithmetic Coding, are described in previous chapters. The entropy coding process produces a sequence of prefix-free, variable-length codewords that can be uniquely decoded. Therefore, only one parameter, a quantizer step size, plus a binary map indicating the above-threshold coefficients have to be sent as overhead for the entire block.

Some systems of image transfom coding use a weighted MSE criterion. The weighting constant $w_{m,n}$ mimics the sensitivity of the eye's response to different spatial frequencies. In order to achieve minimum average weighted squared error, the weighted MSE, $w_{m,n}e^2$, should equal a constant θ for the coded coefficients. Therefore, the step size changes with the coordinates according to

$$\Delta(m, n) = \sqrt{12w_{m,n}^{-1}\theta}. \tag{9.8}$$

The weights $w_{m,n}$ are known beforehand at the encoder and decoder, so again only one parameter, θ, needs to be sent as overhead to reconstruct the values of the above-threshold coefficients indicated by the binary map. The JPEG standard, which we shall visit in the next section, uses uniform quantizers with different step sizes to code coefficients within a transform block.

9.3   The JPEG standard

The JPEG standard for image compression [4] has gained a solid foothold, even an exclusive stronghold, in consumer products and commercial and professional applications. Almost every digital camera features this form of compression in its default storage format. The traffic of images over the Internet contains almost exclusively JPEG images. One of the DICOM standard lossy medical image storage formats uses JPEG compression at file size reductions (compression ratios) of 6:1 or lower. Tele-radiology uses JPEG-compressed images almost exclusively. The JPEG standard is a transform coding method that follows the principles laid out in this chapter. It warrants our attention, not only because of its importance, but also because it serves as a good example of the application of these principles.

The JPEG standard consists of a baseline system, adequate for most coding applications, and a set of extensions that provide special features, such as 12 bit sample inputs, progressive sequential and hierarchical build-up and arithmetic coding. Commercial use

Table 9.1 JPEG quantization table for monochrome images

16	11	10	16	24	40	51	61
12	12	14	19	26	58	60	55
14	13	16	24	40	57	69	56
14	17	22	29	51	87	80	62
18	22	37	56	58	109	103	77
24	35	55	64	81	104	113	92
49	64	78	87	103	121	120	101
72	92	95	98	112	100	103	99

of the baseline system was intended to be cost-free,[1] so that version took off, so to speak, and dominated the market. The exposition of the baseline system entirely suits our purposes here for teaching an understanding of the basic workings of the JPEG system.

9.3.1   The JPEG baseline system

The input image is divided into 8×8 subblocks, every one of which is independently transformed by a two-dimensional DCT, as in Figure 9.1. In the baseline system, the pixel inputs are recorded as 8-bit non-negative integers with values 0 to 255. Prior to transformation of the blocks, the central level of the pixels is shifted to 0 by subtracting 128 from each pixel. Also, if any dimension is not a multiple of eight, the last row or column is replicated to make it so.

The coding method applied to the transform coefficients is uniform step-size quantization followed by entropy coding. A standard 8×8 array of step sizes, called a *quantization table*, determines the relative step sizes of the coefficients in a block. Every block uses the same quantization table. JPEG's standard document contains the example quantization table that appears in Table 9.1. Different code rates are obtained by scaling all the table entries by the same constant. All blocks in the image use the same scaled quantization table.

Let us introduce some notation (reluctantly). Let us designate the coordinate of the linear scan of the subblocks within the image as v, $v = 0, 1, \ldots, N_r N_c$, where N_r and N_c are the number of subblocks in the rows and columns, respectively. $\tilde{x}_v(m, n)$ denotes the transform coefficient at coordinates (m, n) within the v-th subblock. The scaled quantization table assigns step size $\Delta(m, n)$ for this coefficient. The type of uniform quantization is midtread, so that the quantizer bin index for this coefficient is given by

$$q_v(m, n) = \left\lfloor \frac{\tilde{x}_v(m, n)}{\Delta(m, n)} + 1/2 \right\rfloor. \tag{9.9}$$

We shall call these bin indices *quantizer levels*. After decoding the quantizer levels, the dequantization operation

$$\hat{\tilde{x}}_v(m, n) = q_v(m, n)\Delta(m, n), \tag{9.10}$$

Table 9.2 Range categories and bits to represent position within a category

Category	Amplitude interval	Sign bit	Offset bit
0	[0]	no	0
1	[−1], [1]	yes	0
2	[−2, −3], [2, 3]	yes	1
3	[−7, −4], [4, 7]	yes	2
4	[−15, −8], [8, 15]	yes	3
5	[−31, −16], [16, 31]	yes	4
6	[−63, −32], [32, 63]	yes	5
7	[−127, −64], [64, 127]	yes	6
⋮	⋮	⋮	⋮
c	$[-2^c + 1, \ -2^{c-1}], [2^{c-1}, \ 2^c - 1]$	yes	$c - 1$
⋮	⋮	⋮	⋮
15	[−32767, −16384], [16384, 32767]	yes	15

reconstructs the pixel value at the midpoint of the quantization bin. (See Equation (9.6).) For low bit rates, the quantization sets most DCT coefficients to 0, since their magnitudes will be less than their threshold of $\Delta(m, n)/2$.

The different DCT blocks are visited normally in raster scan order (left to right from top to bottom), while the coefficients in each block are scanned in the zigzag order (see Figure 9.2).

The coding of the quantizer levels of the DC (0,0) coefficients of every block $\{q_v(0, 0)\}$ is performed separately from the AC coefficients. Because adjacent blocks are likely to have small differences in DC indices, the successive differences, excepting the first block with index $q_0(0, 0)$, are coded. Thus, the DC difference array takes the mathematical form:

$$\delta q_v(0, 0) = \begin{cases} q_v(0, 0) - q_{v-1}(0, 0), & v \neq 0 \\ q_0(0, 0), & v = 0. \end{cases} \quad (9.11)$$

These differences then replace their original values in the quantizer index array. The addition of the current difference to the reconstruction of the DC of the previous block reconstructs the current quantizer level. The additions shown below are done sequentially in the scan order.

$$q_v(0, 0) = \begin{cases} \delta q_v(0, 0) + q_{v-1}(0, 0), & v \neq 0 \\ \delta q_0(0, 0), & v = 0. \end{cases} \quad (9.12)$$

The recursive addition operations imply that the DC quantizer level of any block can be decoded only if all past blocks in the scan have been received.

The DC level differences form a rather large alphabet (range of integer differences), which presents a hindrance to efficient entropy coding. Therefore, the alphabet is broken into integer amplitude intervals called *categories*. These categories are defined in Table 9.2.

Table 9.3 Example of Huffman DC code table

Category	Codeword	Category	Codeword
0	00	6	1110
1	010	7	11110
2	011	8	111110
3	100	9	1111110
4	101	10	111111110
5	110	11	11111111110

Every number $\delta q_v(0,0)$ is represented by its category and its position within the category. The category alphabet is small, so is conducive to entropy coding. The categories have a non-uniform probability distribution, so they are entropy coded using a fixed Huffman code. The JPEG standard's document [5] presents the code in Table 9.3 for DC differences for luminance (grayscale) images. The position within a category is specified by the difference of its magnitude from the minimum one in the category and its sign. The magnitude difference from the minimum in the set is commonly called the *offset*. The position is represented as a natural (uncoded) binary number in magnitude-sign format. The equivalence of DC difference to the following triplet of category, offset, and sign may be expressed as:

$$\delta q_v(0,0) \equiv (c_v(0,0), o_v(0,0), s_v(0,0)).$$

For example, suppose that the DC level difference $q = -19$. Its representation is the triplet $(5, 3, -1)$ for category, offset, and negative sign, respectively. Table 9.3 points to the 3-bit codeword 110 for Category 5; the offset of 3 in Category 5 requires the 4 bits 0011; and an additional bit (0) specifies the negative sign. So, in general, given category $c \neq 0$, c additional bits are needed to represent the signed offset within this category. The category $c = 0$ uniquely specifies its sole element of 0, so additional bits are unnecessary.

The AC coefficients within a block are coded separately from its DC coefficient. The quantizer levels of the AC coefficients (Equation (9.9)) within a block are scanned in the zigzag order of Figure 9.2. Since all blocks are treated the same, we drop the block index v and denote the quantizer levels of the AC coefficients as $q(m,n)$, $(m,n) \neq (0,0)$. As with DC, every non-zero level is represented by the triplet of (category, offset, sign). That is,

$$q(m,n) \equiv \begin{cases} (c(m,n), o(m,n), s(m,n)), & q(m,n) > 0 \\ 0, & q(m,n) = 0. \end{cases} \tag{9.13}$$

Almost always, the quantization produces a majority of 0's in any block. In progressing along the zigzag scan path through the block, there will occur several runs of 0's. Starting with the first element $q(0,1)$, one counts the number of 0's, denoted Z_{r1}, until the next non-zero element $q(m_1, n_1)$ is encountered. With $q(m_1, n_1)$ represented as

Table 9.4 Excerpt of example of Huffman AC code table

Zero run	Category	Codeword	Zero run	Category	Codeword
0	1	00	2	1	11100
0	2	01	2	2	11111001
⋮	⋮	⋮	⋮	⋮	⋮
0	3	100			
0	4	1011	3	1	111010
0	5	11010	3	2	111110111
0	6	1111000	⋮	⋮	⋮
⋮	⋮	⋮	4	1	111011
1	1	1100	5	1	1111010
1	2	11011	6	1	1111011
1	3	1111001	⋮	⋮	⋮
1	4	111110110	16	–	11111111001
⋮	⋮	⋮	EOB	–	1010

the triplet $(c(m_1, n_1), o(m_1, n_1), s(m_1, n_1))$ of category, offset, and sign, the composite symbol $(Z_{r1}, c(m_1, n_1))$ of run-length and category is formed. The scan continues counting the number of 0's to the next non-zero element at coordinates (m_2, n_2), where is formed the next composite symbol $(Z_{r2}, c(m_2, n_2))$ of run-length and category. The scan continues and repeats this process until the last non-zero element is reached. An end-of-block (EOB) symbol then signifies that the rest of the coefficients along the scan are zero. Each of the composite symbols $(Z_{rk}, c(m_k, n_k))$, $k = 1, 2, \ldots, K$, denoting $(runlength, category)$ is Huffman coded using a Huffman AC table. The standard document [5] contains an example code table that seems to be the one almost universally used. Table 9.4 shows an excerpt from this AC code table. The signs and offsets are written in natural binary format, as before.

In summary, the string of symbols representing a block takes the following form:

$$c_v(0, 0), o_v(0, 0), s_v(0, 0)) | (Z_{r1}, c(m_1, n_1)), s(m_1, n_1), o(m_1, n_1)| \cdots$$
$$|(Z_{rK}, c(m_K, n_K)), s(m_K, n_K), o(m_K, n_K)|\text{EOB.} \tag{9.14}$$

Commas (,) separate the distinct symbols (except within composite ones), and vertical bars (|) separate different coefficients. In words, the block's symbol string is

DC difference| K occurrences of (run-length, category), sign, offset|EOB.   (9.15)

The bitstream generated by this symbol string follows the same order. As mentioned previously, the DC difference category $c_v(0, 0)$ is Huffman coded using the DC Huffman Table 9.3. The composite (run-length, category) $(Z_{rk}, c(m_k, n_k))$ symbols for AC coefficients are encoded using the AC Huffman table, part of which is contained in Table 9.4. The code specifies no single codeword for a run-length greater than 16. So when the run-length exceeds 16, it is split into zero runs of length 15 and a following run-length-category symbol. A special symbol and codeword are reserved for a run-length of 16 zeros. A full AC table lists entries for 16 run-lengths and 10 categories plus 2 special symbols, EOB and run-length 16, giving a total of 162 entries. One bit specifies the sign and $c - 1$ bits specify the offset for category c.

Notice in the decoding of a block's bitstream, the category code precedes the offset code. The category code is prefix-free, so the category is decoded instantaneously and reveals the number of bits in its associated fixed-length offset code. The block is therefore decoded following the same order of the encoder's scan. Decoding a block's bitstream produces the array of its quantizer levels $q(m, n)$. These levels are dequantized according to the operation in (9.10) to reconstruct the DCT coefficients $\hat{\tilde{x}}(m, n)$. Applying the inverse DCT produces the reconstructed image array $\hat{x}(m, n)$ of this block. The blocks are encoded and decoded in raster order, so that the reconstructed image develops from top to bottom.

9.3.2　Detailed example of JPEG standard method

In order to illustrate the baseline JPEG coding method, we present a numerical example of the transform, quantization, and symbol and code streams for a single subblock of an image. Although the standard's block size is 8×8, a 4×4 block suits our purpose of teaching the method just as well. The input block in our example is represented in the image intensity matrix by

$$\mathbf{I} = \begin{bmatrix} 156 & 192 & 144 & 45 \\ 132 & 112 & 101 & 56 \\ 210 & 190 & 45 & 35 \\ 230 & 150 & 70 & 90 \end{bmatrix} \tag{9.16}$$

After subtracting 128 from each pixel and performing the two-dimensional DCT, we obtain the block DCT matrix

$$\tilde{\mathbf{I}} = \begin{bmatrix} -22.5000 & 202.3986 & -12.5000 & -24.8459 \\ -11.6685 & -45.3462 & -81.4960 & 24.9693 \\ 49.0000 & -3.7884 & -5.0000 & 9.1459 \\ 25.3987 & 58.4693 & -20.3628 & -36.1538 \end{bmatrix} \tag{9.17}$$

For quantization of the elements of $\tilde{I}$, we use a haphazardly subsampled form of the standard's example Q-table, Table 9.1, to generate the following step size matrix:

$$\Delta = \begin{bmatrix} 16 & 10 & 24 & 51 \\ 14 & 22 & 51 & 80 \\ 35 & 64 & 104 & 92 \\ 72 & 95 & 112 & 103 \end{bmatrix} \tag{9.18}$$

For this example, we use the Δ matrix without scaling. The elements of $\tilde{I}$ and Δ are substituted into the quantization formula in Equation (9.9) to produce the matrix of quantization levels

$$\mathbf{q} = \begin{bmatrix} -1 & 20 & -1 & 0 \\ -1 & -2 & -2 & 0 \\ 1 & 0 & 0 & 0 \\ 0 & 1 & 0 & 0 \end{bmatrix} \tag{9.19}$$

At this juncture, the level array is scanned in the zigzag manner to yield the linear array

$$\mathbf{q}_{zz} = [\ -1 \ \ 20 \ \ -1 \ \ 1 \ \ -2 \ \ -1 \ \ 0 \ \ -2 \ \ 0 \ \ 0 \ \ 1 \ \ 0 \ \ 0 \ \ 0 \ \ 0 \ \ 0 \ ] \tag{9.20}$$

It is convenient now to count the run-lengths of zeros, starting with the first AC level. This sequence of run-lengths is 0 0 0 0 0 1 2 5.
The first five AC levels are non-zero, so they all receive a run-length of 0.

Then we convert every level (element of $\mathbf{q}_{zz}$) to its triplet of category, sign, and offset. The equivalent array is written as

$$(1, -1, 0), (5, 1, 4), (1, -1, 0), (1, 1, 0), (2, -1, 0), (1, -1, 0), (0, 0, 0), (2, -1, 0),$$
$$(0, 0, 0), (0, 0, 0), (1, 1, 0), (0, 0, 0), (0, 0, 0), (0, 0, 0), (0, 0, 0), (0, 0, 0).$$

Here we use commas to separate triplets. The $(0, 0, 0)$ triplets are redundant representations of 0.

The symbol string for encoding is ready to be synthesized. The DC triplet $(1, -1, 0)$ heads the string followed by the sequence of run-lengths, categories, signs, and offsets of the AC levels. The run-lengths and categories are formed into a single supersymbol. The run-length after the last non-zero level in the block is indicated by the EOB symbol. Executing this procedure, we obtain

$$1, -1, 0|(0, 5), 1, 4|(0, 1), -1, 0|(0, 1), 1, 0|(0, 2), -1, 0|(0, 1), -1, 0|$$
$$(1, 2), -1, 0|(2, 1), 1, 0|EOB. \tag{9.21}$$

Huffman (variable length) codes are reserved for the DC level category difference and the (run-length, AC category) supersymbol. The signs and offsets are expressed as fixed-length codewords with length depending on the category. The category belonging to a level is encoded first to signify the length of the following offset codeword.

The numbers $1, -1, 0$ at the start of the string above represent category 1 with sign -1 and 0 offset. Using Table 9.3, the code is 0100, 010 for category 1, 0 for the negative sign, and nothing for offset in category 1. Next come the AC symbols $(0, 5), 1, 4$. The Huffman AC table (Table 9.4 lists the codeword 11010 for $(0, 5)$. The codeword for the symbols $1, 4$ is 10100, the initial 1 declaring positive sign and 0100 the offset of 4 with 4 bits for category 5. The next $(0, 1)$ points to codeword 00 in the table and the signed offset $-1, 0$ encodes as 0, just one bit for category 1. Continuing in this way produces the following bitstream to represent the symbol string in Equation (9.21):

$$0100|11010, 10100|00, 0|00, 1|01, 00|00, 0|11011, 00|$$
$$11100, 1|1010 \tag{9.22}$$

For tutorial purposes, we have used vertical bars to separate the codewords between different levels and commas to separate the supersymbol and signed offset within each level codeword. Certainly, these bars and commas are not written into the bitstream. The count of the bits in Equation (9.22) that represent these 16 coefficients in the block comes to 44 for a bit rate of $44/16 = 2.75$ bits/pixel.

It is instructive to examine the reconstruction process. The encoding of the quantization levels is lossless, so let us assume that we have decoded the bitstream in Equation (9.22) to obtain the level sequence $\mathbf{q}_{zz}$ in Equation (9.20). Reversing the zigzag scan and dequantizing its elements according to Equation (9.10), produces the reconstructed DCT matrix

$$
\hat{\hat{\mathbf{I}}} = \begin{bmatrix} -16 & 200 & -24 & 0 \\ -14 & -44 & -102 & 0 \\ 35 & 0 & 0 & 0 \\ 0 & 95 & 0 & 0 \end{bmatrix}
\tag{9.23}
$$

After applying the inverse DCT, adding 128 to each element, and rounding to the nearest integer, the reconstucted image array becomes

$$
\hat{\mathbf{I}} = \begin{bmatrix} 152 & 194 & 141 & 26 \\ 111 & 140 & 126 & 77 \\ 239 & 156 & 62 & 11 \\ 232 & 138 & 82 & 97 \end{bmatrix}
\tag{9.24}
$$

To measure the fidelity of this reconstruction, we now calculate the PSNR. The definition of PSNR is

$$
\text{PSNR} \equiv 10 \log_{10} \frac{255^2}{\overline{e^2}},
\tag{9.25}
$$

where $\overline{e^2}$ denotes the MSE, defined as

$$
\overline{e^2} \equiv \frac{1}{M^2} \sum_{i=1}^{M} \sum_{j=1}^{M} (I(i, j) - \hat{I}(i, j))^2
\tag{9.26}
$$

In this case, PSNR calculates to be 22.48 dB. With true JPEG coding of 8×8 blocks of ordinary images, one expects to attain a significantly higher PSNR for the average bit rate of 2.75 bits per pixel (bpp).

The same step size for all DCT coefficients achieves MMSE for a given rate or minimum rate for a given MSE. The JPEG standard's table of different step sizes was obtained through experiments to minimize human visual error. Therefore, we would expect a lower rate than 2.75 bpp to be achieved by constant step size quantization giving the same PSNR as above.

In order to confirm this expectation, we substituted the step size matrix $\boldsymbol{\Delta}$ with the constant step size $\Delta = 73$, which was set by trial and error to give almost the same

PSNR = 22.30 dB. Otherwise, the coding is exactly the same. The symbol string below results:

$$0, 0, 0|(0, 2), 1, 1|(1, 1), 1, 0|(0, 1), -1, 0|(2, 1), -1, 0|(2, 1), 1, 0|EOB. \qquad (9.27)$$

Use of Tables 9.3 and 9.4 results in the corresponding code string

$$0|0111|11001|000|111000|111001|1010 \qquad (9.28)$$

for a total of 29 bits and a bit rate of 1.81 bpp, so the rate is substantially lower for almost the same PSNR, confirming our expectation.

9.4 Advanced image transform coding: H.264/AVC intra coding

The JPEG standard, despite its origin more than 15 years ago, is still the most ubiquitous image compression sytem in use today. Certainly, more efficient and more feature-laden algorithms have evolved, but most applications do not require better efficiency and advanced features. The more advanced techniques often require more active memory and much more computation, making them too costly and power hungry for mobile and hand-held devices.

One example of an advanced image transform coding system is the so-called *intra coder* of the H.264/AVC advanced video coding system [6, 7], the documentation of which is in draft stage at this writing. In a video coding system, the sequence of image frames is coded in groups headed by an intra-coded frame to allow resynchronization of the coder. An intra-coded frame is coded just like a single image, whereas the rest of the frames are coded together using complicated motion compensation together with temporal prediction modes. An intra-coded frame also uses prediction, but only within the frame in the spatial direction, so it can be decoded without use of information from any other frame.

A detailed description of H.264 intra coding is beyond the scope of this chapter, since our purpose is to teach coding principles accompanied by examples of application. The interested reader can consult the books by Sayood [8] and Shi and Sun [9] or the overview paper by Wiegand *et al.* [7] for further details. A skeletal explanation of the differences from the JPEG standard serves to illustrate why the H.264 intra coder achieves better compression. First of all, the subblocks are not encoded independently. The image (frame) is divided into 16×16 squares called *macroblocks*. These macroblocks are further subdivided into two 8×16 or two 16×8, or four 8×8 subblocks, where each subblock may be further split in the same way. The subdividing must stop when a 4×4 subblock is reached. An activity measure determines the subdivision of the macroblock. When the macroblock is subdivided, a subblock to be coded is predicted from its adjacent subblocks; and its prediction residual (the subblock minus its prediction) is transformed and encoded. The various prediction modes are specified in the draft standard document [6]. Although the prediction residual subblocks vary in size, the transform is applied in 4×4 block segments, yielding 16 transform subblocks per macroblock. The transform is not the DCT, but an integer transform with similar properties, given by

$$\mathbf{H} = \begin{bmatrix} 1 & 1 & 1 & 1 \\ 2 & 1 & -1 & -2 \\ 1 & -1 & -1 & 1 \\ 1 & -2 & 2 & -1 \end{bmatrix} \tag{9.29}$$

The integer elements allow integer arithmetic and an option for lossless coding. For lossy coding, the coefficients are quantized using a scaled step size matrix (called a scaling matrix). The quantized DC coefficients undergo a processing step separate from the AC ones. From the 16 4×4 transforms, a 4×4 array of DC values is extracted, transformed by a Hadamard matrix, and then re-inserted into the DC positions from which they were extracted.

The order of visiting the transformed subblocks of dimensions 4×4 follows the indices of these subblocks shown in Figure 9.3. Within every 4×4 array the scan follows the now familiar zigzag path similar to that of JPEG.

There are two methods of entropy coding of the 4×4 arrays: context-adaptive variable length coding (CAVLC) and context-adaptive binary arithmetic coding (CABAC). The former method, CAVLC, is the simpler one and has been adopted in the baseline system. The latter, CABAC, is more complicated and is an option for the enhanced system. We shall describe some aspects of CAVLC here.

The zigzag scan of the 4×4 array of transformed residuals typically yields a linear sequence of small values, with a predominance of zeros. The basis of CAVLC is the so-called exp-Golomb (exponential Golomb) code. Recall that the normal Golomb codeword of a positive value $v = im + r$ consisted of a prefix unary codeword representing the number i of intervals of length m and a suffix codeword representing the remainder r. The exp-Golomb code substitutes fixed intervals of length m for exponential ones, expressing the value as $v = 2^m + r$, where the unary codeword for m is the prefix.[2] Because values in the sequence can be negative, they are all mapped to positive integers with the negative values occupying the odd integers and the positive ones the even integers, according to

0	1	4	5
2	3	6	7
8	9	12	13
10	11	14	15

Figure 9.3   Order of coding of 4×4 transform subblocks of a macroblock.

$$v = \begin{cases} 2c, & c > 0 \\ 2|c| - 1, & c < 0 \end{cases} \tag{9.30}$$

where c is a value in the sequence. The remainders r are encoded using six context-dependent coding tables. The context is the last value of v that was coded.

The coding of the sequence breaks into five parts, one of which is coding of the non-zero values. These parts are as follows.

1. *Number of non-zero coefficients (N) and Trailing 1's.* The number of "Trailing 1's" (T1s) refers to the number of coefficients with absolute value equal to 1 prior to the last non-zero coefficient in the scan. N and T1s are encoded together using VLC (variable length code) lookup tables.
2. *Encode non-zero coefficients with exp-Golomb code lookup tables.* Actually 1 and -1 need not be coded, because once their positions are indicated by the succeeding parts, only sign information needs to be sent. For this step, the coding proceeds in reverse scan order, so that the just-coded likely smaller coefficient (in magnitude) serves as context for the next one.
3. *Sign information.* A single bit indicates the sign of 1 or -1. The sign of the other coefficients is embedded in the exp-Golomb code.
   Positions of non-zero coefficients are conveyed by the coding in the next two parts.
4. *Total zeros.* This codeword indicates the number of zeros (TZ) between the start and last non-zero coefficient of the scan. Since the count of non-zero coefficients (N) is also coded, the range of TZ is limited to $16 - N$. There is a VLC table of TZ codewords for each value of N, except $N = 16$ where there are no zeros.
5. *Run-before:* run-lengths of zeros before non-zero coefficients. Starting at the last non-zero coefficients, count backwards the number of consecutive zeros between non-zero coefficients. The number of non-zero coefficients and the total number of these zeros have been coded, so the number of such zero runs and the individual limits of their ranges can be determined. A suitable VLC code table is accessed to find the codewords for these runs.

These five parts of coding the sequence from the zigzag scan uniquely determine the values and positions of the coefficients.

The H.264/AVC intra coding is superior to JPEG for the following reasons.

1. Coding operates on 4×4 blocks instead of 8×8. For non-stationary sources like images, the pixel values are likely to be more uniform in smaller blocks and hence compress better after a transform operation. The uniformity of small blocks is likely to become even more prevalent, as images are being captured with ever increasing resolution.
2. The statistical dependence of blocks within a macroblock is exploited using predictive coding. The prediction operates in as many as eight directions, including horizontal, vertical, diagonal, and cross-diagonal directions, in order to capture edges and contours. The prediction residuals are likely to be small, because they effectively predict changes in almost any direction, even on edges and contours.

3. A deblocking filter compensates for boundary discontinuities on macroblock edges. The macroblocks are encoded independently, so the boundaries of these 16×16 blocks may start to show at low bit rates. The deblocking filter smooths the transitions across these boundaries, thereby diminishing their prominence.

9.5   Concluding remarks

In this chapter, we described transform coding systems devised for still images. These methods are also pertinent for encoding frames of video. We shall take up the extensions of these methods to video in Chapter 13. The same coding methods can also be applied to one-dimensional sources, where the transform is one-dimensional and there is no need for a zigzag scan. For audio and speech, unlike images, there are good perceptual models for frequency domain weighting. So for these sources, we just insert the proper weighting factors into the coding system. The reigning coding system for home and commercial audio, MP3, and its more advanced extensions are subband coding systems. Chapter 11 will treat subband coding systems. For most biological signal sources, such as electro-encephalograms (EEGs) and electro-cardiograms (ECGs or EKGs), the episodes in their waveforms are analyzed without a perceptual component. Therefore, these sources can be encoded with the one-dimensional transform equivalents of the methods described in this chapter.

Problems

9.1   The 4×4 block of pixels B is cut from an actual image.

$$B = \begin{bmatrix} 90 & 91 & 65 & 42 \\ 86 & 78 & 46 & 41 \\ 81 & 58 & 42 & 42 \\ 82 & 51 & 38 & 43 \end{bmatrix}$$

We wish to encode this block with the methods of the JPEG standard.
(a) Take the two-dimensional (separable) DCT on the block.
(b) Show the sequence of coefficients given from a zigzag scan of the DCT coefficient matrix.
(c) The AC coefficients are now to be encoded. Perform the following steps.
   (i) Quantize every AC coefficient with step size $\Delta = 16$. Use the standard's double-width null zone uniform step size quantizer. Give the resulting sequence of 15 bin indices.
   (ii) For this sequence of bin indices, determine the corresponding sequence of triplets {(runlength, category), sign, offset}
   (iii) Using the tables in this chapter, give the binary code sequence for this sequence of triplets. What is the code rate?

        (iv) Calculate the MSE of the reconstructed image. Why do you not need to reconstruct the coded image for this calculation?

    (d) Repeat the steps from 9.5 on for $\Delta = 32$ and $\Delta = 8$.

9.2 Repeat the previous problem with the same image block B, but with the standard H.264/AVC transform H in Equation (9.29) and $\Delta = 30$ and $\Delta = 70$. In calculating MSE, note that the transform is not unitary.

9.3 Image transform coding methods can also be applied to one-dimensional signals. Let the subject to code be the ECG signal "s110", available via download from http://www.cambridge.org/pearlman/datasets/audio/s110. Extract the first block of 128 samples for encoding.

    (a) Perform a (one-dimensional) $N = 128$ point DCT of this signal block.

    (b) Quantize all (including DC) coefficients with a uniform step size, double width, null-zone quantizer with the same step size Δ. Select Δ such that the code rate is $R = 2.0$ bits per sample by trial and error or the bisection method by repeating the steps below.

        (i) For the sequence of bin indices output from the quantizer, show the corresponding sequence of triplets {(runlength, category), sign, offset}

        (ii) Using the tables in this chapter, determine the binary code sequence for this sequence of triplets.

    (c) Calculate the MSE for $R = 2$ bits per sample.

Notes

1. There have been successful patent infringement law suits in recent years.
2. m is analogous to the category and r the offset in JPEG.

References

1. W. A. Pearlman, "Adaptive cosine transform image coding with constant block distortion," *IEEE Trans. Commun.*, vol. 38, no. 5, pp. 698–703, May 1990.
2. N. O. Anderson, "On the calculation of filter coefficients for maximum entropy spectral analysis," *Geophysics*, vol. 39, pp. 69–72, Feb. 1974.
3. W.-H. Chen and C. H. Smith, "Adaptive coding of color and monochrome images," *IEEE Trans. Commun.*, vol. 25, no. 11, pp. 1285–1292, Nov. 1977.
4. W. B. Pennebaker and J. L. Mitchell, *JPEG Still Image Data Compression Standard*. New York: Van Nostrand Reinhold, 1993.
5. *ISO/IEC 10918-1 and ITU-T Recommendation T.81. Information Technology – Digital Compression and Coding of Continuous-tone Still Images: Requirements and Guidelines*, 1994.
6. *ITU-T Recommendation H.264. Series H: Audiovisual and Multimedia Sytems – Advanced Video coding for Generic Audiovisual Services*, 2009.
7. T. Wiegand, G. J. Sullivan, G. Bjontegaard, and A. Luthra, "Overview of the h.264/avc video coding standard," *IEEE Trans. Circuits Syst. Video Technol.*, vol. 13, no. 7, pp. 560–576, July 2003.
8. K. Sayood, *Introduction to Data Compression*, 3rd edn. San Francisco, CA: Morgan Kaufmann Publishers (Elsevier), 2006.
9. Y. Q. Shi and H. Sun, *Image and Video Compression for Multimedia Engineering*. Boca Raton: CRC Press, Taylor & Francis Group, 2008.

Further reading

W.-H. Chen and W. K. Pratt, "Scene adaptive coder," *IEEE Trans. Commun.*, vol. 32, no. 3, pp. 225–232, Mar. 1984.

W. K. Pratt, W.-H. Chen, and L. R. Welch, "Slant transform image coding," *IEEE Trans. Commun.*, vol. 22, no. 8, pp. 1075–1093, Aug. 1974.

M. Rabbani and P. W. Jones, *Digital Image Compression Techniques*, ser. Tutorial Texts. Bellingham, WA: SPIE Optical Engineering Press, 1991, vol. TT7.

10 Set partition coding

10.1 Principles

The storage requirements of samples of data depend on their number of possible values, called alphabet size. Real-valued data theoretically require an unlimited number of bits per sample to store with perfect precision, because their alphabet size is infinite. However, there is some level of noise in every practical measurement of continuous quantities, which means that only some digits in the measured value have actual physical sense. Therefore, they are stored with imperfect, but usually adequate precision using 32 or 64 bits. Only integer-valued data samples can be stored with perfect precision when they have a finite alphabet, as is the case for image data. Therefore, we limit our considerations here to integer data.

Natural representation of integers in a dataset requires a number of bits per sample no less than the base 2 logarithm of the number of possible integer values.[1] For example, the usual monochrome image has integer values from 0 to 255, so we use 8 bits to store every sample. Suppose, however, that we can find a group of samples whose values do not exceed 15. Then every sample in that group needs at most 4 bits to specify its value, which is a saving of at least 4 bits per sample. We of course need location information for the samples in the group. If the location information in bits is less than four times the number of such samples, then we have achieved compression. Of course, we would like to find large groups of samples with maximum values below a certain low threshold. Then the location bits would be few and the number of bits per sample would be small. The location information of each group together with its size and threshold determine the bit savings of every group and hence the compression compared to so-called raw storage using the full alphabet size for every sample. This is a basic principle behind set partition coding.

The reason why we discussed a group of samples and not individual samples is that location information for single samples would be prohibitively large. Ignoring for now the required location information, we can assess the potential compression gains just by adding the number of bits associated with the actual values of the samples. For example, the 512×512 monochrome image Lena, shown in Figure 10.1, is stored as 8 bits for every sample. A histogram of this image is depicted in Figure 10.2. Note that moderate to large values are relatively frequent. In fact, given these values and their locations, the average number of minimal natural (uncoded) bits is 7.33 bits per sample. (What is meant by minimal natural bits is the base-2

Figure 10.1   Lena image: dimension 512×512 pixels, 8 bits per sample.

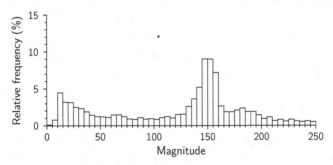

Figure 10.2   Distribution of pixel values for the Lena image.

logarithm of the value rounded up to the nearest integer.) Adding location information would surely exceed the 8 bits per sample for natural or raw storage. Clearly, set partition coding would not be a good method for this image. As this histogram is fairly typical for natural images, set partition coding should not be chosen for direct coding of natural images. For this method to be efficient, the source must contain primarily sets of samples with small maximum values. That is why we seek source transformations, such as discrete wavelet and cosine transformations, that have this property.

In Figure 10.3 is shown a histogram of an integer wavelet transform of the same source image Lena. It is evident that we have a large percentage of wavelet coefficients with small values and a low percentage with large values. In fact, the average number of minimal natural bits for the coefficients of this wavelet transform is 2.62 bits per sample

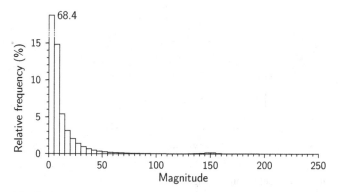

Figure 10.3 Distribution of magnitudes of integer transform coefficients of the Lena image.

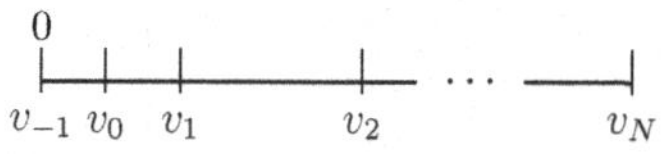

Figure 10.4 Set of thresholds marking ranges of values in dataset.

(for three levels of decomposition). Clearly, here, we have some room for location information to give us compression from the original 8 bits per sample. Set partition coding is a class of methods that can take advantage of the properties of image transforms. The members of this class differ in the methods of forming the partitions and the ways in which the location information of the partitions is represented.

10.1.1 Partitioning data according to value

Let us consider the source as either the original image or an unspecified transform of the image. To treat both image sources and transforms, let us assume that the values of the data samples are their magnitudes.

Let $\mathcal{P} = \{p_1, p_2, \ldots, p_A\}$ be the finite set of all source samples. We postulate a set of increasing thresholds

$$v_{-1} = 0 < v_0 < v_1 < \cdots < v_N,$$
$$\max_k p_k < v_N \le v_{N-1} + 2^{N-1}. \tag{10.1}$$

These thresholds are illustrated in Figure 10.4.

Consider the interval of values p between adjacent thresholds v_{n-1} and v_n and denote them as $\mathcal{P}_n$, where

$$\mathcal{P}_n = \{p : v_{n-1} \le p < v_n\}. \tag{10.2}$$

Let us define the set of indices of the samples in an interval $\mathcal{P}_n$ as

$$S_n = \{k : p_k \in \mathcal{P}_n\} \tag{10.3}$$

Assume a procedure whereby the indices of the samples in $\mathcal{P}$ are sorted into a number of these index sets, $S_0, S_1, S_2, \ldots, S_N$. That is, indices of samples with values in

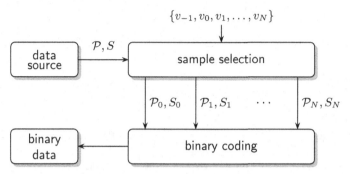

Figure 10.5   Separation of values in dataset before coding.

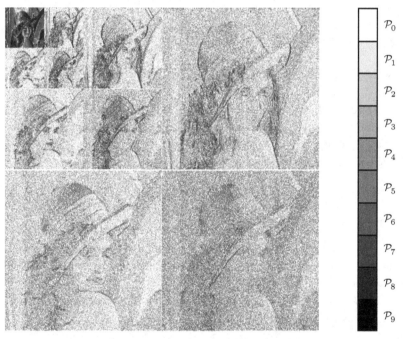

Figure 10.6   Distribution of magnitudes of integer-wavelet-transform coefficients. Logarithmic scale, darker pixels representing larger values. ($\mathcal{P}_n$ notation is explained in Section 10.1.1.)

interval $\mathcal{P}_0$ are put into S_0, indices of samples with values in interval $\mathcal{P}_1$ are put into S_1, and so forth. This procedure is what we call *set partitioning* of the data. A block diagram of a system that separates samples of the source into these sets is shown in Figure 10.5. Figure 10.6 shows an example of this type of data partitioning. In this figure, the magnitudes of coefficients of an integer wavelet transform, applied to image Lena (Figure 10.1) minus its mean, are shown with pixel gray value defined by the set $\mathcal{P}_n$ to which the coefficient magnitudes belong. For instance, all pixels in $\mathcal{P}_0$ are shown in white, and those in $\mathcal{P}_9$ are shown in black.

Using $|S_n|$ to represent the number of elements (size) in the set S_n, it is easy to conclude that the minimal number of natural (uncoded) bits B to represent all these source values is

$$B = \sum_{n=0}^{N} |S_n| \left\lceil \log_2 (v_n - v_{n-1}) \right\rceil. \tag{10.4}$$

If we use entropy coding, we may achieve lossless representation of the source with fewer than B bits, but we concentrate here on the principles to keep the discussion uncluttered. Therefore, in all that follows, except where especially noted, the code is natural binary (raw) representation.

We remind that B includes only the bits used to code the values of the samples. The information needed to determine the location of samples is conveyed by the indices in the sets S_n.

The coding process is simplified and more efficient when the intermediate thresholds are powers of two, i.e.,

$$v_n = 2^n, \quad n = 0, 1, \ldots, N - 1 \tag{10.5}$$

with

$$N - 1 = \left\lfloor \max_{k \in \mathcal{P}} \{ \log_2 p_k \} \right\rfloor \equiv n_{\max}. \tag{10.6}$$

We note that

$$v_n - v_{n-1} = 2^{n-1}$$

for $n \neq 0$ using these thresholds. The exponent of two (n) is called the bitplane index, and the largest index $(n_{\max})$ is called the most significant bitplane. It is possible to represent every data element perfectly using $n_{\max} + 1$ bits.[2] However, with the knowledge that $p_k \in \mathcal{P}_n$, $n \neq 0$, the number of bits to represent p_k perfectly is just $n - 1$, because there are 2^{n-1} integer members of $\mathcal{P}_n$. The only value in $\mathcal{P}_0$ is 0, so knowing that an element is in $\mathcal{P}_0$ is sufficient to convey its value. Thus, Equation (10.4) for the minimum number of natural bits becomes

$$B = \sum_{n=1}^{N} |S_n|(n - 1). \tag{10.7}$$

The indexing of thresholds and sets, although adopted for simplicity and consistency, may cause some confusion, so we illustrate with the following example.

Example 10.1 Suppose that the maximum value in our data set $\max_k p_k = 9$. Then $n_{\max} = 3$ and $N = 4$. Our intermediate thresholds are 2^n, $n = 0, 1, 2, 3$. The following Table 10.1 shows the association of indices with the bitplanes, thresholds, and sets.

Table 10.1 Association of indices with various quantities

n	Bit plane	v_n	2^n	$\mathcal{P}_n$
−1	–	0	–	–
0	0	1	1	{0}
1	1	2	2	{1}
2	2	4	4	{2, 3}
3	3	8	8	{4, 5, 6, 7}
4	4	9–16	16	{8, 9, 10, . . . , 15}

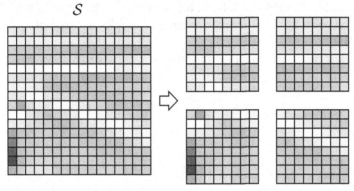

Figure 10.7 Partitioning of an array of pixels $\mathcal{S}$ into four equal-sized subsets. Gray values are used for representing pixel values.

Three bits are needed to code $\max_k p_k = 9$, because eight elements are in its containing set $\mathcal{P}_4$.

As remarked earlier, the location information of the samples is contained in the elements of the sets S_n. We now turn our attention to methods of set partitioning that lead to efficient conveyance of these sets.

10.1.2   Forming partitions recursively: square blocks

An efficient method for generating partitions is through recursion. Here, we shall present a method where the partitions consist of square blocks of contiguous data elements. Since these elements are arranged as a two-dimensional array, we shall call them *pixels* and suppose that we have a square $2^m \times 2^m$ array of pixels.

First, the square array of source data is split into four $2^{m-1} \times 2^{m-1}$ quadrants, pictured in Figure 10.7. At least one of those quadrants contains an element $p_j \geq 2^{n_{\max}}$. Label those quadrants containing elements $p_j \geq 2^{n_{\max}}$ with "1," otherwise label those having no such elements with "0." The data elements in these quadrants labeled with "0" are seen to require, at most, $n_{\max}$ bits for lossless representation. Now we split the

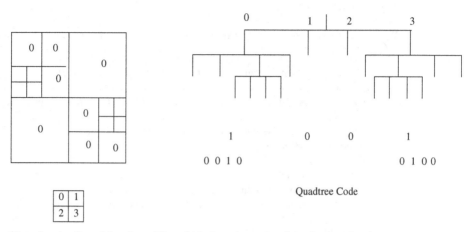

Figure 10.8 Three levels of quadrisection of 8 × 8 block and associated quadtree and code.

"1" labeled quadrants into four $2^{m-2} \times 2^{m-2}$ element quadrants and test each of these four new quarter-size quadrants, whether or not all of its elements are smaller than $2^{n_{max}}$. Again, we label these new quadrants with "1" or "0," depending whether any contained an element $p_j \geq 2^{n_{max}}$ or not, respectively. Again any "0" labeled quadrant requires n_{max} bits for lossless representation of its elements. Any quadrant labeled "1" is split again into four equal parts (quadrisected) with each part again tested whether its elements exceed the threshold $2^{n_{max}}$. This procedure of quadrisection and testing is continued until the "1"-labeled quadrants are split into four single elements, whereupon all the individual elements greater than or equal to $2^{n_{max}}$ are located. These elements are known to be one of the $2^{n_{max}}$ integers from $2^{n_{max}}$ to $2^{n_{max}+1} - 1$, so their differences from $2^{n_{max}}$ are coded with n_{max} bits and inserted into the bitstream to be transmitted. The single elements less than $2^{n_{max}}$ can be coded now with n_{max} bits. What also remains are sets of sizes 2×2 to $2^{m-1} \times 2^{m-1}$ labeled with "0" to indicate that every element within these sets is less than $2^{n_{max}}$. An illustration of three levels of splitting and labeling is shown in Figure 10.8.

We wish to do better than find sets requiring just one less bit for representation of its elements, so we lower the threshold by a factor of 2 to $2^{n_{max}-1}$ and repeat the above procedure of quadrisection and labeling on the "0"-labeled sets already found. However, since we wish to identify and code individual elements as early as possible (the reason will become clearer later), these sets are tested in increasing order of their size. The result will emulate that of the previous higher threshold: individual elements labeled with "1" or "0" coded with n_{max-1} bits each and "0"-labeled sets of sizes 2×2 up to $2^{m-2} \times 2^{m-2}$. We then iterate this procedure on the previously found "0"-labeled sets by successively lowering the threshold by a factor of 2, quadrisecting when the maximum element in the set equals or exceeds the threshold, and labeling the new quadrants with "0" or "1" through the last threshold $v_1 = 2^0 = 1$.

The tests of comparing the maximum element of a set to a threshold are called *significance* tests. When the maximum is greater than or equal to a given threshold,

ALGORITHM 10.1 _____

Algorithm for recursive set partitioning to encode a $2^m \times 2^m$ block of non-negative source elements

1. **Initialization:**
   (a) Find n_{max}, the most significant bit of the largest magnitude of the source.
   (b) Create a LIS, initially empty, to contain "0"-labeled (called insignificant) sets.
   (c) Put descriptor of $2^m \times 2^m$ source block onto LIS (upper left corner coordinates, plus size m).
   (d) Set $n = n_{max}$. Encode n_{max} and write to codestream buffer.
2. **Testing and partitioning:**
   (a) For each set on the LIS, do
      (i) If maximum element in set is less than 2^n, write "0" to codestream buffer.
         A. if set has more than one element, return to 2(a).
         B. if set has one element, write value using n bits to codestream buffer. Remove set from LIS.
      (ii) Otherwise write "1" to codestream buffer and do the following:
         A. if set has more than one element, divide the set into four equal quadrants, and add each quadrant to the end of the LIS.
         B. if set has one element, write its value minus 2^n using n bits to codestream buffer.
         C. remove set from LIS.
3. If $n > 0$, set $n = n - 1$ and return to 2(a). Otherwise stop.

we say the set is *significant*, with it being understood that it is with respect to this threshold. Otherwise, when the maximum is less than this threshold, the set is said to be *insignificant* (with respect to this threshold). The algorithm requires that sets that test insignificant for a certain threshold be tested again at successively lower thresholds until they become significant. Therefore, we need a way to keep track of the insignificant sets that are constantly being created, so that they can be retested at lower thresholds. One way is to put them immediately onto an ordered list, which we call the list of insignificant sets (LIS). The coordinates of the top left corner of a set suffice to locate the entire set, since the size of the set is determined by the image size and the outcomes of the significance tests. Individual insignificant elements are also put onto the LIS through their coordinates. The steps are presented in detail in Algorithm 10.1.

In order to decode this bitstream, n_{max} and the labels "0" and "1" that are the outcomes of the threshold tests must also be sent to the coded bitstream. The decoder can thereby follow the same procedure as the encoder as it receives bits from the codestream. The order of visiting the quadrants and the "0"-labeled sets are understood by prior agreement at the encoder and decoder. To describe the decoding algorithm, the word "encode" is replaced by "decode" and the words "write to" are replaced by "read from" and the phrases in which they appear are reversed in order. (First "read" and then "decode".) In the last line, when $n = 0$, the "0"-labeled sets are decoded and all their elements are set to 0 value. Algorithm 10.2 shows the decoding algorithm in its entirety.

Clearly, this recursive set partitioning algorithm achieves the objective of locating different size sets of elements with upper limits of its elements in a defined

ALGORITHM 10.2 _____

Algorithm for recursive set partitioning to decode a $2^m \times 2^m$ block of source elements.

1. Initialization:
    (a) Create a LIS, initially empty, to contain "0"-labeled (called insignificant) sets.
    (b) Put descriptor of $2^m \times 2^m$ source block onto LIS (upper left corner coordinates, plus size m).
    (c) Read from codestream buffer and decode n_{max}. Set $n = n_{max}$.
2. **Partitioning and recovering values:**
    (a) For each set on the LIS, do
        (i) Read next bit from codestream. If "0", do
            A. if set has more than one element, return to 2a.
            B. if set has one element, read next n bits from codestream and decode value. Remove set from LIS.
        (ii) Otherwise, if "1", do the following:
            A. if set has more than one element, divide the set into four equal quadrants, and add each quadrant to the end of the LIS.
            B. if set has one element, then read n bits from codestream to decode partial value. Value of "1"-associated element is decoded partial value plus 2^n.
            C. remove set from LIS.
3. If $n > 0$, set $n = n - 1$ and return to 2(a).
4. Otherwise, set the value of all elements belonging to all sets remaining in the LIS to zero, and stop.

range of values, so that they can be represented with fewer than n_{max} bits. The location information is comprised of the 4-bit binary masks that specifies whether or not a given set is quadrisected. These masks can be mapped onto a quadtree structure, as shown in the example of Figure 10.8. Therefore, this type of coding is often called quadtree coding. Note that the threshold lowering can be stopped before $2^0 = 1$, if we are content to use the smaller number of bits corresponding to a higher threshold.

The performance of the algorithm in terms of compression ratio or bits per sample reduction depends very much on the characteristics of the source data. The input data must be a series of integers, so they can represent images, quantized images and transforms, and integer-valued transforms. Whatever the integer input data, they are encoded without loss by the algorithm. The coefficients of transforms can be negative, so they are represented with their magnitudes and signs. The threshold tests are carried out on the magnitudes, and when the individual elements are encoded, one more bit is added for the sign. We have programmed this algorithm and run it to compress an integer wavelet transform of a few images to assess its performance. For our purposes now, we can just consider that the original image array has been transformed to another array of integers with a different distribution of amplitudes.[3] The results for this perfectly lossless compression are presented in Table 10.2.

So we do get compression, since the amount of location information of the masks is less than the number of bits needed to represent the elements in the partitioned input. We can get further compression, if we entropy encode the masks, the individual coefficients,

Table 10.2 Recursive block partitioning in lossless coding of images

Image	Attributes	Rate in bpp
Lena	512 × 512, 8 bpp	4.489
Goldhill	720 × 576, 8 bpp	4.839
Bike	2048 × 2560, 8 bpp	4.775
Barbara	512 × 512, 8 bpp	4.811

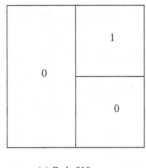   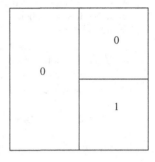

(a) Code 010   (b) Code 00

Figure 10.9   Bisection codes of significance patterns: (a) 0100 and (b) 0001.

or the signs or any combination thereof. Of course, that requires gathering of statistics and substantial increase in computation. One can usually get results comparable to the state-of-the-art by adaptive entropy coding of the masks only.

10.1.3   Binary splitting

Quadrisection is not the only means of partitioning a square block. Another potentially effective procedure is bisection or binary splitting. The idea is to split into two in one dimension followed by a second split into two in the second dimension of any significant half. An example is shown in Figure 10.9(a) in which the significant square set is split into two vertically and the right hand significant half is split horizontally. We adopt the usual convention that a "0" signifies that a subset is insignificant and "1" that it is significant and the order is left to right in the vertical split and top to bottom in the horizontal split. Following this convention, the output bit sequence using binary splitting is 0110 for the significance pattern 0100 found by scanning the four quadrants in raster order. However, since we only split significant sets, the first "0" also signifies that the right half is significant, so the following "1" is redundant. Therefore, without loss of information, we can reduce the output bit sequence to 010. In quadsplitting, we would need four bits to represent the significance pattern 0100.

The conventional order for encoding bits in quadsplitting is also the raster order (left to right and then top to bottom), so that the significance pattern 0001 can be encoded

Table 10.3 Bisection versus quadrisection of significant square array

Significance pattern (raster order)	Quadrisection code	Bisection code
0001	000	00
0010	0010	100
0011	0011	1100
0100	0100	010
0101	0101	011
0110	0110	11010
0111	0111	11011
1000	1000	1010
1001	1001	11100
1010	1010	1011
1011	1011	11110
1100	1100	111010
1101	1101	111011
1110	1110	111110
1111	1111	111111

using quadrisection to 000 without loss of information. In fact, that is the only reduction possible in quadsplitting. More are possible in binary splitting. It may take up to three binary splits to deduce the significance pattern of the four quadrants. Whenever the raw pattern "01" occurs in any of these binary splits, it can be reduced to "0." In the one case of the significance pattern 0001 (raster order), depicted in Figure 10.9(b), the raw output pattern "0101" reduces to only the two bits "00." However, there are significance patterns needing more than two binary splits to encode. For example, take the significance pattern and quadsplit code 1101. Both the left and right halves of the vertical split are significant, giving 11. The left half horizontal split yields 10 and the right one 11. This is an example of when both halves of the first split are significant. Then we must split each of these subsets horizontally and send the additional bit patterns 10, 11, or 0 (for 01). (Pattern "00" cannot occur.) In Table 10.3 we list the bisection and quadrisection codes for all significant quadrant patterns. In order to determine whether bisection or quadrisection is preferable, one must take into account the relative frequencies of the occurrences of the patterns of significance of the four quadrants. In other words, if the data favors more frequent occurrences of the significance patterns giving shorter bisection codewords, then bisection would produce a shorter average codeword length than would the quadrisection code. Entropy coding would yield the same codes with both methods, since the codewords then would be a function only of the probabilities of the significance patterns.

In practice, bisection and quadrisection seem to give nearly identical results for ordinary images. However, bisection seems to have an advantage in adaptive coding of non-rectangular regions and generalizes more easily to higher dimensions [2, 3]. The latter is accomplished simply by bisecting each dimension in turn from lowest to highest.

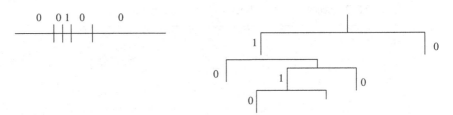

Figure 10.10 Bisection of signal: left, splitting pattern 10011001; right, corresponding binary tree (bintree) with bisection code 100100 labeled on branches.

10.1.4   One-dimensional signals

We have been using two-dimensional data arrays as our targets for encoding by set partitioning. However, bisection is a natural and efficient choice for encoding one-dimensional digital signal waveforms. A finite length signal then would be encoded by recursive binary splitting of segments that are significant for the given threshold until the significant single elements are located. The encoding algorithm is the same as Algorithm 10.1, except that there are 2^m single coordinates, sets are divided equally into two subsets, and the significance labeling of these subsets creates a 2-bit mask. The same changes obtain in the corresponding decoding algorithm, Algorithm 10.2. The splitting sequence can be mapped onto a binary tree, as shown in Figure 10.10. Note that when "01" reduces to "0," it results in a collapse of the "1" branch. This kind of binary tree is often called a *bintree*.

10.2   Tree-structured sets

The previously described sets were blocks of contiguous elements that form a partition of the source. Another kind of partition is composed of sets in the form of trees linking their elements. These so-called tree-structured sets are especially suitable for coding the discrete wavelet transform (DWT) of an image. The DWT is organized into contiguous groups of coefficients called subbands that are associated with distinct spatial frequency ranges. One can view the DWT as a number of non-overlapping spatial orientation trees. These trees are so called, because they branch successively to higher frequency subbands at the same spatial orientation. An example of a single SOT in the wavelet transform domain is in Figure 10.11. Note that the roots are the coefficients in the lowest frequency subband and branch by three to a coefficient in the same relative position in each of the other three lowest level subbands. All subsequent branching of each node is to four adjacent coefficients in the next higher level subband in the same relative position. The set of three or four nodes branching from a single node are called the *offspring* set. The set of all descendants from nodes in the offspring set is called the *grand-descendant* set. The SOTs are non-overlapping and taken altogether include all the transform coefficients. Therefore, the set of all SOTs partition the wavelet transform.

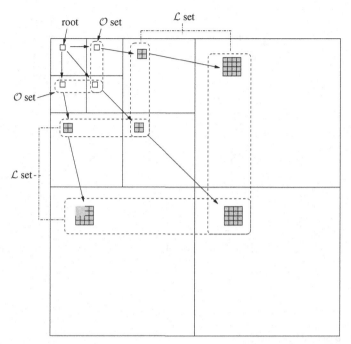

Figure 10.11 A SOT of a discrete wavelet transform. $\mathcal{O}$ denotes offspring set and $\mathcal{L}$ denotes grand-descendant set. The full descendant set $\mathcal{D} = \mathcal{O} \cup \mathcal{L}$.

Although tree-structured partitions are used mostly with wavelet transforms, they can be used for the DCT [4] or even for the source itself, when the data organization seems appropriate. The DWT is used here to present motivation and an example of tree-structured sets. The focus here is the formation of trees and coding of the elements that are the nodes in these trees. It is immaterial as to how the dataset or transform was generated. Within these trees, we wish to locate subsets whose coefficients have magnitudes in a known range between two thresholds. We take our hint from the recursive block procedure just described.

Here we start with a number of these tree-structured sets, one for each coefficient in the lowest frequency subband. However, now we rearrange the coefficients of each tree into a block, as shown in Figure 10.12. Let us call each such block a tree-block. Each tree-block is coded in turn as in Algorithm 10.1. However, we have the option of using either a single LIS and passing through all the blocks in turn at the same threshold or using an LIS for each block and staying within the same block as the threshold is successively lowered. For the latter option, we obtain essentially a separate codestream for each block, while in the former we encode all higher value coefficients across the blocks before the lower value ones.

The procedure that uses an LIS for each tree-block can be summarized as follows. Assume the DWT is a square $M \times M$ array of coefficients $c_{i,j}, i, j = 1, 2, \ldots, M$. Let $\mathcal{H}$ denote the set of coefficients in the lowest spatial frequency subband. Every $c_{i,j}$ in $\mathcal{H}$ is the root of a SOT. Group the coefficients of each SOT with root in $\mathcal{H}$ into a tree-block.

tree   tree-block

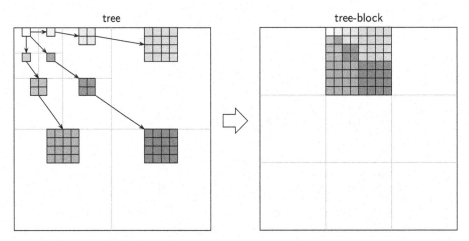

Figure 10.12   Rearranging a spatial orientation tree into a tree-block placed in the position of the image region it represents.

ALGORITHM 10.3 _____

Algorithm for encoding tree-blocks

1. Form the tree-blocks $\mathcal{B}_1, \mathcal{B}_2, \ldots, \mathcal{B}_K$.
2. Create a LIS initially empty. LIS will contain "0"-labeled (called insignificant) sets.
3. Put coordinates of roots of tree-blocks on the LIS.
4. Calculate $n_{\max}$, the highest bitplane index among all the blocks.
5. Set $n = n_{\max}$.
6. For $k = 1, 2, \ldots, K$, set m=D and execute Step 10.1 of Algorithm 10.1 on $\mathcal{B}_k$.
7. If $n > 0$ and there are multiple element sets on the LIS, set $n = n - 1$ and return to 6.
8. If $n > 0$ and all the LIS sets comprise single elements, encode each with n bits, write to codestream, and stop.
9. If $n = 0$, stop.

For a dyadic DWT with D levels of decomposition, there will be $M^2/2^{2D}$ tree-blocks with dimensions $2^D \times 2^D$ elements. Denote these tree-blocks $\mathcal{B}_1, \mathcal{B}_2, \ldots, \mathcal{B}_K$, $K = M^2/2^{2D}$. They are usually ordered either by a raster scan or zigzag scan of their roots in $\mathcal{H}$. For each tree-block $\mathcal{B}_k$, encode with the set partitioning algorithm, Algorithm 10.1, with $m = D$ while generating a corresponding LIS_k. Note that each tree-block has its own most significant bitplane $n_{\max}$. The codestream is a concatenation of the separate codestreams for each $\mathcal{B}_k$.

The alternative procedure uses a single LIS and passes through all the tree-blocks at the same threshold. It starts as above with the tree-blocks $\mathcal{B}_1, \mathcal{B}_2, \ldots, \mathcal{B}_K$, $K = M^2/2^{2D}$. When a tree-block is fully tested and partitioned at a certain threshold, the next tree-block is entered for testing and partitioning at the same threshold. Insignificant sets are put on the common LIS. The steps of this encoding algorithm are delineated in Algorithm 10.3. The decoding algorithm follows in the usual obvious way.

10.2.1 A different wavelet transform partition

10.2.1.1 Spatial orientation tree encoding

Another method for partitioning the wavelet transform is the one introduced in the SPIHT coding algorithm [5], which will be visited later. Here, we describe essentially the same method, but on a different configuration of the spatial orientation trees than defined in SPIHT. We consider here the same SOTs as before with the same significance test. However, we do not gather the coefficients of the trees into blocks. We test a given tree for significance at the current threshold and if significant, we divide the tree into its immediate offspring (child) coefficients and the set of all coefficients descending from the child coefficients, as shown in Figure 10.11. The latter set is called the grand-descendant set. We start with a full tree rooted at a coefficient in $\mathcal{H}$, the lowest frequency subband. When this tree tests significant, it is split into child nodes (coefficients) and the grand-descendant set. The child coefficients are individually tested for significance, followed by testing of the grand-descendant test for significance at the same threshold. If the grand-descendant set is significant, it is split into its component subtrees descended from the child coefficients, again depicted in Figure 10.11. Now each of these subtrees (minus their roots) is tested for significance and split as before into its child coefficients and their grand-descendant set, the set of all coefficients descended from these child coefficients. The testing of the individual child coefficients and the grand-descendant set of each of these subtrees continue in the same manner until there are empty grand-descendant sets at the terminal nodes of the trees.

In order to keep track of the various sets, we use here two lists: one for keeping track of single coefficients, called list of insignificant pixels (LIP); and LIS, for keeping track of sets of more than one coefficient. All single coefficients not yet tested at the current threshold are put onto the LIP. All sets not tested at the current threshold are put onto the LIS. Therefore, when a tree is split, the coordinates of the child nodes (three for the first split in a tree, four thereafter) go to the LIP and the locator of a grand-descendant set goes to the LIS. The coordinates of the tree root together with a Type B tag, indicating a grand-descendant set, constitute the locator. Now the LIP coefficients are tested at the same threshold. If an LIP member is significant, it is removed from the LIP, and a "1" and the code of its value (magnitude and sign) are written to the codestream.[4] If insignificant, it remains on the LIP and a "0" is written to the codestream. After all the LIP coefficients are tested, the LIS is tested. If insignificant, it remains on the LIS and a "0" is written to the codestream. If significant, it is removed from the LIS and partitioned into the subtrees descending from the child nodes (see Figure 10.11). The subtrees are put onto the LIS for immediate testing. They are designated by their roots, the child node coordinates, together with a Type A tag, to indicate all descendents of the child node root, but not including this root.

We see that there are two kinds of splitting, one that produces Type A LIS entries, indicating all descendents of the root, and one producing Type B LIS entries, indicating grand-descendant sets. The initial tree and the descendants of a child node are Type A sets. The use of an additional list, the LIP, for single insignificant coefficients, is convenient in this kind of splitting procedure that explicitly splits off single coefficients whenever any multiple element set is split. As with the previous procedures, the single elements on the LIP are always tested before the multiple element ones on the LIS. The justification is that the coefficients in a SOT generally decrease in magnitude from top to bottom and that the children of a significant root are highly likely to be significant either at the same or the next lower threshold.

10.2.1.2 Encoding SOTs together

When we search through one tree at a time for significant pixels and sets, according to the procedure above, it is called a *depth-first* search. When we search across the trees for significant pixels and sets, it is called a *breadth-first* search. In such a search procedure, all the significant pixels at a given threshold are determined before re-initiating the search at the next lower threshold. Therefore, we achieve an encoding that builds value in each threshold pass as quickly as possible. We now combine the above splitting and list management procedures to create the breadth-first SOT encoding algorithm. The following sets of coordinates are used to present this coding algorithm:

- $\mathcal{O}(i, j)$: set of coordinates of all offspring of the node (i, j);
- $\mathcal{D}(i, j)$: set of coordinates of all descendants of the node (i, j);
- $\mathcal{H}$: set of coordinates of all spatial orientation tree roots (nodes in the lowest spatial frequency subband);
- $\mathcal{L}(i, j) = \mathcal{D}(i, j) - \mathcal{O}(i, j)$, the grand-descendant set.

For instance, except at the highest and lowest tree levels, we have

$$\mathcal{O}(i, j) = \{(2i, 2j), (2i, 2j + 1), (2i + 1, 2j), (2i + 1, 2j + 1)\}. \tag{10.8}$$

We also formalize the significance test by defining the following function

$$\Gamma_n(\mathcal{T}) = \begin{cases} 1, & \max_{(i,j) \in \mathcal{T}} \{|c_{i,j}|\} \geq 2^n, \\ 0, & \text{otherwise,} \end{cases} \tag{10.9}$$

to indicate the significance of a set of coordinates $\mathcal{T}$. To simplify the notation of single pixel sets, we write $\Gamma_n(\{(i, j)\})$ as $\Gamma_n(i, j)$.

As stated previously, we use parts of the spatial orientation trees as the partitioning subsets in the sorting algorithm. The set partitioning rules are simply:

1. the initial partition is formed with the sets $\{(i, j)\}$ and $\mathcal{D}(i, j)$, for all $(i, j) \in \mathcal{H}$;
2. if $\mathcal{D}(i, j)$ is significant, then it is partitioned into $\mathcal{L}(i, j)$ plus the four single-element sets with $(k, l) \in \mathcal{O}(i, j)$.
3. if $\mathcal{L}(i, j)$ is significant, then it is partitioned into the four sets $\mathcal{D}(k, l)$, with $(k, l) \in \mathcal{O}(i, j)$.

As explained before, to maintain proper order for testing significance of sets, the significance information is stored in two ordered lists, LIS, and LIP. In these lists each entry is identified by a coordinate (i, j), which in the LIP represents individual pixels, and in the LIS represents either the set $\mathcal{D}(i, j)$ or $\mathcal{L}(i, j)$. To differentiate between them,

ALGORITHM 10.4 _____

SOT encoding algorithm: encoding an image wavelet transform along SOTs

1. **Initialization:** output $n = \lfloor \log_2 \left(\max_{(i,j)} \{|c_{i,j}|\} \right) \rfloor$; add the coordinates $(i, j) \in \mathcal{H}$ to the LIP, and also to the LIS, as type A entries.
2. **Sorting pass:**
   (a) for each entry (i, j) in the LIP do:
      (i) output $\Gamma_n(i, j)$;
      (ii) if $\Gamma_n(i, j) = 1$ then output (write to codestream) magnitude difference from 2^n (n bits) and sign of $c_{i,j}$;
   (b) for each entry (i, j) in the LIS do:
      (i) if the entry is of type A then
         - output $\Gamma_n(\mathcal{D}(i, j))$;
         - if $\Gamma_n(\mathcal{D}(i, j)) = 1$ then
            - for each $(k, l) \in \mathcal{O}(i, j)$ do:
               · output $\Gamma_n(k, l)$;
               · if $\Gamma_n(k, l) = 1$, output the magnitude difference from 2^n (n bits) and sign of $c_{k,l}$;
               · if $\Gamma_n(k, l) = 0$ then add (k, l) to the end of the LIP;
            - if $\mathcal{L}(i, j) \neq \emptyset$ then move (i, j) to the end of the LIS, as an entry of type B; otherwise, remove entry (i, j) from the LIS;
      (ii) if the entry is of type B then
         - output $\Gamma_n(\mathcal{L}(i, j))$;
         - if $\Gamma_n(\mathcal{L}(i, j)) = 1$ then
            - add each $(k, l) \in \mathcal{O}(i, j)$ to the end of the LIS as an entry of type A;
            - remove (i, j) from the LIS.
3. **Threshold update:** decrement n by 1 and go to Step **2**.

_____

we say that a LIS entry is of type A if it represents $\mathcal{D}(i, j)$, and of type B if it represents $\mathcal{L}(i, j)$.

All the pixels in the LIP – which were insignificant in the previous threshold pass – are tested in turn, and those that become significant are encoded and outputted. Similarly, sets are sequentially evaluated following the LIS order, and when a set is found to be significant, it is removed from the list and partitioned. The new subsets with more than one element are added back to the LIS, while the single-coordinate sets are added to the end of the LIP or encoded, depending whether they are insignificant or significant, respectively. Algorithm 10.4 delineates the entire procedure.

One important characteristic of the algorithm is that the entries added to the end of the LIS in Step 2(b)(ii) are evaluated before that same sorting pass ends. So, when we say "for each entry in the LIS," we also mean those that are being added to its end. Note that in this algorithm, all branching conditions based on the significance test outcomes Γ_n – which can only be calculated with the knowledge of $c_{i,j}$ – are output by the encoder. Thus, the decoder algorithm duplicates the encoder's execution path as it receives the significance test outcomes. Thus, to obtain the decoder algorithm, we simply have to replace the words *output* by *input*. The other obvious difference is that the value $n_{\max}$, which is received by the decoder, is not calculated.

10.2.2   Data-dependent thresholds

Using a set of fixed thresholds obviates the transmission of these thresholds as side information. Choices of thresholds and their number have an impact on compression efficiency. One wants to identify as many newly significant sets as possible between successive thresholds, but at the same time have these sets be small and identify large insignificant sets. Choosing too many thresholds may result in sending redundant information, since too few changes may occur between two thresholds. Choosing too few thresholds will inevitably lead to large significant sets that require many splitting steps to locate the significant elements. The best thresholds for a given source are necessarily data-dependent and must decrease as rapidly as possible without producing too many large significant sets.

Consider the thresholds to be the maximum (magnitude) elements among the sets split at a given stage. For example, say we have found the maximum value in a set and split this set into four subsets. Find the maximum elements in each of these four subsets. Split into four any subset having an element whose value equals the set maximum. Such a subset is said to be significant. Now repeat the procedure for each of these four new subsets and so on until the final splits that produce four single elements. These single elements are then encoded and outputted to the codestream. Clearly, we are creating subsets, the ones not further split, whose elements are capped in value by the full set maximum. These may be called the insignificant sets. The maximum of each subset that is not further split is less than the previous maximum. Then each of these subsets is split, just as the full set, but with a lesser maximum value to guide further splitting. These subsets should be tested from smallest to largest, that is, in reverse of the order they were generated. The reason is that we can more quickly locate significant single elements, thereby finding earlier value information, when the sets are smaller and closer to those already found. Clearly, we are producing a series of decreasing thresholds, as so-called insignificant sets are generated. The steps in the modification of the fixed threshold algorithm, Algorithm 10.1, to an adaptive one are presented in detail in Algorithm 10.5.

One major difference from the previous preset thresholds is that the thresholds generated here depend on the original data, so have to be encoded. Prior to splitting of any set, its maximum is found. An efficient scheme is to generate a series of 4-bit binary masks, where a "1" indicates a significant set (having an element equal to the maximum) and "0" an insignificant set (no element within having a value equal to the maximum). A "1" in one of these masks specifies that the maximum is to be found in this set. A "0" indicates that the maximum element is not in the associated set and that a new (lesser) maximum has to be found and encoded. The masks themselves have to be encoded and outputted, so that the decoder can follow the same splitting procedure as the encoder. The maxima can be coded quite efficiently through differences from the global maximum or a similar differencing scheme. Each 4-bit mask can be encoded with the natural binary digits or with a simple, fixed Huffman code of 15 symbols (four "0"s cannot occur).

The above scheme works well when there are relatively large areas with 0-value elements, as found naturally in the usual image transforms. The compression results are comparable to those yielded from fixed thresholds. If the probabilities of values are

ALGORITHM 10.5 ──

Algorithm for recursive set partitioning with adaptive thresholds to encode a $2^m \times 2^m$ block of non-negative source elements.

1. **Initialization:**
   (a) Find v_{max}, the largest magnitude among the source elements.
   (b) Create a LIS initially empty. LIS will contain "0"-labeled (called insignificant) sets.
   (c) Put coordinates of $2^m \times 2^m$ source block onto LIS. (Only upper-left corner coordinates are needed.) Label set with "0".
   (d) Set threshold $t = v_{max}$ and bitplane index $n = \lfloor \log_2 t \rfloor$. Encode t and write to codestream buffer.

2. **Testing and partitioning:**
   (a) For each multiple element set on the LIS (all "0"-labeled sets with more than one element), do
     (i) if maximum element less than t, write "0" to codestream buffer, else write "1" to codestream buffer and do the following:
       A. Remove set from LIS and divide the set into four equal quadrants.
       B. Find maximum element of each quadrant $v_q, q = 1, 2, 3, 4$. For each quadrant, label with "1", if $v_q = t$, otherwise label with "0" if $v_q < t$. (This step creates a 4-bit binary mask.)
       C. Add coordinates of "0"-labeled quadrants onto LIS. Encode 4-bit binary mask and write to codestream buffer.
       D. For each quadrant of size 2×2 or greater labeled with "1", go to 2(a)(i)A.
       E. When quadrants have been split into single elements, encode "0"-labeled elements with $n + 1$ bits and write to codestream buffer. ("1"-labeled single elements have values equal to the known t.)

3. If $t > 0$ and there are multiple element sets on the LIS, reset t equal to maximum of all the maxima of the LIS sets, encode t, set $n = \lfloor \log_2 t \rfloor$, and return to 2(a).

4. If $t = 0$, stop.

──

approximately or partially known beforehand, it is advisable to index the values, so that the highest probability value has 0 index, the next highest 1, and so forth. Then this set partitioning and encoding procedure operates on these indices. The association of the indices to the actual values must be known beforehand at the decoder to reconstruct the source data.

10.2.3 Adaptive partitions

The methods described above always split the set into fixed-shape subsets, such as quadrants of a block or children and grand-descendant sets of a tree. Also, the iterative splitting into subsets numbering four seems somewhat arbitrary and other numbers may prove better. Since the attempt is to find clusters of insignificant (or significant) elements, why not adjust the partition shape to conform better to the outer contours of the clusters? As stated before, methods whose sets are exactly these clusters need too much location information as overhead. But there are methods that try to vary the number of clusters and the splitting shape in limited ways that prove to be quite effective for compression.

One such method is a form of *group testing*. The basic procedure is to start with a group of k linearly ordered elements in the set K. We wish to test this set as before to

Table 10.4 Group testing of binary sequence 00000110

Step number	Test elements (in braces)	Output code
1	⌈00000110⌉	1
2	⌈0000⌉0110	0
3	0000⌈01⌉10	1
4	0000⌈0⌉110	0
5	000001⌈10⌉	1
6	000001⌈1⌉0	1
7	0000011⌈0⌉	0

locate significant elements and clusters of insignificant elements. Again, we envision a threshold test of a set that declares "1" as significant when there is a contained element whose value passes the test and declares "0" as insignificant when all contained elements fail the test. If the set K is significant, it is then split into two nearly equal parts, K_1 and K_2 of k_1 and k_2 elements, respectively. The smaller part, say K_1, is first tested for significance. If significant, K_1 is further split into two nearly equal subsets. If insignificant, then the other part, K_2, must be significant, so it is split into two nearly equal parts. This testing and splitting is repeated until the significant single elements are located. Whenever the threshold test is enacted, a "1" or "0" is emitted to indicate a significant or insignificant result, respectively. We call this procedure a group iteration, where K is the group.

Consider the example of coding the binary sequence $K = \{00000110\}$ of $k = 8$ elements in Table 10.4. Here the threshold is 1, so "0" denotes an insignificant element and "1" a significant one. The test of K produces an output "1," because there is at least one "1" in the sequence. Therefore, it is split into the subsequences 0000 and 0110, whereupon the test of the first produces a "0" output, because it is insignificant. The second subsequence must be significant, so it is split into the dibits 01 and 10. The first digit is significant, so a "1" is output, and it is split into 0 and 1. Since the 0 is insignificant, the second element must be significant, so only a "0" is output for 0 and 1. The 10 is significant, so a "1" is output and it is likewise split into single digits 1 and 0. Since the first digit is 1, a "1" is output, followed by the second digit's "0," since it too has to be tested. The resulting code is 1010110, seven digits compared to the original eight in this example.

Now when we have a large set of elements, we do not want to start the above iteration with the full set, because there is an optimum group size for the above iteration, based on the probability θ of a "0" or insignificant element. It turns out there is an equivalence between the encoding procedure above and the Golomb code [6], where it has been shown that a good rule for choosing group size k is k that satisfies[5]

$$\theta^k \approx \frac{1}{2},$$

but not less than $\frac{1}{2}$.

You usually do not know a priori the value of θ and it is often too inconvenient or practically impossible to pass through all the data to estimate it. Furthermore, this value may vary along the data set. Therefore, an adaptive method, based on a series of estimates of θ is a possible solution. One method that seems to work well is to start with group size $k = 1$ and double the group size on the next group iteration if no significant elements are found on the current iteration. Once the significant elements are identified, we can estimate the probability θ as the number of insignificant elements identified so far divided by the total of elements thus far identified. For example, in the single group iteration of Table 10.4, the probability estimate would be $\frac{6}{8} = 0.75$. Therefore, we would take the next group size $k = 2$, according to the approximate formula above with $\theta = 0.75$. We would accumulate the counts of insignificant and total elements thus far identified and update these estimates as we pass through the data with these varying group sizes.

10.3 Progressive transmission and bitplane coding

All the coding techniques so far described have in common that higher magnitude elements are encoded before lower value ones. Therefore, they provide a form of progressive transmission and recovery, since the reconstruction fidelity is increasing at the fastest possible rate as more codewords are received. One might call such encoding progressive in value. Another form of progressive encoding of finer granularity is achieved by progressive bitplane coding, when the same bitplanes of currently significant elements are sent together from higher to lower bitplanes. In order to realize such encoding, the significant elements cannot be immediately outputted, but instead are put onto another list, the list of significant pixels (LSP). These elements are represented by magnitude in natural binary and sign. After the next lower threshold, the intermediate bitplanes of all previously significant elements are output to the codestream. When the thresholds are successive integer powers of 2, then only the current bitplane bits of previously significant elements are outputted. An illustration of the LSP with progressive bitplane coding appears in Figure 10.13 and 10.14 for cases of successive and non-successive integer powers of two thresholds.

In the non-successive case, where decreasing group maxima are the thresholds, it may be more efficient to set the thresholds to the largest power of two in the binary representation of magnitude, in other words, to the highest non-zero bit level. That is, for a given maximum V, set the threshold to 2^n, where $n = \lfloor \log_2 V \rfloor$. In this way, the thresholds will decrease more rapidly, giving fewer passes and fewer and more compact transmission of thresholds that are represented with small integers.

One of the benefits of progressive bitplane coding is that the passes through bitplanes can be stopped before the end of the bottom bitplane to produce a lossy source code that has close to the smallest squared error for the number of bits encoded up to the stopping point.[6] This means of lossy coding is called *implicit quantization*, because its effect is a quantization step size one-half of the threshold corresponding to the bitplane where the coding stops. The partial ordering of LSP coefficients by bitplane significance

S	S	S	S	S	S	S	S	S	S	S	S	S	S	S
msb 5: 1	1	0	0	0	0	0	0	0	0	0	0	0	0	0
4: →	→	1	1	0	0	0	0	0	0	0	0	0	0	0
3: →	→	→	→	1	1	1	1	0	0	0	0	0	0	0
2: →	→	→	→	→	→	→	→	1	1	1	1	1	1	1
1: →	→	→	→	→	→	→	→	→	→	→	→	→	→	→
lsb 0: →	→	→	→	→	→	→	→	→	→	→	→	→	→	→

Figure 10.13 Progressive bitplane coding in the LIS for successive power of 2 thresholds. S above the most significant bit (msb) stands for the bit indicating the algebraic sign.

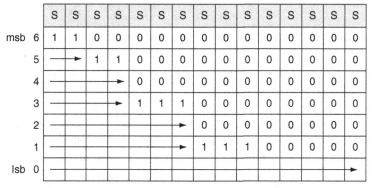

S	S	S	S	S	S	S	S	S	S	S	S	S	S	S
msb 6: 1	1	0	0	0	0	0	0	0	0	0	0	0	0	0
5: →	→	1	1	0	0	0	0	0	0	0	0	0	0	0
4: →	→	→	→	0	0	0	0	0	0	0	0	0	0	0
3: →	→	→	→	1	1	1	0	0	0	0	0	0	0	0
2: →	→	→	→	→	→	→	→	0	0	0	0	0	0	0
1: →	→	→	→	→	→	→	→	1	1	1	0	0	0	0
lsb 0: →	→	→	→	→	→	→	→	→	→	→	→	→	→	→

Figure 10.14 Progressive bitplane coding in the LIS for general thresholds. S above the most significant bit (msb) stands for the bit indicating the algebraic sign.

and successive refinement from highest to lowest bitplanes gives almost the smallest distortion possible for the bit rate corresponding to the stopping point. For a codestream of a given rate (number of bits), all the best lower rate codes can be extracted just by cutting the end of the codestream. Such a codestream is said to be *embedded* and, in this case, is said to be *bit embedded* or *finely embedded,* because the stream can be cut to any lower number of bits.

10.4   Applications to image transform coding

Set partition coding methods are especially suitable for coding transforms of images. The reason is that transforms compact the energy of the source into a relatively small percentage of the transform coefficients, leaving relatively large regions of zero or small values. The two most prevalent transforms, the DCT and the DWT, are clear examples of this phenomenon. The typical DCT of an image contains large values at low spatial frequency coefficients and small or zero values otherwise. A DWT produces coefficients belonging to spatial frequency ranges called *subbands*. The size (number of coefficients) of subbands increases in octaves from low to high frequency. The typical DWT of an image contains coefficients of large values in the low spatial frequency subbands and decreasing values as the spatial frequency of its subbands increases. For a stationary Gauss–Markov image source, the DCT coefficients are Gaussian and asymptotically statistically independent [8]. Subbands of transforms generated by ideal ("brick-wall") filtering of a stationary Gaussian source are statistically independent, because they have no overlap in frequency. These asymptotic or ideal statistical properties of transforms have motivated much of the original coding methodology in the literature. Images, like most real-world sources, are neither stationary nor Gaussian and realizable filters are not ideal, so non-statistically-based techniques, such as set partition coding, have arisen to treat these more realistic sources whose transforms have these energy clustering characteristics.

In this section, we shall describe the application of methods of set partition coding to image coding. However, because our focus is the particular coding algorithm in its use to take advantage of certain data characteristics, we deliberately do not describe here the mathematical operations or implementation of these transforms. They are amply described in Chapter 7. We describe here just the characteristics of the output of these transforms that are needed for proper application of the coding algorithms.

10.4.1 Block partition coding and amplitude and group partitioning (AGP)

We consider, for the sake of simplicity, a square $M \times M$ image that has been transformed either blockwise by a $2^n \times 2^n$, $n = 2, 3, 4$ DCT or by a DWT yielding subbands of sizes $M2^{-m} \times M2^{-m}$, $m = 1, 2, \ldots, D$, where D is the number of stages of decomposition. For example, the JPEG standard calls for 8×8 DCT blocks and D is typically 3 to 6 in wavelet transform coders. In either case, we are faced with square blocks of transform coefficients.

One of the problems confronting us is that of large alphabet size. For almost any method of entropy coding, large alphabet means large complexity. For the block partition method (Algorithm 10.1) which uses group maxima as thresholds, this complexity is reflected in many closely spaced maxima and therefore many passes through the data, so before we enact this method, we reduce the alphabet size substantially using *alphabet partitioning*. Alphabet partioning consists of forming groups of values and representing any value by two indices, one for the group, called the *group index*, and one for the position or rank within the group, called the *symbol index*. We itemize the steps in the explanation of alphabet partitioning below.

- The source alphabet is partitioned, before coding, into a relatively small number of groups;
- Each data symbol (value) is coded in two steps: first the group to which it belongs (called *group index*) is coded; followed by the rank of that particular symbol inside that group (the *symbol index*);

- When coding the pair (group index, symbol index) the group index is entropy-coded with a powerful and complex method, while the symbol index is coded with a simple and fast method, which can be simply the binary representation of that index.

A theory and procedure for calculating optimal alphabet partitions has been presented by Said [9]. The advantage of alphabet partitioning comes from the fact that it is normally possible to find partitions that allow large reductions in the coding complexity and with a very small loss in the compression efficiency.

In the current case, the powerful entropy coding method is block partitioning as in Algorithm 10.1. The particular alphabet partitioning scheme to be used in the image coding simulations is shown in Table 10.5. The groups are adjacent ranges of magnitudes (magnitude sets) generally increasing in size and decreasing in probability from 0 to higher values. The groups are indexed so that 0 has the highest probability, 1 the next lower, and so forth, until the lowest probability has the highest index. These group indices are the symbols coded by block partitioning. The table indicates whether a sign bit is needed and lists the number of bits needed to code the values inside each group under the column marked "magnitude-difference bits." Knowledge of the group index

Table 10.5 Example of an effective alphabet partitioning scheme for subsequent block partition coding

Magnitude set (MS)	Amplitude intervals	Sign bit	Magnitude-difference bits
0	[0]	no	0
1	[−1], [1]	yes	0
2	[−2], [2]	yes	0
3	[−3], [3]	yes	0
4	[−5, −4], [4, 5]	yes	1
5	[−7, −6], [6, 7]	yes	1
6	[−11, −8], [8, 11]	yes	2
7	[−15, −12], [12, 15]	yes	2
8	[−23, −16], [16, 23]	yes	3
9	[−31, −24], [24, 31]	yes	3
10	[−47, −32], [32, 47]	yes	4
11	[−63, −48], [48, 63]	yes	4
12	[−127, −64], [64, 127]	yes	6
13	[−255, −128], [128, 255]	yes	7
14	[−511, −256], [256, 511]	yes	8
15	[−1023, −512], [512, 1023]	yes	9
16	[−2047, −1024], [1024, 2047]	yes	10
17	[−4095, −2048], [2048, 4095]	yes	11
18	[−8191, −4096], [4096, 8191]	yes	12
19	[−16383, −8192], [8192, 16383]	yes	13
20	[−32767, −16384], [16384, 32767]	yes	14
21	[−65535, −32768], [32768, 65535]	yes	15
⋮	⋮	⋮	⋮

reveals the smallest magnitude in the group and the symbol index is just the difference from this smallest magnitude. Notice that there are only 22 group indices associated with a 2^{16} range of magnitudes. This range is normally sufficient for a DCT or DWT of images with amplitude range 0–255.

The raw binary representation of the symbol indices results only in a small loss of coding efficiency and no loss at all when the values inside the groups are uniformly distributed.

10.4.2   Enhancements via entropy coding

We have already mentioned that the 4-bit masks can be simply and efficiently encoded with a fixed 15-symbol Huffman code. Another simple entropy coding can be enacted once the block partitioning reaches the final 2×2 group. If the maximum of this group is small, then a Huffman code on the extension alphabet is feasible. For example, suppose that the maximum is 2, then the extension alphabet has $3^4 - 1$ possible 4-tuples ((0,0,0,0) cannot occur), so is feasible for coding. For higher maxima, where the extension alphabet is considerably larger, the elements may be encoded individually, say with a simple adaptive Huffman code. This strategy of switching between Huffman codes of different alphabet sizes is made possible by calculating and sending the maximum values of the groups formed during group partitioning.

10.4.3   Traversing the blocks

The coding of the blockwise DCT of an image usually proceeds in the usual raster scan of the blocks from top left to bottom right, where each DCT block is independently coded. Independent coding of the DCT blocks favors no scan mode over another. For the DWT, it matters how one scans the subbands when coding, because the subbands have different statistical properties, even for a stationary source. One should start with subbands of lowest spatial frequency, where the energy is highest and move next to a subband of the same or higher spatial frequency. One should never move next to a subband of lower spatial frequency. The usual such scan is shown in Figure 10.15 for a three-level wavelet decomposition.

An inefficiency of the scanning mode in Figure 10.15 is that each subband has to be tested individually for significance, so that areas larger than a single subband can never be insignificant and signified with a single 0 bit. To remedy this situation, another layer of partitioning, called *octave band partitioning* precedes the partitioning of the coding procedure. The idea is to consider the DWT as initially consisting of two sets, one being the LL band marked as an S set and the remainder of the DWT marked as an $\mathcal{I}$ set, as depicted in Figure 10.16. The S set is coded by set partitioning, after which the $\mathcal{I}$ set is tested. If not significant, a "0" is output and the threshold is lowered or a new maximum is calculated, depending on whether fixed or data-dependent thresholds are used. If $\mathcal{I}$ is significant, it is partitioned into three subbands marked as S at the same level adjacent to the previous S set and the remainder set marked as $\mathcal{I}$. This partitioning of $\mathcal{I}$ is illustrated in Figure 10.17. Testing and partitioning of $\mathcal{I}$ continues in the same

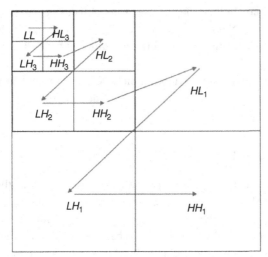

Figure 10.15 Scanning order of subbands in a 3-level wavelet decomposition. Subbands formed are named for horizontal and vertical low or highpassband and level of decomposition, e.g., LH_2 is horizontal lowpassband and vertical highpassband at second recursion level.

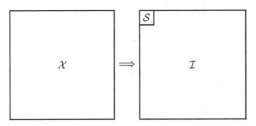

Figure 10.16 Partitioning of image $\mathcal{X}$ into sets $\mathcal{S}$ and $\mathcal{I}$.

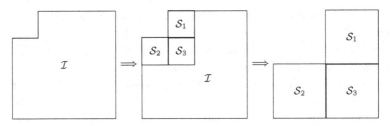

Figure 10.17 Partitioning of set $\mathcal{I}$.

manner until $\mathcal{I}$ becomes empty, as indicated in Figure 10.17. Octave band partitioning was first used in wavelet transform coding by Andrew [10] in the so-called SWEET method and then utilized in the embedded, set partition coding method SPECK [11], which will be described in the next section.

Algorithm 10.6

The SPECK algorithm

1. **Initialization**
   - Partition image transform $\mathcal{X}$ into two sets: $\mathcal{S} \equiv$ root, and $\mathcal{I} \equiv \mathcal{X} - \mathcal{S}$ (Figure 10.16).
   - Output $n_{max} = \lfloor \log_2(\max_{(i,j) \in \mathcal{X}} |c_{i,j}|) \rfloor$
   - Add $\mathcal{S}$ to LIS and set LSP $= \phi$
2. **Sorting pass**
   - In increasing order of size $|\mathcal{S}|$ of sets (smaller sets first)
     - for each set $\mathcal{S} \in$ LIS do ProcessS($\mathcal{S}$)
   - if $\mathcal{I} \neq \emptyset$, ProcessI()
3. **Refinement pass**
   - for each $(i, j) \in$ LSP, except those included in the last sorting pass, output the n-th MSB of $|c_{i,j}|$
4. **Threshold update**
   - decrement n by 1, and go to step 2

10.4.4 Embedded block coding of image wavelet transforms

Using successive fixed powers of two thresholds, the octave band and block partitioning methods just described can be utilized together with progressive bitplane coding to encode an image (integer) wavelet transform. When a significant coefficient is found, instead of outputting a code for its value, only its sign is output and its coordinates are put onto a LSP. The procedure of significance testing to sort pixels and sets into the categories identified with the LIP, LIS, and LSP is called the *sorting pass*. When a sorting pass at a given n is completed, the LSP is visited to output the bits of previously significant coefficients in bitplane n. This step is just the progressive bitplane coding depicted in Figure 10.13. This step is called the *refinement pass*, because its output bits refine the values of the previously significant coefficients. The LSP is populated and the bits are sent to the codestream sequentially from the highest to lowest bitplane. We see that this codestream is *embedded*, because truncation at any point gives the smallest distortion code corresponding to the rate of the truncation point.

We can state the steps of the full algorithm with the pseudo-code in Algorithm 10.6. The functions called by this algorithm are explained in the pseudo-code of Algorithm 10.7. The full algorithm is called SPECK for Set Partioning Embedded bloCK.

After the splitting into the sets $\mathcal{S}$ and $\mathcal{I}$, the function ProcessS tests $\mathcal{S}$ and calls CodeS to do the recursive quadrisection of $\mathcal{S}$ to find the significant elements to put onto the LSP and populate the LIS with the insignificant quadrant sets of decreasing size. ProcessI tests $\mathcal{I}$ and calls CodeI to split $\mathcal{I}$ into three $\mathcal{S}$ sets and one new $\mathcal{I}$ set, whereupon ProcessS and CodeS are called to test and code the $\mathcal{S}$ sets as above.

10.4.5 A SPECK coding example

In order to clarify the SPECK coding procedure, we present an example of encoding data of the type resulting from an 8×8 two-level wavelet transform. Figure 10.18

ALGORITHM 10.7 ──

The functions used by the SPECK algorithm
Procedure ProcessS(S)

   1. output $\Gamma_n(S)$
   2. if $\Gamma_n(S) = 1$
     • if S is a pixel, then output sign of S and add S to LSP
     • else CodeS(S)
     • if $S \in$ LIS, then remove S from LIS
   3. else
     • if $S \notin$ LIS, then add S to LIS
   4. return

Procedure CodeS(S)

   1. Partition S into four equal subsets $\mathcal{O}(S)$ (see Figure 10.7)
   2. for each set $S_i \in \mathcal{O}(S)$ ($i = 0, 1, 2, 3$)
     • output $\Gamma_n(S_i)$
     • if $\Gamma_n(S_i) = 1$
       – if S_i is a pixel, output its sign and add S_i to LSP
       – else CodeS(S_i)
     • else
       – add S_i to LIS
   3. return

Procedure ProcessI()

   1. output $\Gamma_n(\mathcal{I})$
   2. if $\Gamma_n(\mathcal{I}) = 1$
     • CodeI()
   3. return

Procedure CodeI()

   1. Partition $\mathcal{I}$ into four sets—three S_i and one $\mathcal{I}$ (see Figure 10.17)
   2. for each of the three sets S_i ($i = 0, 1, 2$)
     • ProcessS(S_i)
   3. ProcessI()
   4. return

depicts this data in the usual pyramidal subband structure. This is the same data example used by Shapiro to describe his EZW image coding algorithm [12]. A Powerpoint tutorial on SPECK that includes an animation of this example is available via the link: http://www.cambridge.org/pearlman/animations/Vortrag_SPECK.ppt.

    The output bits, actions, and population of the lists are displayed in Tables 10.6 and 10.7 for two full passes of the algorithm at bitplanes $n = 5$ and $n = 4$. The following explains the notational conventions.

• i is row index and j is column index in coordinates (i, j).
• $S^k(i, j)$ under Point or Set denotes $2^k \times 2^k$ set with (i, j) upper-left corner coordinate.

	0	1	2	3	4	5	6	7
0	63	–34	49	10	7	13	–12	7
1	–31	23	14	–13	3	4	6	–1
2	15	14	3	–12	5	–7	3	9
3	–9	–7	–14	8	4	–2	3	2
4	–5	9	–1	47	4	6	–2	2
5	3	0	–3	2	3	–2	0	4
6	2	–3	6	–4	3	6	3	6
7	5	11	5	6	0	3	–4	4

Figure 10.18   Example of coefficients in an 8×8 transform used by example. The numbers outside the box are vertical and horizontal coordinates.

- (i, j)**k** under Control Lists denotes $2^k \times 2^k$ set with (i, j) upper-left corner coordinate.
- (i, j) in LSP is always a single point.

The maximum magnitude of the full transform is 63, so $n = 5$ is the initial bitplane significance level. The initial S set is the top left 2×2 subband, so $S = S^1(0, 0)$ and $(0,0)$**1** initializes the LIS, and LSP is initially empty. The set $S^1(0, 0)$ is tested and is significant, so it is quadrisected into four singleton sets added to the LIS and a "1" is output to the bitstream. These singleton sets (pixels) are tested in turn for significance. The point $(0, 0)$ with magnitude 63 is significant, so a "1" designating "significant" and a "+" indicating its sign are sent to the bitstream and it is moved to the LSP. Likewise, $(0, 1)$ is significant and negative, so a "1–" is output and its coordinate is moved to the LSP. Both $(1, 0)$ and $(1, 1)$ are insignificant with magnitudes below 32, so a "0" is output for each and they stay in the LIS. Next, the remainder set $\mathcal{I}$ is tested for significance, so a "1" is sent and it is partitioned into three new S sets and a new $\mathcal{I}$. Each of these new S sets, $S^1(0, 2)$, $S^1(2, 0)$, and $S^1(2, 2)$ are processed in turn as was $S^1(0, 0)$. Of the three, only the first tests significant and is further quadrisected to four points, one of which, $(0, 2)$, moves to the LSP. The other three points, $(0, 3)$, $(1, 2)$, and $(1, 3)$ stay in the LIS and three "0"s are output to indicate their insignificance. The other two insignificant S sets, $S^1(2, 0)$ and $S^1(2, 2)$, are added to the LIS after the single point sets with bold suffixes **1**, since they are size 2×2, larger than the single point sets. Three "0"s are also output to indicate the insignificance of each.

The algorithm continues in this way until the $n = 5$ sorting pass is complete. Note that SPECK puts out 29 raw (uncoded) bits in this first ($n = 5$) pass. Subsequent entropy coding can reduce this number of bits.

The significance threshold is now lowered to $n = 4$. The LIS at the end of the previous pass is the initial list for this pass. It transpires that only two single points are significant for $n = 4$, $(1, 0)$ and $(1, 1)$. They are accordingly moved to the LSP and removed from the LIS. Outputs are "1–" and "1+," respectively. All other LIS sets are insignificant, so stay in the LIS with "0" emitted for each to indicate insignificance. Note that there is

Table 10.6 Example of SPECK coding of wavelet transform, from bitplane $n = 5$

Comment	Point or set	Output bits	Action	Control lists
$n = 5$ Sorting $S = S^1(0, 0)$, $\mathcal{I} =$ rest				LIS = {(0, 0)**1**} LSP = ϕ
	$S^1(0, 0)$	1	quad split, add to LIS(0)	LIS = { (0, 0)**0**, (0, 1)**0**,(1, 0)**0**,(1, 1)**0**} LSP = ϕ
	(0, 0)	1+	(0, 0) to LSP	LIS = {(0, 1)**0**, (1, 0)**0**, (1, 1)**0**} LSP = {(0, 0)}
	(0, 1)	1−	(0, 1) to LSP	LIS = {(1, 0)**0**, (1, 1)**0**} LSP = {(0, 0), (0, 1)}
	(1, 0)	0	none	
	(1, 1)	0	none	
Test $\mathcal{I}$	$S(\mathcal{I})$	1	split to 3 S's, new $\mathcal{I}$	
	$S^1(0, 2)$	1	quad split, add to LIS(0)	LIS = {(1, 0)**0**, (1, 1)**0**}, (0, 2)**0**, (0, 3)**0**}(1, 2)**0**, (1, 3)**0**
	(0, 2)	1+	(0, 2) to LSP	LSP = {(0, 0), (0, 1),(0, 2)} LIS = {(1, 0)**0**, (1, 1)**0**}, (0, 3)**0**}(1, 2)**0**, (1, 3)**0**}
	(0, 3)	0	none	
	(1, 2)	0	none	
	(1, 3)	0	none	
	$S^1(2, 0)$	0	add to LIS(1)	LIS = {(1, 0)**0**, (1, 1)**0**}, (0, 3)**0**}(1, 2)**0**, (1, 3)**0**, (2, 0)**1**}
	$S^1(2, 2)$	0	add to LIS(1)	LIS = {(1, 0)**0**, (1, 1)**0**}, (0, 3)**0**}(1, 2)**0**, (1, 3)**0**, (2, 0)**1**, (2, 2)**1**}

Table 10.6 (cont.)

Comment	Point or set	Output bits	Action	Control lists
Test $\mathcal{I}$	$S(\mathcal{I})$	1	split to 3 S's	
	$S^2(0, 4)$	0	add to LIS(2)	LIS = {(1, 0)**0**, (1, 1)**0**}, (0, 3)**0**}(1, 2)**0**, (1, 3)**0**, (2, 0)**1**, (2, 2)**1**, (0, 4)**2**}
	$S^2(4, 0)$	1	quad split, add to LIS(1)	LIS = {(1, 0)**0**, (1, 1)**0**, (0, 3)**0**, (1, 2)**0**, (1, 3)**0**, (2, 0)**1**, (2, 2)**1**, (4, 0)**1**, (4, 2)**1**, (6, 0)**1**, (6, 2)**1**, (0, 4)**2**}
	$S^1(4, 0)$	0	none	
	$S^1(4, 2)$	1	quad split, add to LIS(0)	LIS = {(1, 0)**0**, (1, 1)**0**, (0, 3)**0**, (1, 2)**0**, (1, 3)**0**, (4, 2)**0**, (4, 3)**0**, (5, 2)**0**, (5, 3)**0**, (2, 0)**1**, (2, 2)**1**, (4, 0)**1**, (4, 2)**1**, (6, 0)**1**, (6, 2)**1**, (0, 4)**2**}
	(4, 2)	0	none	
	(4, 3)	1+	move (4, 3) to LSP	LSP = {(0, 0), (0, 1), (0, 2), (4, 3)} LIS = {(1, 0)**0**, (1, 1)**0**, (0, 3)**0**}(1, 2)**0**, (1, 3)**0**, (4, 2)**0**, (4, 2)**0**, (5, 2)**0**, (5, 3)**0**, (2, 0)**1**, (4, 2)**1**, (6, 0)**1**, (6, 2)**1**, (0, 4)**2**}
	(5, 2)	0	none	
	(5, 3)	0	none	
	$S^1(6, 0)$	0	none	
	$S^1(6, 2)$	0	none	
	$S^2(4, 4)$	0	add to LIS(2)	LIS = {(1, 0)**0**, (1, 1)**0**, (0, 3)**0**, (1, 2)**0**, (1, 3)**0**, (4, 2)**0**, (5, 2)**0**, (5, 3)**0**, (2, 0)**1**, (2, 2)**1**, (4, 0)**1**, (6, 2)**1**, (0, 4)**2**, (4, 4)**2**}
End $n = 5$ Sorting				LSP = {(0, 0), (0, 1), (0, 2), (4, 3)}

Table 10.7 Example of SPECK coding of wavelet transform, continued to bitplane $n = 4$

Comment	Point or set	Output bits	Action	Control lists
$n = 4$ sorting				LIS = {(1, 0)**0**, (1, 1)**0**, (0, 3)**0**, (1, 2)**0**, (1, 3)**0**, (4, 2)**0**, (5, 2)**0**, (5, 3)**0**, (2, 0)**1**, (2, 2)**1**, (4, 0)**1** , (4, 2)**1**, (6, 0)**1**, (6, 2)**1**, (0, 4)**2**, (4, 4)**2**} LSP = {(0, 0), (0, 1), (0, 2), (4, 3)}
Test LIS (0)	(1, 0)	1−	(1, 0) to LSP	
	(1, 1)	1+	(1, 1) to LSP	LIS = {(0, 3)**0**, (1, 2)**0**, (1, 3)**0**, (4, 2)**0**, (5, 2)**0**, (5, 3)**0**, (2, 0)**1**, (2, 2)**1**, (4, 0)**1**, (4, 2)**1**, (6, 0)**1**, (6, 2)**1**, (0, 4)**2**, (4, 4)**2**} LSP = {(0, 0), (0, 1), (0, 2), (4, 3), (1, 0), (1, 1)}
	(0, 3)	0	none	
	(1, 2)	0	none	
	(1, 3)	0	none	
	(4, 2)	0	none	
	(5, 2)	0	none	
	(5, 3)	0	none	
Test LIS (1)	$S^1(2, 0)$	0	none	
	$S^1(2, 2)$	0	none	
	$S^1(4, 0)$	0	none	
	$S^1(6, 0)$	0	none	
	$S^1(6, 2)$	0	none	
Test LIS (2)	$S^2(0, 4)$	0	none	
	$S^2(4, 4)$	0	none	
Refinement	(0, 0)	1	decoder adds 2^4	
	(0, 1)	0	decoder subtracts 0	
	((0, 2)	1	decoder adds 2^4	
	(4, 3)	0	decoder adds 0	
End $n = 4$				

no octave band partitioning for this nor any other lower threshold. The LSP coefficients significant for the $n = 5$ pass are now visited and the $n = 4$ bits in the binary expansion of their magnitudes are now sent to the code bitstream.

The decoder will duplicate the action of the encoder when receiving the code bitstream. In fact, if you replace the words in the column "Output bits" by "Input bits" in Tables 10.6 and 10.7, the same exact table will be built from the codestream.

10.4.6 Embedded tree-based image wavelet transform coding

In previous sections, we have described coding of the wavelet transform of an image by set partitioning along SOTs and progressive bitplane coding of significant coefficients. When putting these two coding procedures together, we obtain what is essentially the SPIHT coding algorithm [5].

We start with a SOT partition of the wavelet transform, such as that shown in Figure 10.11 and initialize, test, and partition the SOTs using the SOT encoding algorithm 10.4. However, instead of immediately outputting the values of coefficients just found significant, we merely output their signs and put their coordinates on the LSP. When a pass at a given n is completed, the LSP is visited to output the bits of previously significant coefficients in bitplane n. This step is just the progressive bitplane coding depicted in Figure 10.13. This step is called the *refinement pass*, because its output bits refine the values of the previously significant coefficients. Therefore, the bits are sent to the codestream sequentially from the highest to lowest bitplane. We see that this codestream is *embedded*, because truncation at any point gives the smallest distortion code corresponding to the rate of the truncation point.

The detailed description of the embedded SOT encoding algorithm is the same as the SOT encoding algorithm, except for the additions of an LSP list and *refinement pass*, and changes to two places, Steps 2(a)(ii) and 2(b)(i), when significant coefficients are found. The details of the algorithm are given in Algorithm 10.8.

The steps of the embedded SOT algorithm are exactly those of the SPIHT algorithm. In the initialization step, all coordinates in $\mathcal{H}$, the lowest frequency LL subband, are put on the LIP and only those with descendants are tree roots to be put onto the LIS. For trees as defined in Figure 10.11, all coordinates in $\mathcal{H}$ are tree roots, so are put onto the LIS. However, in SPIHT, the trees are configured slightly differently. For each 2×2 block in the LL subband, the upper-left point has no descendant tree. Other points in each 2×2 block are roots of trees branching along the corresponding spatial orientations. The first branching is four to the same resolution level and then also four to each subsequent level until the leaves (termini) at the highest resolution level. The branching rules or parent–offspring dependencies are illustrated in Figure 10.19. Therefore, in the initialization step for these trees, only 3/4 of the points in the LL subband are put onto the LIS.

This algorithm provides very efficient coding with low computational complexity. Small improvements in efficiency may be achieved through adaptive entropy coding of the binary significance decisions (called the *significance map*), and the refinement and sign bits. In this chapter, our objective is to present fundamental principles of theory

ALGORITHM 10.8

Embedded SOT encoding algorithm: embedded coding in spatial orientation trees – the SPIHT algorithm

1. **Initialization:** output $n = \lfloor \log_2 (\max_{(i,j)}\{|c_{i,j}|\}) \rfloor$; set the LSP as an empty list, and add the coordinates $(i, j) \in \mathcal{H}$ to the LIP, and only those with descendants also to the LIS, as type A entries.
2. **Sorting pass:**
   (a) for each entry (i, j) in the LIP do:
      (i) output $\Gamma_n(i, j)$;
      (ii) if $\Gamma_n(i, j) = 1$ then move (i, j) to the LSP and output the sign of $c_{i,j}$;
   (b) for each entry (i, j) in the LIS do:
      (i) if the entry is of type A then
         • output $\Gamma_n(\mathcal{D}(i, j))$;
         • if $\Gamma_n(\mathcal{D}(i, j)) = 1$ then
            • for each $(k, l) \in \mathcal{O}(i, j)$ do:
               • output $\Gamma_n(k, l)$;
               • if $\Gamma_n(k, l) = 1$ then add (k, l) to the LSP and output the sign of $c_{k,l}$;
               • if $\Gamma_n(k, l) = 0$ then add (k, l) to the end of the LIP;
            • if $\mathcal{L}(i, j) \neq \emptyset$ then move (i, j) to the end of the LIS, as an entry of type B; otherwise, remove entry (i, j) from the LIS;
      (ii) if the entry is of type B then
         • output $\Gamma_n(\mathcal{L}(i, j))$;
         • if $\Gamma_n(\mathcal{L}(i, j)) = 1$ then
            • add each $(k, l) \in \mathcal{O}(i, j)$ to the end of the LIS as an entry of type A;
            • remove (i, j) from the LIS.
3. **Refinement pass:** for each entry (i, j) in the LSP, except those included in the last sorting pass (i.e., with same n), output the n-th most significant bit of $|c_{i,j}|$;
4. **Threshold update:** decrement n by 1 and go to Step 2.

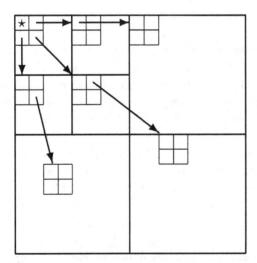

Figure 10.19 Examples of parent–offspring dependencies in the spatial-orientation trees. Coefficients in the LL band marked "*" have no descendants.

and practice in set partition coding. Therefore, we leave to a later chapter the discussion of complete coding systems.

10.4.7 A SPIHT coding example

We describe now in detail an example of SPIHT coding of the 8×8, three-level wavelet transform in Figure 10.18. A Powerpoint tutorial on SPIHT that includes an animation of this example is available via the link:

http://www.cambridge.org/pearlman/animations/Animated_SPIHT.ppt.

We applied the SPIHT algorithm to this set of data, for one pass. The results are shown in Table 10.8, indicating the data coded and the updating on the control lists (to save space only the modifications are shown). The notation is defined in the previous description of the algorithm. For a quick reference, here are some of the important definitions.

LIS List of insignificant sets: contains sets of wavelet coefficients which are defined by tree structures, and which had been found to have magnitude smaller than a threshold (are insignificant). The sets exclude the coefficient corresponding to the tree or all subtree roots, and have at least four elements.

LIP List of insignificant pixels: contains individual coefficients that have magnitude smaller than the threshold.

LSP List of significant pixels: pixels found to have magnitudes larger than or equal to the threshold (are significant).

$\mathcal{O}(i, j)$ in the tree structures, Set of offspring (direct descendants) of a tree node defined by pixel location (i, j).

$\mathcal{D}(i, j)$ Set of descendants of node defined by pixel location (i, j).

$\mathcal{L}(i, j)$ Set defined by $\mathcal{L}(i, j) = \mathcal{D}(i, j) - \mathcal{O}(i, j)$.

The following refer to the respective numbered entries in Table 10.8:

(1) These are the initial SPIHT settings. The initial threshold is set to 32. The notation (i, j)A or (i, j)B, indicates that an LIS entry is of type "A" or "B," respectively. Note the duplication of coordinates in the lists, as the sets in the LIS are trees without the roots. The coefficient $(0, 0)$ is not considered a root.

(2) SPIHT begins coding the significance of the individual pixels in the LIP. When a coefficient is found to be significant, it is moved to the LSP, and its sign is also coded. We used the notation 1+ and 1− to indicate when a bit 1 is immediately followed by a sign bit.

(3) After testing pixels, it begins to test sets, following the entries in the LIS (active entry indicated by bold letters). In this example $\mathcal{D}(0, 1)$ is the set of 20 coefficients $\{(0, 2), (0, 3), (1, 2), (1, 3), (0, 4), (0, 5), (0, 6), (0, 7), (1, 4), (1, 5), (1, 6), (1, 7), (2, 4), (2, 5), (2, 6), (2, 7), (3, 4), (3, 5), (3, 6), (3, 7)\}$. Because $\mathcal{D}(0, 1)$ is

Table 10.8 Example of image coding using the SPIHT method

Comm.	Pixel or set tested	Output bit	Action	Control lists
(1)				LIS = {(0, 1)A, (1, 0)A, (1, 1)A}
				LIP = {(0, 0), (0, 1), (1, 0), (1, 1)}
				LSP = ∅
(2)	(0, 0)	1+	(0, 0) to LSP	LIP = {(0, 1), (1, 0), (1, 1)}
				LSP = {(0, 0)}
	(0, 1)	1−	(0, 1) to LSP	LIP = {(1, 0), (1, 1)}
				LSP = {(0, 0), (0, 1)}
	(1, 0)	0	none	
	(1, 1)	0	none	
(3)	$\mathcal{D}$(0, 1)	1	test offspring	LIS = {**(0, 1)A**, (1, 0)A, (1, 1)A}
	(0, 2)	1+	(0, 2) to LSP	LSP = {(0, 0), (0, 1), (0, 2)}
	(0, 3)	0	(0, 3) to LIP	LIP = {(1, 0), (1, 1), (0, 3)}
	(1, 2)	0	(1, 2) to LIP	LIP = {(1, 0), (1, 1), (0, 3), (1, 2)}
	(1, 3)	0	(1, 3) to LIP	LIP = {(1, 0), (1, 1), (0, 3), (1, 2), (1, 3)}
(4)			type changes	LIS = {(1, 0)A, (1, 1)A, (0, 1)B}
(5)	$\mathcal{D}$(1, 0)	1	test offspring	LIS = {**(1, 0)A**, (1, 1)A, (0, 1)B}
	(2, 0)	0	(2, 0) to LIP	LIP = {(1, 0), (1, 1), (0, 3), (1, 2), (1, 3), (2, 0)}
	(2, 1)	0	(2, 1) to LIP	LIP = {(1, 0), (1, 1), (0, 3), (1, 2), (1, 3), (2, 0), (2, 1)}
	(3, 0)	0	(3, 0) to LIP	LIP = {(1, 0), (1, 1), (0, 3), (1, 2), (1, 3), (2, 0), (2, 1), (3, 0)}
	(3, 1)	0	(3, 1) to LIP	LIP = {(1, 0), (1, 1), (0, 3), (1, 2), (1, 3), (2, 0), (2, 1), (3, 0), (3, 1)}
			type changes	LIS = {(1, 1)A, (0, 1)B, (1, 0)B}

Table 10.8 (cont.)

Comm.	Pixel or set tested	Output bit	Action	Control lists
(6)	$\mathcal{D}(1, 1)$	0	none	LIS = {(**1, 1**)**A**, (0, 1)B, (1, 0)B}
(7)	$\mathcal{L}(0, 1)$	0	none	LIS = {(1, 1)A, (**0, 1**)**B**, (1, 0)B}
(8)	$\mathcal{L}(1, 0)$	1	add new sets	LIS = {(1, 1)A, (0, 1)B, (2, 0)A, (2, 1)A, (3, 0)A, (3, 1)A}
(9)	$\mathcal{D}(2, 0)$	0	none	LIS = {(1, 1)A, (0, 1)B, (**2, 0**)**A**, (2, 1)A, (3, 0)A, (3, 1)A}
(10)	$\mathcal{D}(2, 1)$	1	test offspring	LIS = {(1, 1)A, (0, 1)B, (2, 0)A, (**2, 1**)**A**, (3, 0)A, (3, 1)A}
	(4, 2)	0	(4, 2) to LIP	LIP = {(1, 0), (1, 1), (0, 3), (1, 2), (1, 3), (2, 0), (2, 1), (3, 0), (3, 1), (4, 2)}
	(4, 3)	1+	(4, 3) to LSP	LSP = {(0, 0), (0, 1), (0, 2), (4, 3)}
	(5, 2)	0	(5, 2) to LIP	LIP = {(1, 0), (1, 1), (0, 3), (1, 2), (1, 3), (2, 0), (2, 1), (3, 0), (3, 1), (4, 2), (5, 2)}
	(5, 3)	0	(5, 3) to LIP	LIP = {(1, 0), (1, 1), (0, 3), (1, 2), (1, 3), (2, 0), (2, 1), (3, 0), (3, 1), (4, 2), (5, 2), (5, 3)}
(11)			(2, 1) removed	LIS = {(1, 1)A, (0, 1)B, (2, 0)A, (3, 0)A, (3, 1)A}
(12)	$\mathcal{D}(3, 0)$	0	none	LIS = {(1, 1)A, (0, 1)B, (2, 0)A, (**3, 0**)**A**, (3, 1)A}
	$\mathcal{D}(3, 1)$	0	none	LIS = {(1, 1)A, (0, 1)B, (2, 0)A, (3, 0)A, (**3, 1**)**A**}
(13)				LIS = {(1, 1)A, (0, 1)B, (2, 0)A, (3, 0)A, (3, 1)A}
				LIP = {(1, 0), (1, 1), (0, 3), (1, 2), (1, 3), (2, 0), (2, 1), (3, 0), (3, 1), (4, 2), (5, 2), (5, 3)}
				LSP = {(0, 0), (0, 1), (0, 2), (4, 3)}

significant, SPIHT next tests the significance of the four offspring $\{(0, 2), (0, 3), (1, 2), (1, 3)\}$.

(4) After all offspring are tested, $(0, 1)$ is moved to the end of the LIS, and its type changes from "A" to "B," meaning that the new LIS entry meaning changed from $\mathcal{D}(0, 1)$ to $\mathcal{L}(0, 1)$ (i.e., from set of all descendants to set of all descendants minus offspring).

(5) The same procedure as in comments (3) and (4) applies to set $\mathcal{D}(1, 0)$. Note that even though no offspring of $(1, 0)$ is significant, $\mathcal{D}(1, 0)$ is significant because $\mathcal{L}(1, 0)$ is significant.

(6) Since $\mathcal{D}(1, 1)$ is insignificant, no action need be taken. The algorithm moves to the next element in the LIS.

(7) The next LIS element, $(0, 1)$, is of type "B," and thus $\mathcal{L}(0, 1)$ is tested. Note that the coordinate $(0,1)$ was moved from the beginning of the LIS in this pass. It is now tested again, but with another interpretation by the algorithm.

(8) The same as above, but $\mathcal{L}(1, 0)$ is sigificant, so the set is partitioned into $\mathcal{D}(2, 0)$, $\mathcal{D}(2, 1)$, $\mathcal{D}(3, 0)$, and $\mathcal{D}(3, 1)$, and the corresponding entries are added to the LIS. At the same time, the entry $(1, 0)$B is removed from the LIS.

(9) The algorithm keeps evaluating the set entries as they are appended to the LIS.

(10) Each new entry is treated as in the previous cases. In this case the offspring of $(2, 1)$ are tested.

(11) In this case, because $\mathcal{L}(2, 1) = \emptyset$ (no descendant other than offspring), the entry $(2, 1)$A is removed from the LIS (instead of having its type changed to "B").

(12) Finally, the last two entries of the LIS correspond to insignificant sets, and no action is taken. The sorting pass ends after the last entry of the LIS is tested.

(13) The final list entries in this sorting pass form the initial lists in the next sorting pass, when the threshold value is 16.

Without using any other form of entropy coding, the SPIHT algorithm used 29 bits in this first pass, exactly the same number of bits as SPECK in its first pass for the same data. The initial lists for the second pass of SPIHT at $n = 4$ are those at the end of the first pass in the last line (13) of Table 10.8.

10.4.8   Embedded zerotree wavelet (EZW) coding

An embedded SOT coding method that is not strictly a set partioning coder is the embedded zerotree wavelet (EZW) coder [12]. It is described in this chapter, because it inspired the creation of SPIHT and shares many of its characteristics. It operates on SOTs of a wavelet transform such as SPIHT, but locates only significant coefficients and trees, all of whose coefficients, including the root, are insignificant. When the thresholds are integer powers of 2, insignificant trees mean that all the bits in the corresponding bit-plane located at the coordinates of the tree nodes are 0's. Hence arises the term *zerotree*. SPIHT has also been called a *zerotree* coder, but it locates zerotrees in a broader sense than the EZW zerotrees [13]. For our purposes here, it is better to view SPIHT as a set partitioning coder.

The structure of a full spatial orientation tree used in EZW is shown in Figure 10.11. The EZW algorithm determines certain attributes of these trees and their subtrees. Given a significance threshold, EZW visits a wavelet coefficient and makes the following determinations about its significance and the subtree rooted at that coefficient. Subtrees, whose coefficients (including the root) are all insignificant, are called *zerotrees* and its root is deemed a *zerotree root*. When the subtree root is insignificant and some of its descendants are significant, it is called an *isolated zero*. When the root of a subtree is significant, it is noted whether it is positive or negative. The descendants in its subtree are not characterized in this case. Therefore, when the algorithm comes to a certain coefficient having descendants in its wavelet transform subtree, it emits one of four symbols:

- ZTR: zerotree root
- IZ: isolated zero
- POS: significant and positive
- NEG: significant and negative

When a coordinate has no descendants (i.e., it is contained within a level 1 subband or is a leaf node), then, if insignificant, a zero symbol (Z) is emitted. Note that raw encoding each of the four symbols for non-leaf nodes requires 2 bits, while encoding of the Z symbol for a leaf node sends a single "0" to the codestream. For coefficients associated with leaf nodes, the symbol alphabet changes to Z, POS, NEG, corresponding to the raw bits 0, 11, and 10, respectively. The encoder and decoder will know when the leaf node set has been reached, so can adjust its symbol alphabet accordingly.

Having characterized the symbols used in EZW, we can now describe this algorithm in more detail. Since the trees or subtrees in EZW include the root, we introduce the notation for a tree (or subtree) rooted at coordinates (i, j) as

$$\mathcal{T}(i, j) = (i, j) \bigcup \mathcal{D}(i, j), \tag{10.10}$$

the union of the root coordinate and the set of its descendants. The branching policy is the one shown in Figure 10.11. If (i, j) has descendants, we test both the root coefficient $c(i, j)$ and all the coefficients in $\mathcal{T}(i, j)$ for significance. Referring to (10.9), the outcomes are as follows.

$$\Gamma_n(\mathcal{T}(i, j)) = 0, \text{ output ZTR}$$
$$\Gamma_n(\mathcal{T}(i, j)) = 1 \text{ AND } \Gamma_n(i, j) = 0, \text{ output IZ}$$
$$\Gamma_n(i, j) = 1 \text{ AND } c(i, j) > 0, \text{ output POS}$$
$$\Gamma_n(i, j) = 1 \text{ AND } c(i, j) < 0, \text{ output NEG} \tag{10.11}$$

If (i, j) has no descendants, the outcomes are:

$$\Gamma_n(i, j) = 0, \text{ output Z}$$
$$\Gamma_n(i, j) = 1 \text{ AND } c(i, j) > 0, \text{ output POS}$$
$$\Gamma_n(i, j) = 1 \text{ AND } c(i, j) < 0, \text{ output NEG} \tag{10.12}$$

We maintain two ordered control lists, the dominant list (DL) and the subordinate list (SL).[7] The dominant list keeps track of coordinates of subtree roots that are to be examined at the current and lower thresholds. The SL stores coordinates of significant coefficients, so is identical to the LSP in the previous algorithms. The DL is initialized with the coordinates in the lowest frequency subband. The algorithm first tests entries in order from the DL. When the outcome is not ZTR and there are offspring, the coordinates of the offspring are added to the end of the list DL. When the coefficient for the entry is significant, its coordinates are added to the end of the list SL. These actions for populating the lists may be summarized as follows.

for (i, j) in DL,
– if (i, j) not ZTR AND not Z, add (k, l) in $\mathcal{O}(i, j)$ to end of DL.
– if $\Gamma_n(i, j) = 1$, add (i, j) to end of SL.

Once a coefficient tests significant, it should not be tested again. It is customary to keep it on the DL, but mark it with an internal flag to skip over it. Putting the previous procedures together leads us to the full description of the EZW algorithm. As before, we calculate n_{max}, initialize DL with the coordinates of $\mathcal{H}$, and set SL to empty. The order of scanning of the subbands follows the zigzag shown in Figure 10.15. Coordinates within a subband are visited in raster scan order from the top to the bottom row. We then visit the coordinates in the DL in order starting at $n = n_{max}$. We test these coordinates and output the symbols according to (10.11) and repopulate the DL and SL by the rules above. We continue through the subbands in the prescribed scan order at the same threshold. When we reach DL coordinates belonging to the leaf node set in the level 1 subbands, we test and output symbols by (10.12), but nothing more is added to the DL, since there are no offspring. The so-called *dominant pass* is complete, once all the DL coordinates have been visited at a given threshold n. Then the SL is entered to output the n-th bit of coefficients found significant at previous higher values of n. This pass through the SL, called in EZW the *subordinate pass*, which is identical to the *refinement pass* previously described for progressive bitplane coding, is omitted for coefficients found significant on the current pass, because the output symbols, POS and NEG, already signify that the n-th bit is 1. When the subordinate pass is complete, the threshold is lowered by a factor of 2 (n is decremented by 1) and the dominant pass at this new threshold is executed on the coordinates in the DL left from the previous pass. The coordinates belonging to coefficients that have become significant on previous passes are removed or marked or set to 0 to be skipped. This process continues either until completion of the $n = 0$ passes or the bit budget is exhausted.

10.4.8.1   An EZW coding example

We now present an example of EZW coding of the same 8×8 wavelet transform shown in Figure 10.18. The notation of **F** following a coordinate on the dominant list means that an internal flag is set to indicate "significant" on that pass and its magnitude on the dominant list is set to 0 for subsequent passes. Recall that i signifies row index

Table 10.9 Example of image coding using Shapiro's EZW method

Tree root	Output symbol	DL: dominant list SL: subordinate list
		DL = {(0,0)} SL = ∅
(0,0)	POS	DL = {(0,0)**F**,(0,1),(1,0),(1,1)} SL = {(0,0)}
(0,1)	NEG	DL = {(0,0)**F**,(1,0),(1,1),(0,2),(0,3),(1,2),(1,3),(0,1)**F**} SL = {(0,0),(0,1)}
(1,0)	IZ	DL = {(0,0)**F**,(1,0),(1,1),(0,2),(0,3),(1,2),(1,3),(0,1)**F**,(2,0),(2,1),(3,0),(3,1)}
(1,1)	ZTR	
(0,2)	POS	DL = {(0,0)**F**,(1,0),(1,1),(0,3),(1,2),(1,3),(0,1)**F**,(2,0),(2,1),(3,0),(3,1),(0,4),(0,5),(1,4), (1,5),(0,2)**F**} SL = {(0,0),(0,1),(0,2)}
(0,3)	ZTR	
(1,2)	ZTR	
(1,3)	ZTR	
(2,0)	ZTR	
(2,1)	IZ	DL = {(0,0)**F**,(1,0),(1,1),(0,3),(1,2),(1,3),(0,1)**F**, (2,0),(2,1),(3,0),(3,1),(0,4),(0,5),(1,4),(1,5),(0,2)**F**,(4,2),(4,3),(5,2),(5,3)}
(3,0)	ZTR	
(3,1)	ZTR	
(0,4)	Z	
(0,5)	Z	
(1,4)	Z	
(1,5)	Z	
(4,2)	Z	
(4,3)	POS	DL = {(0,0)**F**,(1,0),(1,1),(0,3),(1,2),(1,3),(0,1)**F**,(2,0),(2,1),(3,0), (3,1),(0,4),(0,5),(1,4),(1,5),(0,2)**F**,(4,2),(5,2),(5,3),(4,3)**F**} SL = {(0,0),(0,1),(0,2),(4,3)}
(5,2)	Z	
(5,3)	Z	
		DL = {(0,0)**F**,(1,0),(1,1),(0,3),(1,2),(1,3),(0,1)**F**,(2,0),(2,1),(3,0), (3,1),(0,4),(0,5),(1,4),(1,5),(0,2)**F**,(4,2),(5,2),(5,3),(4,3)**F**} SL = {(0,0),(0,1),(0,2),(4,3)}

and j column index. We display only the actions and outputs for the first dominant pass at the top bitplane $n = 5$. Table 10.9 shows the results.

Assuming that the EZW uses (at least initially) two bits to code symbols in the alphabet {POS, NEG, ZTR, IZ}, and one bit to code the symbol Z, the EZW algorithm used $26 + 7 = 33$ bits in the first pass. Since the SPECK, SPIHT, and EZW methods are coding the same bitplane defined by the threshold 32, both find the same set of significant coefficients, yielding images with the same MSE. However, SPIHT and SPECK used 29 bits, about 10% fewer bits to obtain the same results, because they coded different symbols.

The final lists of EZW and SPIHT may have some equal coordinates, but as shown in the examples, the interpretation and use of those coordinates by the two methods are quite different. Also, in the following passes they grow and change differently.

10.4.9   Group testing for image wavelet coding

Group testing has been applied successfully to embedded coding of image wavelet transforms. The framework is the same as that in the embedded SOT algorithm 10.8, but the sorting pass is replaced by the method of group testing. Specifically, one encodes SOTs and starts testing for significance at threshold $2^{n_{max}}$, successively lowers the thresholds by factors of 2, and sends significant coefficients to the LSP, where they are refined using progressive bitplane coding (Figure 10.13). One might consider ordering the coefficients in a SOT from lowest to highest frequency subband and use the adaptive method of setting the group sizes in a series of group iterations. However, underlying assumptions in group testing are independence and equal probabilities of significance among coefficients in the group. These assumptions are not valid for coefficients in a SOT. The solution is to divide the SOT coefficients into different classes, within which independence and equal probabilities of significance are approximately true. The details of how these classes are formed and other aspects of applying group testing in this framework are described in [15]. Here, we just want to indicate how group testing can be applied to wavelet transform coding to produce an embedded codestream.

10.5   Conclusion

In this part, we have explained principles and described some methods of set partition coding. We have also shown how these methods naturally extend to produce embedded codestreams in applications to transform coding. Although we have described a single partitioning method to a given transform, different methods can be mixed within such a transform. For example, subbands of a wavelet transform have different statistical characteristics, so that different partitioning methods may be utilized for different subbands of the same transform. For example, the SOT partitioning becomes inefficient for transforms with substantial energy in the highest spatial or temporal frequency subbands, because it produces lots of LIP entries at several thresholds as it proceeds toward the significant coefficients at the end of the tree. Once a coefficient is listed on the LIP, one bit for each threshold pass is required to encode it. Therefore, you want an LIP entry to become significant as early as possible. One solution is to terminate the trees before the highest frequency subbands. Then the descendant sets are no longer significant, so they are not further partitioned and are put onto the LIS and represented with a single "0" bit. To maintain embedded coding, for a given threshold, one could encode these high frequency subbands by block partitioning, using quadrisection or bisection, after the pass through the SOTs at the same threshold is complete.

No doubt there are other scenarios when a hybrid of two (or more) partitioning methods becomes potentially advantageous. In the next part, we shall explain further the advantageous properties of set partition coding as it is used in modern coding systems.

Problems

10.1 Consider the following 8×8 array of magnitudes of transform coefficients.

$$
\begin{bmatrix}
63 & 34 & 49 & 10 & 7 & 13 & 12 & 7 \\
31 & 23 & 14 & 13 & 3 & 4 & 6 & 1 \\
15 & 14 & 3 & 12 & 5 & 7 & 3 & 9 \\
9 & 7 & 14 & 8 & 4 & 2 & 3 & 2 \\
5 & 9 & 1 & 47 & 4 & 6 & 2 & 2 \\
3 & 0 & 3 & 2 & 3 & 2 & 0 & 4 \\
2 & 3 & 6 & 4 & 3 & 6 & 3 & 6 \\
5 & 11 & 5 & 6 & 0 & 3 & 4 & 4
\end{bmatrix}
$$

Split this array into partititions by means of recursive quadrisection governed by testing against power of 2 thresholds. Stop after tests on three different thresholds. The significance of the partitions of the first quadrisection should be tested against n_{max}.

(a) What is n_{max} in this array?

(b) Give the sequence of significance bits and draw the quadtree corresponding to the partitioning of this array.

10.2 Continuing with the previous problem, determine the number of uncoded bits needed to represent the coefficient magnitudes in each partition.

Give the code rate in bits per sample, counting the number of significance bits.

10.3 The AGP method can also be used to convey the partitioning path and coding of the partitions of the array in Problem 10.1. The thresholds for the significance tests are no longer powers of 2, but maxima of the four sets produced after a quadrisection.

(a) Perform quadrisection for three thresholds using the AGP method. Show the corresponding quadtree.

(b) Count the number of raw (uncoded) bits needed to convey the execution path and the maxima.

(c) Give the number of raw bits needed to code the coefficient magnitudes in each partition.

(d) What is the rate in bits per sample of the code for this arrray?

10.4 Consider the following one-dimensional array, which is formed by a zigzag scan of the array in Problem 10.1.

$$[63 \;\; 34 \;\; 31 \;\; 15 \;\; 23 \;\; 49 \;\; 10 \;\; 14 \;\; 14 \;\; 9 \;\; 5 \;\; 7 \;\; 3 \;\; 13 \;\; 7 \;\; 13 \;\; 3 \;\; 12$$
$$14 \;\; 9 \;\; 3 \;\; 2 \;\; 0 \;\; 1 \;\; 8 \;\; 5 \;\; 4 \;\; 12 \;\; 7 \;\; 6 \;\; 7 \;\; 4 \;\; 47 \;\; 3 \;\; 3 \;\; 5$$
$$11 \;\; 6 \;\; 2 \;\; 4 \;\; 2 \;\; 3 \;\; 1 \;\; 9 \;\; 3 \;\; 6 \;\; 3 \;\; 4 \;\; 5 \;\; 6 \;\; 3 \;\; 2 \;\; 2 \;\; 2$$
$$2 \;\; 0 \;\; 6 \;\; 0 \;\; 3 \;\; 3 \;\; 4 \;\; 6 \;\; 4 \;\; 4]$$

Partition this array using bisection governed by testing against thresholds that are powers of 2. Stop after four levels of partitioning (four thresholds).

(a) Give the sequence of significance bits and draw the binary tree corresponding to the partitioning of this array.

(b) Determine the number of uncoded bits needed to represent the coefficient magnitudes in each partition.

(c) Give the code rate in bits per sample, counting the number of significance bits.

10.5 We now identify our magnitude array as belonging to a two-level wavelet transform with seven subbands as shown in Figure 10.20. We wish to organize this data array into tree blocks prior to set partition coding.

(a) Organize the data into four tree blocks. See Figure 10.12.

(b) Partition each tree block by recursive quadrisection. Test set significance against three thresholds. Show the four quadtrees corresponding to the significance bits from each block.

(c) Determine the number of uncoded bits needed to represent the coefficient magnitudes in each partition.

(d) Give the code rate in bits per sample, counting the number of significance bits.

10.6 In this continuation of Problem 10.1, perform progressive bitplane coding to encode the top three bitplanes of significant coefficients left after three threshold passes.

(a) What is the sequence of bits from the top three bitplanes. Identify the bitplane number of these bits.

63	34	49	10	7	13	12	7
31	23	14	13	3	4	6	1
15	14	3	12	5	7	3	9
9	7	14	8	4	2	3	2
5	9	1	47	4	6	2	2
3	0	3	2	3	2	0	4
2	3	6	4	3	6	3	6
5	11	5	6	0	3	4	4

Figure 10.20   Array of magnitudes of coefficients of the two-level wavelet transform having seven subbands in Problem 10.5.

86	67	4	−1
48	41	9	0
−5	21	−1	−1
5	−8	16	2

Figure 10.21   Two-level, seven subband wavelet transform of a 4 × 4 image.

(b) What is the code rate counting these bitplane bits and the significance map bits?

(c) Stopping the procedure after three bitplanes implies quantization of the coefficients. Reconstruct the array represented by the bits from the three bitplanes and the significance map bits.

(d) Determine the MSE between the original array and the reconstructed array.

10.7 In Figure 10.21 appears an integer wavelet transform of 4 × 4 image. The transform has two levels of wavelet decompositions and seven subbands.

(a) Use the SPIHT algorithm to generate a lossless bitstream.

(b) Decode your bitstream to show that the original transform is reproduced.

(c) Cut the lossless bitstream just after the second threshold pass and decode. Show the reconstructed wavelet transform and calculate its MSE.

10.8 Repeat the previous problem using the SPECK algorithm.

10.9 Consider the one-dimensional S+P integer wavelet transform consisting of three levels of dyadic decomposition of a speech sequence of $N = 32$ samples shown below.

$$\left| 6 \ 14 \ 10 \ 8 \ \right| 0 \ -1 \ 2 \ 0 \ \left| 4 \ 1 \ -3 \ -3 \ 1 \ 0 \ 0 \ 0 \right|$$

$$\left| 1 \ -2 \ 2 \ -2 \ 0 \ -4 \ -3 \ -3 \ -2 \ 1 \ 1 \ 3 \ -2 \ 2 \ 1 \ 3 \right|$$

The vertical bars mark the boundaries of subbands from low frequency to the left to high frequency at the right. The second row contains the highest frequency band.

(a) Construct a binary branching, spatial orientation tree analogous to the two-dimensional one used by SPIHT.

(b) Give the coordinates of the initial LIP and tree roots in $\mathcal{H}$ of the SPIHT algorithm.

(c) Use the SPIHT algorithm to produce a lossless codestream for the given wavelet transform.

(d) Cut the codestream just after the second threshold pass and decode to reconstruct the transform. Calculate the MSE.

Notes

1. When the lowest possible value in the dataset is not zero, then this offset from zero is stored as overhead.

2. Sometimes there is confusion about why n_{max} is defined in terms of the floor ($\lfloor \ \rfloor$) operator. ($\lfloor x \rfloor$ is defined as the largest integer not exceeding x.) If one recalls representation of an integer in a binary expansion, then n_{max} is the highest power of 2 in that expansion starting from 2^0.

3. For these tests, we have used the integer S+P filters [1] in five levels of wavelet decomposition.

4. The "1" indicates a value equal or above a known threshold, so only the code of the difference from the threshold needs to be sent to the codestream.

5. A more precise statement of the rule is $\theta^k + \theta^{k+1} < 1 < \theta^k + \theta^{k-1}$ [7].

6. The smallest squared error is obtained either when the stopping point is the end of a bitplane or the coefficients happen to be fully ordered by magnitude.

7. These lists keep track of the actions of the algorithm and point to future actions. Other means that do not require these lists can be employed to achieve the same actions. For example, see Taubman and Marcellin [14]. SPIHT can also be implemented without lists.

References

1. A. Said and W. A. Pearlman, "An image multiresolution representation for lossless and lossy compression," *IEEE Trans. Image Process.*, vol. 5, no. 9, pp. 1303–1310, Sept. 1996.
2. J. E. Fowler, "Shape-adaptive coding using binary set splitting with k-d trees," in *Proceedings of the IEEE International Conference on Image Processing*, Singapore, Oct. 2004, pp. 1301–1304.
3. J. B. Boettcher and J. E. Fowler, "Video coding using a complex wavelet transform and set partitioning," *IEEE Signal Process Lett.*, vol. 14, no. 9, pp. 633–636, Sept. 2007.
4. Z. Xiong, O. G. Guleryuz, and M. T. Orchard, "A DCT-based embedded image coder," *IEEE Signal Process. Lett.*, vol. 3, no. 11, pp. 289–290, Nov. 1996.
5. A. Said and W. A. Pearlman, "A new, fast, and efficient image codec based on set partitioning in hierarchical trees," *IEEE Trans. Circuits Syst. Video Technol.*, vol. 6, no. 3, pp. 243–250, June 1996.
6. S. W. Golomb, "Run-length encodings," *IEEE Trans. Inf. Theory*, vol. IT-12, no. 3, pp. 399–401, July 1966.
7. R. G. Gallager and D. van Voorhis, "Optimal source codes for geometrically distributed integer alphabets," *IEEE Trans. Inf. Theory*, vol. 21, no. 2, pp. 228–230, Mar. 1975.
8. Y. Yemini and J. Pearl, "Asymptotic properties of discrete unitary transforms," *IEEE Trans. Pattern Anal. Mach. Intell.*, vol. 1, no. 4, pp. 366–371, Oct. 1979.
9. A. Said and W. A. Pearlman, "Low-complexity waveform coding via alphabet and sample-set partitioning," in *Proc. SPIE Vol. 3024: Visual Commun. Image Process.*, San Jose, CA, Feb. 1997, pp. 25–37.
10. J. Andrew, "A simple and efficient hierarchical image coder," in *Proc. IEEE Int. Conf. Image Process.*, vol. 3, Santa Barbara, CA, Oct. 1997, pp. 658–661.
11. A. Islam and W. A. Pearlman, "Embedded and efficient low-complexity hierarchical image coder," in *Proceedings of the SPIE Vol. 3653: Visual Communications Image Processing*, San Jose, CA, Jan. 1999, pp. 294–305.
12. J. M. Shapiro, "Embedded image coding using zerotrees of wavelet coefficients," *IEEE Trans. Signal Process.*, vol. 41, no. 12, pp. 3445–3462, Dec. 1993.
13. Y. Cho and W. A. Pearlman, "Quantifying the coding performance of zerotrees of wavelet coefficients: degree-k zerotree," *IEEE Trans. Signal Process.*, vol. 55, no. 6, pp. 2425–2431, June 2007.
14. D. S. Taubman and M. W. Marcellin, *JPEG2000: Image Compression Fundamentals, Standards, and Practice*. Norwell, MA: Kluwer Academic Publishers, 2002.
15. E. S. Hong and R. E. Ladner, "Group testing for image compression," *IEEE Trans. Image Process.*, vol. 11, no. 8, pp. 901–911, Aug. 2002.

Further reading

C. Chrysafis, A. Said, A. Drukarev, A. Islam, and W. A. Pearlman, "SBHP – a low complexity wavelet coder," in *Proceedings of the IEEE International Conference on Acoustics Speech Signal Processing*, vol. 4, Istanbul, Turkey, June 2000, pp. 2035–2038.

S.-T. Hsiang and J. W. Woods, "Embedded image coding using zeroblocks of subband/wavelet coefficients and context modeling," in *Proceedings of the IEEE International Symposium on Circuits Systems*, vol. 3, Geneva, Switzerland, May 2000, pp. 662–665.

Information Technology – JPEG2000 Image Coding System, Part 1: Core Coding System, ISO/IEC Int. Standard 15444-1, Geneva, Switzerland, 2000.

Information Technology – JPEG2000 Extensions, Part 2: Core Coding System, ISO/IEC Int. Standard 15444-2, Geneva, Switzerland, 2001.

A. Munteanu, J. Cornelis, G. V. der Auwera, and P. Cristea, "Wavelet image compression – the quadtree coding approach," *IEEE Trans. Inf. Technol. Biomed.*, vol. 3, no. 3, pp. 176–185, Sept. 1999.

W. A. Pearlman, A. Islam, N. Nagaraj, and A. Said, "Efficient, low-complexity image coding with a set-partitioning embedded block coder," *IEEE Trans. Circuits Syst. Video Technol.*, vol. 14, no. 11, pp. 1219–1235, Nov. 2004.

11 Subband/wavelet coding systems

In this chapter, we shall describe coding systems, primarily for images, that use the principles and algorithms explained in previous chapters. A complete coding system uses a conjunction of compression algorithms, entropy coding methods, source transformations, statistical estimation, and ingenuity to achieve the best result for the stated objective. The obvious objective is compression efficiency, stated as the smallest rate for a given distortion for lossy coding or smallest rate or compressed file size in lossless coding. However, other attributes may be even more important for a particular scenario. For example, in medical diagnosis, decoding time may be the primary concern. For mobile devices, small memory and low power consumption are essential. For broadcasting over packet networks, scalability in bit rate and/or resolution may take precedence. Usually to obtain other attributes, some compression efficiency may need to be sacrificed. Of course, one tries to obtain as much efficiency as possible for the given set of attributes wanted for the system. Therefore, in our description of systems, we shall also explain how to achieve other attributes besides compression efficiency.

11.1 Wavelet transform coding systems

The wavelet transform consists of coefficients grouped into subbands belonging to different resolutions or scales with octave frequency separation. As such, it is a natural platform for producing streams of code bits (hereinafter called *codestreams*) that can be decoded at multiple resolutions. Furthermore, since the coefficients are the result of short FIR filters acting upon the input data, they retain local characteristics of the data.[1] Most natural images show wavelet transforms with magnitudes of their coefficients generally decreasing as the subband scale or resolution increases. In other words, most of the energy of these images is packed into the lower frequency subbands. Furthermore, there are intra-subband and interscale statistical dependencies that can be exploited for compression. A typical wavelet transform with three scales, that of the 512×512 Lena image, is displayed in Figure 11.1. The values of the coefficients may be negative, except in the lowest frequency subband in the top-left corner, and require precision exceeding the eight bits of most displays. Therefore, for display purposes, the coefficients in all subbands were scaled to the range of 256 gray levels. In all but the lowest frequency subband, the zero value corresponds to the middle gray level of 128.

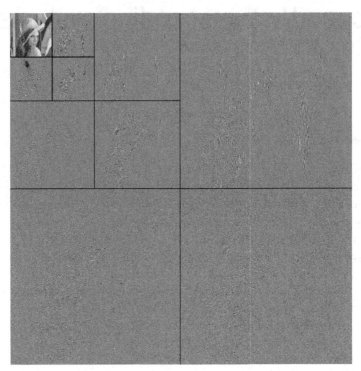

Figure 11.1   Display of image subbands of a 3-level, dyadic wavelet transform. Middle gray level of 128 corresponds to 0 value of a coefficient in all subbands, excluding the lowest frequency one in the top-left corner.

One can clearly see the edge detail propagating across scales to the same relative locations in the higher frequency subbands as one moves from the top left to the top right, bottom left, or bottom right of the transform. However, this detail becomes quite faint in the highest frequency subbands, where the coefficients are predominantly middle gray with actual values of zero. Notice also that within subbands, the close neighbors of a coefficient do not change much in value, except across edges. All these characteristics make the wavelet transform an effective platform for efficient compression with the attributes of resolution scalability and random access decoding. The arrangement of the coefficients into subbands belonging to different scales or resolutions makes possible encoding or decoding different scales. Random access to selected regions of interest in the image is possible, because of the local nature of the transform, since one can select regions of subbands at different scales to decode a particular region of an image.

The subband labeling diagram for the wavelet transform in Figure 11.1 is depicted in Figure 11.2. This subband arrangement was produced by three stages of alternate low and highpass horizontal and vertical filterings of the resulting low horizontal and low vertical subbands followed by 2:1 downsampling. (The first stage input is just the source image itself.) We show the analysis and synthesis for two stages explicitly in Figures 11.3 and 11.4. The LL2 (low horizontal, low vertical, 2nd stage) subband is just a coarse, factor of 2^2 reduction (in both dimensions) of the original image. Upsampling

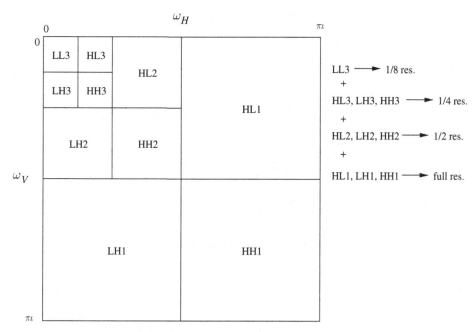

Figure 11.2   Subbands of a 3-level, dyadic wavelet transform.

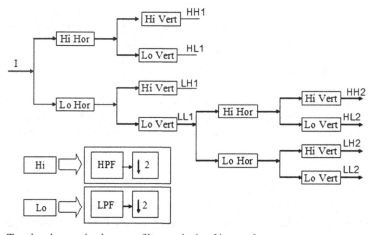

Figure 11.3   Two-level recursive lowpass filter analysis of image I.

and filtering of LL2, LH2, HL2, and HH2 subbands yields the LL1 subband, which is two times the scale of LL2, but still scaled down by a factor of 2^1 from the original. And so it repeats on the LL1 subband for one more stage to obtain the full-scale reconstruction of the original input. Therefore, at each synthesis stage, we can obtain a reduced scale version of the original image.

In order to synthesize just any given (rectangular) region of an image, one needs only to locate the coefficients in the corresponding regions in the subbands of the wavelet transform and apply them to the same synthesis filter bank. These regions are located in

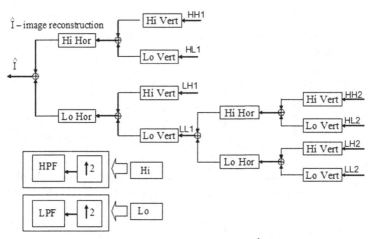

Figure 11.4   Two-level recursive lowpass filter synthesis of image $\hat{I}$.

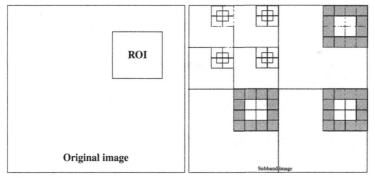

Figure 11.5   Subband regions in the wavelet transform corresponding to an image region. The inner rectangles correspond to the exact corresponding fractional area of the image region of interest (ROI). The areas between the inner and outer rectangles contain coefficients needed to reconstruct ROI exactly.

the subbands in the same relative position as in the image, as illustrated in Figure 11.5 for an image and its two-level wavelet transform. The fractional area in these subbands is slightly larger than that in the image, because coefficients outside the designated region result from filtering of image samples inside the region only near the boundaries, due to the finite length of the filters. That is why the rectangles in the subbands that exactly correspond to the image rectangle are shown inside larger rectangles. Again, the region can be reconstructed at different resolutions, if desired.

The best filters to use from the standpoint of achieving the best compression efficiency or highest coding gain have real number tap values, represented as floating point numbers that are precise only within the limits of the computer. These filters produce floating point wavelet coefficients. Hence, the inverse transform, done again with floating point filters, may not produce an exact replica of the source. For example, if a filter tap value contains a factor of $1 + \sqrt{3}$, then the floating point representation cannot be exact and all subsequent mathematical operations will propagate this inexactness. Coding the wavelet coefficients means converting them to a compact integer representation. Therefore, even if the coding is perfectly lossless, the inverse wavelet transform may reconstruct the image with error from the original. Therefore, one can not guarantee perfectly lossless image compression in a wavelet transform system using floating point filters. One achieves what is often called *virtually lossless* reconstruction. In most practical applications, that is not a hindrance, but sometimes, often for legal reasons that arise in certain fields such as diagnostic medicine, it is crucial to achieve perfectly lossless compression. Therefore, for perfectly lossless image compression in a wavelet-based coding system, one must use filters that operate with integer arithmetic to produce integer transform coefficients. This also assures that there are no errors due to limited precision in the synthesis stage. There will be a small, usually tolerable, reduction in potential coding gain from the use of these integer-to-integer filters for wavelet transformation.

11.2 Generic wavelet-based coding systems

The generic wavelet transform coding system, regardless of the source, normally starts with subband/wavelet transformation of the source input. There is often then an optional preprocessing step that consists of statistical estimation, segmentation, weighting, and/or classification of the transform coefficients. Then the coefficients are subjected to quantization and/or set partitioning, usually followed by entropy coding, such as Huffman or arithmetic coding. Along with overhead information generated by the preprocessor, the encoded transform coefficients are written to the output codestream. The block diagram of this generic system is shown in Figure 11.6.

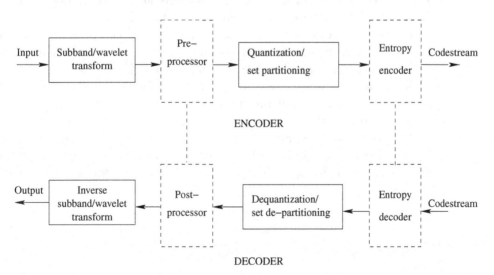

Figure 11.6 Encoder and decoder of subband/wavelet transform coding system. The boxes with dashed lines denote optional actions.

The decoding system, also shown in Figure 11.6, reverses the process by decoding the codestream to reconstruct the wavelet transform, post-processing this transform, according to the preprocessor's actions, and then inverting the wavelet transform to reconstruct the source. One simple example of preprocessing is weighting transform coefficients to affect the distribution of code bits, in order to enhance visual quality of an image or aural quality of audio. The post-processing step must do inverse weighting, or the reconstruction will be distorted. The pairs of the pre- and post-processors and the entropy encoder and decoder are optional for some coding systems, so are depicted in dashed line boxes in Figure 11.6.

11.3   Compression methods in wavelet-based systems

We now describe various methods to compress and decompress coefficients of a wavelet transform of a source, such as the image shown in Figure 11.1. We shall describe methods referring specifically to images, although these methods can almost always be applied to one-dimensional or higher-dimensional wavelet transforms with obvious modifications. We shall start with some of the set-partition coding methods of the previous chapter.

Within the quantization/set partitioning system block in Figure 11.6, the quantization is necessary for floating point transform coefficients and optional for integer ones. Any quantization will result in reconstruction error, which is unavoidable for floating point coefficients. For embedded coding enabled through coding of bitplanes, the quantization may be just a mere truncation of every coefficient to the nearest integer. Deeper quantization is implicit when the coding stops above the least significant $n = 0$ bitplane. Suppose that coding stops just after completing bitplane n. Then every coefficient has an indeterminate value between 0 and 2^{n-1}, corresponding to a quantizer interval. The decoder then assigns a reconstruction point within this interval.

For explicit quantization, the coefficients are quantized with a particular type of uniform quantizer, called a uniform, null-zone quantizer. For this quantizer, thresholds are uniformly spaced by step size Δ, except for the interval containing zero, called the null-(or dead-)zone, which extends from $-\Delta$ to $+\Delta$. Figure 11.7 illustrates the input–output characteristic of a uniform null-zone quantizer with seven quantizer levels and midpoint reconstruction values.

This quantization is enacted by scaling by the step size and truncating to integers to produce the indices of the decision intervals (often called quantization bins). The mathematical operation upon the input x to produce a bin index q, given a quantizer step size Δ is

$$q = \text{sign}(x)\lfloor |x|/\Delta \rfloor \tag{11.1}$$

The reconstruction (dequantization) $\hat{x}$ is given by

$$\hat{x} = \begin{cases} (q + \xi)\Delta, & q > 0 \\ (q - \xi)\Delta, & q < 0 \\ 0, & q = 0 \end{cases} \tag{11.2}$$

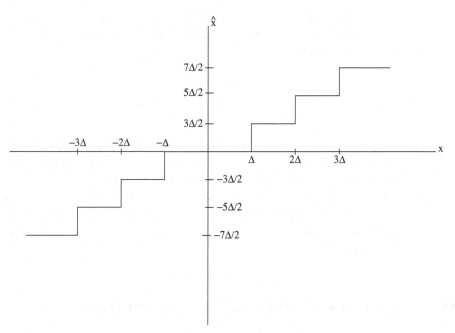

Figure 11.7   Input–output characteristic of a 7-level, uniform null-zone quantizer.

where $0 \leq \xi < 1$. Henceforth, the bin index q will be called the *quantizer level*. The parameter ξ is often set to place the reconstruction value at the centroid of the decision interval. It has been derived through a model and confirmed in practice that $\xi \approx 0.38$ usually works well. In many cases, $\xi = 0.5$ is used, which places the reconstruction at the interval's midpoint. It is important to realize that when $-\Delta < x < \Delta$, the quantizer level and reconstruction value are both 0. For a subband or linear transform, there may be many coefficients, belonging especially to higher frequencies, that are set to 0. The array of quantizer levels (bin indices) q are further encoded losslessly. As we have seen, clusters of zeros can be represented with very few bits.

For optimal coding that minimizes MSE for a given number of bits, statistically independent subbands encoded with non-zero rate should exhibit the same MSE. Although not statistically independent, the subbands are often modeled as independent and are encoded independently. In such circumstances, use of the same step size for all subbands minimizes the mean-squared reconstruction error for high rates and orthonormal synthesis filters. In accordance with this usual model, notwithstanding less than optimal coding, all subbands encoded with non-zero rates are quantized with the same step size Δ.

Often, the preprocessor function of weighting subbands to enhance perceptual quality or to compensate for scaling of non-orthonormal synthesis filters is combined with quantization by using different step sizes among the subbands. In this case, there is posited a set of step sizes, $\{\Delta_m\}$, $m = 1, 2, \ldots, M$, with Δ_m being the step size for subband m. The quantization follows the same formulas as (11.1) and (11.2).

In almost all wavelet transform coding methods, the quantizer levels are represented by their sign and magnitude. In the sections that follow, we shall describe specifically some of the many methods used to code the quantizer levels of wavelet transforms of images. For images that are represented by 8 bits per pixel per color, the quality criterion is PSNR, defined by

$$\text{PSNR} = \frac{255^2}{\frac{1}{MN} \sum_{i=1}^{M} \sum_{j=1}^{N} (x[i, j] - \hat{x}[i, j])^2} \tag{11.3}$$

where $x[i, j]$ and $\hat{x}[i, j]$ denote the original and reconstructed image values, respectively, at coordinates (i, j), and M and N the total row and column elements, respectively. The denominator is just the MSE per pixel. Usually PSNR is expressed in dB, so that it expresses the dB difference between the peak value and root mean squared (RMS) error. We remind the readers that MSE is exactly the same whether calculated in the source or transform domain, if the transform or filters are orthonormal.

11.4   Block-based wavelet transform set partition coding

In this section, we shall describe block-based techniques, presented in detail in Chapter 10, for set partition coding of the quantizer levels. The principles of providing resolution and quality scalable coding are explained first, followed by the descriptions of the specific techniques of SBHP [1], the JPEG2000 Standard, Part 1 [2], and EZBC [3, 4].

The quantizer levels are represented in sign-magnitude format and tests for significance are enacted on the magnitudes. There are various choices for setting the order of coding of the subbands, coding and/or partitioning within subbands, setting the thresholds for significance, coding the results of significance tests, and coding the coefficient values. We shall describe coding systems that utilize some of these possible combinations of choices.

First, we consider the subbands as the blocks to be coded. The order in which the subbands are coded follows the indicated zigzag path from lowest to highest frequency subband, as illustrated in Figure 11.8. The advantages of following this particular path is that the image is encoded and decoded progressively in resolution or in scale. For example, referring to Figure 11.8, decoding just the LL_3 subband produces just a 1/8 scale reconstruction. Decoding LL_3, HL_3, LH_3, and HH_3 produces a 1/4 scale reconstruction, and so forth. One also follows this path in the search related to quality progressive coding, because the subband energy tends to decrease along this path for most natural images. The general framework is delineated in Algorithm 11.1.

Any one of the methods described in the previous chapter can be used for coding of the subbands in Step 4 of Algorithm 11.1. We shall describe below how some of these methods fit in this framework.

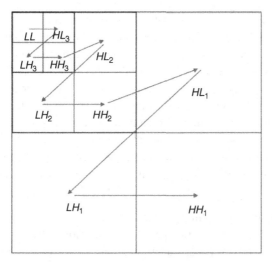

Figure 11.8 Scanning order of subbands in a 3-level wavelet decomposition. Subbands formed are named for horizontal and vertical low- or highpassband and level of decomposition, e.g., LH_2 is horizontal lowpassband and vertical highpassband at second recursion level.

ALGORITHM 11.1

Framework for progressive-resolution wavelet transform coding

1. Start with M subbands $\{SB_k\}$ having quantization levels of their coefficients denoted by $q_{i,j}$.
2. Calculate top threshold in each subband, i.e., calculate $n_k = \log_2 \lfloor s_k \rfloor$, $s_k = \max_{(i,j) \in SB_k} |q_{i,j}|$. Order subbands SB_k according to the progressive-resolution zigzag scan, as shown in Figure 11.8.
3. Encode $n_1, n_2, \ldots, n_M$.
4. For $k = 1, 2, \ldots, M$, if $n_k > 0$, encode SB_k.

11.4.1 Progressive resolution coding

First, we consider the (quantized) subbands to be coded progressively from lower to higher resolution by the fixed-threshold, recursive quadrisection procedure in Algorithm 10.1. Recall that this algorithm produces (approximate) value-embedded code, but not bit-embedded code. The thresholds are fixed to be successive integer powers of two and tests for significance are enacted on the magnitudes of the coefficients, as defined in Equation (10.9). The maximum threshold, $2^{n_{max}}$, is calculated for each subband. We deposit these thresholds into the header of the codestream. We scan the subbands in order from lowest to highest as illustrated in Figure 11.8 and partition each one in turn by the recursive quadrisection procedure of Algorithm 10.1. Briefly, we start with a LIS for each subband, each initialized with the coordinates of its top-left corner. Each LIS set of a subband is tested at the current threshold, say 2^n, and if not significant, a "0" is sent to the codestream. But, if significant, the set is split into four equal quadrants,[2] each of which is tested at threshold 2^n for significance. Significant quadrants are labeled with "1" and insignificant ones are labeled with "0," thereby creating a 4-bit binary mask that is encoded and sent to the codestream. The insignificant quadrants are appended to the bottom of their subband LIS, represented by their top-left corner coordinates. Significant sets continue to be split and labeled in the same way until all significant single elements are located and encoded. Then we lower the threshold from 2^n to 2^{n-1} and test the LIS sets in order of increasing size. For each subband, we start with its initial threshold $2^{n_{max}}$, lower it by a factor of 2 in each succeeding pass, and continue through the last pass at threshold 2^0. The description of this procedure is Algorithm 11.1 with SB_k in Step 4 encoded using Algorithm 10.1. The following example illustrates use and displays results from this resolution scalable algorithm.

Example 11.1 Resolution scalable coding.

The source image is the 512×512, 8 bits per pixel, gray Goldhill image, shown in Figure 11.9. This image is then wavelet transformed and its wavelet coefficients are then quantized through scaling by a factor of 0.31 (step size of 1/0.31) and truncating to integers. The resolution scalable coding algorithm described above encodes the wavelet transform's quantization bin indices losslessly to a rate of 1.329 bpp (codestream size of 43 538 bytes). Portions of this codestream corresponding to the desired resolution are then decoded, dequantized, and inverse wavelet transformed, to produce the reconstructed images. Figure 11.10 shows the reconstructed images for full, one-half, and one-quarter resolutions.

Figure 11.9   Original 512×512, 8 bits per pixel Goldhill image.

Figure 11.10 Reconstructions from codestream of Goldhill coded to rate 1.329 bpp, quantizer step size = 1/0.31 at full, 1/2, and 1/4 resolutions.

The aggregate of 4-bit binary masks corresponds to a quadtree code, as noted in Chapter 10. It is advantageous to entropy code these 4-bit masks instead of sending the raw quadrisection codes to the codestreams. (See Table 10.3.) Even a simple fixed Huffman code of 15 symbols (0000 cannot occur) shows improvement. A more sophisticated fixed Huffman code can be conditioned on the size of the block and/or the particular subband. Another level of sophistication is to adapt the code by estimating and updating the significance state probabilities as coding proceeds. Certainly, arithmetic coding may be substituted for Huffman coding in the above scenarios for entropy coding.

The decoder has prior knowledge of the wavelet transform decomposition and scanning order of its subbands, and receives the image storage parameters and subbands' maximal thresholds from the header of the incoming codestream. It therefore knows which subbands to skip in every scan corresponding to the current threshold. It initializes the LIS of every subband with its corner coordinates. It reads the significance decisions, so is able to build the same LIS populations. It will then know when it is reading the code of significant coefficients and will decode them.

11.4.2 Quality-progressive coding

Encoding every subband completely in the zigzag order above produces a resolution-progressive codestream. Furthermore, each subband's codestream is approximately progressive in quality or value, because coefficients with larger most significant bits (MSBs) precede those with smaller MSBs. Figure 11.11 illustrates the coding order of the magnitude bits of coefficients in a value-progressive scan. The coefficients are ordered by the level of their MSBs and every coefficient's bits are coded from its MSB to its LSB (least significant bit). Although the subbands individually are progressive in quality, the composite codestream is not. Bits from larger coefficients in a subband scanned later will follow those of smaller coefficients from a subband scanned earlier.

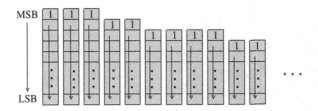

Progressive Value Scan

Figure 11.11 Coding order for a progressive value codestream.

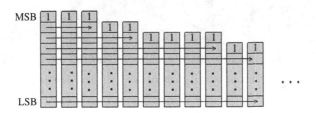

Progressive Bitplane Scan

Figure 11.12 Coding order for a bit-embedded codestream.

One can reorganize this codestream to be progressive in value, if we insert indicators[3] to separate the bits from coding passes at different thresholds. The code bits of coefficients significant for the same threshold are kept together in the codestream. With the aid of these indicators, we can gather together from different subbands the code bits of coefficients significant for the same threshold. Therefore, in the decoder, the coefficients with larger significance are decoded before those of lesser significance. This reorganization scheme does not order the values within the same threshold, so is only partially progressive in value. Furthermore, this reorganization may vitiate the progressiveness in resolution. One or more lower resolution subbands may not have significant coefficients at the current threshold, while higher resolution ones do. If we continue to collect the bits from these higher resolution significant coefficients at the current threshold, then we will not be keeping together code bits from the same resolution.

One can reorganize this codestream not only to be partially value-progressive, but also to be bit-embedded. Again note that bits belonging to the same threshold (with same MSB) are kept together in the codestream. The value-progressive scan reads code bits from the most significant bit downward to the least significant before proceeding to the next coefficient with the same most significant bit. We can produce a bit-embedded codestream, if starting with the largest most significant bitplane, we always move to the same bitplane of the next coefficient in the scan, whether in the same or a different subband. Referring to Figure 11.12, this path corresponds to scanning a bitplane from extreme left to extreme right at the same horizontal (bitplane) level, dropping down one level, and returning to the extreme left to repeat the rightward scan holding the same level. The general framework for value-progressive coding is delineated in Algorithm 11.2.

ALGORITHM 11.2 _____

Framework for value-progressive wavelet transform coding.

1. Start with M subbands $\{SB_k\}$ having quantization levels of their coefficients denoted by $q_{i,j}$.
2. Calculate top threshold in each subband, i.e., calculate $n_k = \log_2 \lfloor s_k \rfloor$, $s_k = \max_{(i,j) \in SB_k} |q_{i,j}|$. Order subbands SB_k according to the progressive-resolution zigzag scan, as shown in Figure 11.8.
3. Encode $n_1, n_2, \ldots, n_M$.
4. Initialize LIS with top-left corner coordinates of subbands in order of $SB_1, SB_2, \ldots, SB_M$.
5. Let $k = 1$.
   (a) Let $n = n_k$. If $n > 0$,
       (i) do single pass of set partition coding of SB_k at threshold 2^n until all single significant elements are located;
       (ii) encode significant single elements with (no more than) $n + 1$ bits.
   (b) Move to next subband, i.e., let $k = k + 1$;
   (c) If $k \le M$, go to Step 5(a).
   (d) If $k > M$, decrement n, i.e., let $n = n - 1$.
   (e) If $n > 0$, return to Step 5(a)(i); if $n = 0$, stop.

_____

The coding passes in Step 5(a)(i) may be accomplished either by the recursive quadrature splitting of Step 10.1 of Algorithm 10.1 or octave band partitioning and recursive quadrature splitting of the SPECK algorithm 10.6. The initial S sets for the subbands would be the 2×2 blocks in their top-left corners. Instead of maintaining a LSP and executing a refinement pass, the single elements are encoded immediately and written to the codestream when found to be significant, as in Step 5(a)(ii) in Algorithm 11.2.

As before, if there are one or more subbands with empty layers at the given bit-plane, the codestream will lose its progressiveness in resolution. Then higher bitplane bits from higher resolution subbands will reside among those from the lower resolution ones. Therefore, a truncated codestream cannot contain purely bits from a certain scale.

Another way to obtain a bit-embedded codestream is to keep the same threshold in the zigzag scan across the subbands and maintain a list of significant elements, called the list of significant points (LSP), initially empty. The starting threshold is the largest among all the subbands and is almost always that of the lowest frequency subband. When the current threshold exceeds the maximal threshold of a given subband, that subband is skipped. For the threshold 2^n, we code a subband by the same quadrisection procedure described above until all the single elements are located. The coordinates of the significant single elements are added to the LSP. The next subband in the scan order is visited and encoded in the same way, until all the subbands have been visited at the threshold 2^n. The bits in the same bit-plane n from coefficients found significant at higher thresholds are then read from the LSP and written to the codestream. The threshold is then lowered by a factor of 2 to 2^{n-1} and the scan returns to its first subband to test its sets in the LIS left after the just-completed threshold pass. The LIS sets within a subband decrease in size from top to bottom and are visited in reverse order from bottom to top, so that the smallest sets are tested first. In this way, we obtain a bit-embedded codestream, because code bits from the higher bitplanes always precede those from the lower bitplanes.

Example 11.2 Reconstructions of different quality from a bit-embedded codestream. The wavelet transform of the Goldhill image is encoded, but this time using a quality scalable algorithm giving a bit-embedded codestream of size 65 536 bytes (rate of 2.00 bpp). (The transform coefficients are not scaled and are truncated to integers, if necessary.) The full codestream and its truncation to sizes of 32 768 bytes (1.00 bpp), 16 384 bytes (0.50 bpp), and 8192 bytes (0.50 bpp) are then decoded, dequantized, and inverse wavelet transformed. The reconstructions and their PSNRs are shown in Figure 11.13.

11.4.3   Octave band partitioning

Instead of a predetermined scan order of subbands, we could use octave band partitioning of the wavelet transform, as is done in conjunction with the SPECK algorithm described in the last chapter. We would need only to calculate and send the maximum threshold for the entire transform and have the potential to locate larger insignificant sets. Recalling briefly this procedure, the initial S set is the lowest frequency subband and the remainder of the transform comprises the initial I set. These sets are depicted in Figure 11.14. The set S is tested for significance at the current threshold and, if significant, is coded as above using recursive quadrisection. If not significant, the I set is then tested for significance and, if significant, it is partitioned into three S sets and an I set. These three S sets are exactly the three subbands that complete the next higher resolution scale. If I is not significant, then the threshold is lowered by a factor of 2, the octave band partitioning is re-initialized, and only the resident LIS entries left from the last threshold are tested when the S set for a subband becomes significant. As the threshold becomes lower, more sets of three subbands that comprise the next higher resolution are encoded. Therefore, this scheme of octave band partitioning is naturally progressive in resolution.

A binary digit is sent to the codestream after every significance test. An insignificant subband then is indicated by a single "0" at the given threshold. So the maximal threshold of a subband is conveyed by a succession of "0"s and a single "1" when it becomes significant. Sometimes, a "0" indicating insignificance is shared among insignificant subbands, as in an I set. The previous method sent the maximal threshold of every subband to the codestream. Although it did not require another bit to skip insignificant subbands, more bits are needed for conveying the maximal threshold of every subband than in the method of octave band partitioning.

Figure 11.13 Reconstructions of Goldhill from same codestream by a quality scalable coding method.

(a) 2.00 bpp, 42.02 dB	(b) 1.00 bpp, 36.55 dB
(c) 0.50 bpp, 33.13 dB	(d) 0.25 bpp, 30.56 dB

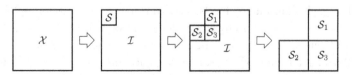

Figure 11.14 Partitioning of image $\mathcal{X}$ into sets $\mathcal{S}$ and $\mathcal{I}$, and subsequent partitioning of set $\mathcal{I}$.

Figure 11.15   Uncoded (left) and coded (right) reconstructions of 512 × 512 Lena image with identical quantizer step sizes. Both have PSNR = 37.07 dB; rate of uncoded (left) = 0.531 bpp and rate of coded (right) is 0.500 bpp.

Example 11.3 Effect of mask coding. As an example, we performed one experiment to compare an uncoded and coded same step size quantization of the wavelet transform (4 levels, 9/7 biorthogonal filters) of the Lena (512 × 512) image. The quantization was done according to Equation (11.1) with the same step size (Δ) for all subbands and the coded one used fixed Huffman coding of the masks only. We used octave band partitioning for determining the scanning order of the subbands. No overhead marker bits were in the codestream, as there was no attempt to produce progressiveness in quality. As expected, the identical uncoded and coded reconstructions showed the same PSNR of 37.07 dB. However, coding the masks resulted in an average bit rate of 0.500 bpp, whereas without coding the average bit rate was 0.531 bpp, an increase of 6%. These reconstructions are displayed in Figure 11.15.

11.4.4   Direct bit-embedded coding methods

The utilization of SPECK just described achieves complete coding of a lower resolution level before the next higher one. As explained in the previous chapter, SPECK realizes bit-embedded coding when the threshold is held at the same value across subbands and through all resolution levels before it is lowered by a factor of 2 for the next pass through the LIS. Such a procedure violates resolution progressiveness, because bits from different resolution levels are intermingled in the codestream. The benefit is the approximate realization of the optimal rate allocation that prescribes that the most important of the code bits, that is, the ones contributing to the largest reduction in distortion, are sent to the codestream. The coding can then be terminated when the bit budget is reached with the confidence that the lowest distortion for this budget is achieved for the corresponding reconstruction. One does not need to do any codestream reorganization or Lagrangian optimization, as required when certain portions of the wavelet transform are independently and completely coded.

The coding procedure of SPECK (or SPIHT) will produce optimal rate allocation when the probability density function of the wavelet coefficients is symmetric about zero and monotonically non-increasing with increasing magnitude. Then we are assured that the distortion reduction caused by a newly found significant coefficient is more than any refinement bit in the same bitplane of a previously significant coefficient. It is relatively rare when the opposite phenomenon is true.

11.4.5   Lossless coding of quantizer levels with adaptive thresholds

The coding paradigm remains the same, except now we decide to use adaptive thresholds rather than fixed ones. The coding mechanism using adaptive thresholds was described in the AGP method in the last chapter. We take the subbands as the blocks to be coded and test them starting as usual with the lowest frequency subband. Briefly, the alphabet is partitioned by symbol grouping and representing each coefficient by its group index and symbol index. A typical alphabet partitioning scheme appears in Table 10.5. The group indices are coded via set partitioning. The symbol indices are represented by their magnitude range and sign in raw binary.

The subbands are now encoded by the adaptive-threshold, quadrature splitting procedure described in Algorithm 10.5. The values to be coded are the group indices. The initial thresholds of the subbands are the maxima of their respective group indices. Briefly, starting with the lowest frequency subband, the subband to be encoded is split into four nearly equal size sets. The maxima of these sets are compared to the subband maximum. Those with lower maxima are labeled with "0" and those equal are labeled with "1," thereby creating a four bit mask that is sent to the codestream. The "1"-labeled (significant) sets are recursively split into four quadrants which are labeled with "0" or "1," depending whether the quadrant maximum is lower or equal to the subband maximum. When the recursive splitting reaches single elements of value equal to the threshold, it is encoded and sent to the codestream. The maxima of the "0"-labeled (insignificant) quadrants remaining comprise new lower thresholds. The new threshold for the next pass is the maximum among these thresholds, which is the largest group index magnitude among these remaining insignificant sets. These insignificant sets are now recursively split and tested against this new lower threshold in the same way until all coefficients in the subband are encoded. These insignificant sets are always visited in order of smallest to largest, so that significant single elements are located as early as possible. When "0"-labeled sets are 2×2 blocks having small maxima, they can be coded together using an extension alphabet to obtain coding gains. Once a subband is encoded, the same processing moves to the non-visited subband with the largest maximum group index magnitude greater than zero. A subband with such maximum equaling 0 has all its elements 0 in value. The 0 in the four-bit mask sent to the codestream conveys this information.

ALGORITHM 11.3 ——————————————————————————————

Wavelet transform coding by set partitioning with adaptive thresholds

1. Start with M subbands $\{SB_k\}$ and represent quantized coefficients $\hat{c}_{i,j}$ with group index, symbol index pairs $(g_{i,j}, r_{i,j})$, where $r_{i,j}$ includes sign.
2. Calculate maximum group index in each subband and order subbands by decreasing maxima. That is, let $s_k = \max_{(i,j)\in SB_k} g_{i,j}$. Order subbands SB_k either by progressive resolution zigzag scan or by decreasing maxima, $s_1 \geq s_2 \geq \cdots \geq s_M$.
3. Encode $s_1, s_2, \ldots, s_M$.
4. For $k = 1, 2, \ldots, M$, if $s_k > 0$, do set partitioning coding of SB_k using Algorithm 10.5.

ALGORITHM 11.4 ——————————————————————————————

Value-progressive, wavelet transform coding by set partitioning with adaptive thresholds

1. Start with M subbands $\{SB_k\}$ and represent quantized coefficients $\hat{c}_{i,j}$ with group index, symbol index pairs $(g_{i,j}, r_{i,j})$, where $r_{i,j}$ includes sign.
2. Calculate maximum group index in each subband and order subbands by decreasing maxima. That is, let $s_k = \max_{(i,j)\in SB_k} g_{i,j}$ and order subbands SB_k so that $s_1 \geq s_2 \geq \cdots \geq s_M$.
3. For every $k = 1, 2, \ldots, M$,
   (a) encode s_k
   (b) set initial thresholds $t_k = s_k$
   (c) initialize a list LIS_k with top-left corner coordinates of subband $\{SB_k\}$.
4. Let $k = 1$.
   (a) if $t_k > 0$, do single pass of set partition coding of SB_k in Step 2 of Algorithm 10.5.
   (b) If $t_k > 0$ and there are multiple element sets on LIS_k,
       (i) reset t_k equal to largest maximum among the sets of LIS_k;
       (ii) encode t_k.
       (iii) Move to subband with largest current threshold, i.e., SB_{k^*}, where $k^* = \arg\max_{\ell=1,2,\ldots,k,k+1} t_\ell$.
       (iv) Set $k = k^*$ and return to Step 4(a).
   (c) If $t_k > 0$ and all the LIS_k sets comprise single elements, encode them with no more than $\lfloor \log_2 t_k \rfloor + 1$ bits, write to codestream buffer, and stop.
   (d) If $t_k = 0$, stop.

The steps of this coding procedure are delineated in Algorithm 11.3. There, the subbands are ordered by decreasing maximum value (group index). For images, SB_1 is almost always the lowest frequency subband and the ordering by maximum decreasing value is often progressive in resolution. But for other sources, such as audio, the largest maximum may belong to another subband. This prior ordering, whether by maximum value or by increasing resolution, simplifies the step of moving to the next subband to be coded. Moreover, for ordering by decreasing maximum, once we come to a subband with maximum value of 0, we can stop the coding. For some applications, it might be advantageous to produce a value-progressive codestream. Achieving such property requires crossing subbands to code an LIS with the next lower threshold, instead of executing the next pass at the threshold that is the largest maximum of the LIS sets of the current subband. This means that a separate LIS must be maintained for every subband. Algorithm 11.4 presents the steps of this value-progressive coding method.

The new lower thresholds in either mode, value-progressive or not, are encoded efficiently using differential coding. The thresholds are always decreasing and the successive differences are small. Recall that these thresholds are group indices, the largest of which is under 32. A unary code for these differences is a good choice.

Modest gains in coding efficiency may be obtained through entropy coding of significant coefficients either singly or in 2×2 blocks when the threshold is small. Recall that a small threshold allows extension alphabets of small size. In the quadrisection steps (see Step 4(a) in Algorithm 11.4), 4-bit binary masks are generated to convey whether the quadrant maxima are below or equal to the testing threshold. These masks are encoded with a fixed 15-symbol Huffman code. We carried out the experiment of encoding the same quantized wavelet transform of the Lena image, but now used Algorithm 11.3 with entropy coding of masks and small-threshold 2×2 blocks. The resulting code rate was 0.479 bits/pixel for the same 37.07 dB PSNR. Therefore, even with the extra overhead of coding the thresholds, the combination of adaptive thresholds and simple entropy coding of these blocks saved about 4% in rate or file size.

11.4.6 Tree-block coding

In Chapter 10, we introduced a partition of the wavelet transform into spatial orientation trees (SOTs) and formed these trees into blocks, called *tree-blocks*, as depicted in Figure 10.12. We illustrated the order of coding these tree-blocks in Algorithm 10.3 by the fixed-threshold, quadrature splitting method of Algorithm 10.1. However, any block coding method can be chosen to code the tree-blocks. In fact, the adaptive-threshold set partition coding algorithms can be used for coding the tree-blocks, instead of the subbands taken as blocks. All it requires is the substitution of tree-blocks for subbands in the input to these algorithms. More specifically, suppose we denote the tree-blocks as $S_1, S_2, \ldots, S_T$, where T is the number of these blocks, which is usually equal to number of coefficients in the lowest frequency subband. Then in Algorithms 11.3 and 11.4, we replace subband SB_k with S_k and subband count M with tree-block count T.

Again, we performed the analogous experiment of coding the same quantized wavelet transform of the Lena image, but now coded tree-blocks instead of subbands. The code rate turned out to be 0.485 bpp, which is slightly higher than the 0.479 bpp resulting from coding subbands as blocks. There are two factors contributing to the superiority of the subband block coding. First, the inter-dependence of coefficients in subbands may be stronger than the interdependence in the SOTs. Second, the subbands constitute mainly bigger blocks, so that larger insignificant sets and zero blocks are located. In a 5-level wavelet decomposition of a 512×512 image, there are 256 tree-blocks each with 1024 coefficients, but only 16 subbands having from 256 to 65 536 coefficients. However, coding in tree-blocks provides natural random access to regions directly in the codestream, since each tree-block corresponds to a different image region.

11.4.7   Coding of subband subblocks

Coding blocks that are contained within subbands, unlike direct coding of blocks of source samples, will not produce blocking artifacts at low code rates, because the filtering in the inverse transform performs a weighted average of the reconstructed coefficients that crosses block boundaries.[4] This realization led to systems that divide the subbands into square blocks, typically 32×32 or 64×64 in their dimensions, and that code these blocks. The subbands of a 3-level wavelet transform divided into these so-called *subblocks* are depicted in Figure 11.16. The smallest subbands are not divided, if their size is comparable to that of the subblock. The figure therefore shows no division of the subbands at the coarsest level.

These subblocks may be encoded by any of the means previously presented for coding of blocks that are full subbands. All the subblocks within a given subband will be encoded before transition to the next subband. The order of encoding subblocks within a subband is predetermined. Usually it is raster scan order, that is, horizontal from left to right and then vertical from top to bottom, but it could be the zigzag or some other space-filling scan. The advantages of coding subblocks are small memory usage and random access or ROI capability. The independent coding of subblocks allows the latter capability, because subblocks forming a group of spatial orientation trees can be selected from the image or the codestream for encoding or decoding a certain spatial region of the image.

The coding of subblocks of subbands entails extra codestream overhead in the form of the initial (largest) threshold for every subblock and markers to delineate boundaries between subblocks.[5] The thresholds are either group indices or bitplane maxima and can be compressed, if necessary. For example, consider again the 5-level wavelet decomposition of a 512×512 image, for a subblock size of 32×32. There are 64 subblocks in each of the 3 level-1 subbands, 16 subblocks in each of the 3 level-2 subbands, 4 subblocks in each of the level-3 subbands, 1 subblock in each of the 3 level-4 subbands, and 1 subblock making up remaining subbands for a total of 256 blocks. Assuming that the thresholds are maximal bitplane levels, no threshold can exceed 13 for an original 8-bit image. Therefore, 4 bits for each threshold gives an overhead of 1024 bits or 0.004 bits per pixel. At very low bit rates, such overhead may be problematical. These thresholds, especially those associated with subblocks within a subband, are statistically dependent and can be encoded together to save overhead rate.

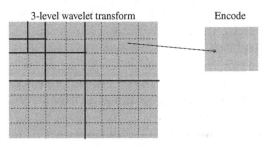

Figure 11.16   Division of wavelet subbands into subblocks. Note subbands of coarsest level are too small to be divided.

11.4.8   Coding the initial thresholds

The initial thresholds that are bitplane maxima of subblocks contained within a subband are often not too different, especially for the lower level, high frequency subbands. Therefore, it is often advantageous to encode the initial thresholds in each level-1 subband separately and group the initial thresholds in the remaining higher level subbands together. For the 5-level wavelet decomposition of the 512×512 image, there would be coded four 8×8 arrays of initial thresholds, three belonging to the level-1 subbands and one belonging to all the remaining subbands of the higher levels.

Let us assume that these initial subblock thresholds are a square array of integers $n[i, j]$, $i, j = 1, 2, \ldots, K$, representing the bitplane maxima, the $n_{\max}$'s of these subblocks. Let $n_M = \max_{i,j=1,2,\ldots,K} n[i, j]$ be the maximum of these integers. Because the thresholds within a block are dependent, they are likely to be close in value to n_M. Because small values require fewer code bits than large ones, it will be more efficient to encode the differences from the maximum, that is, $\bar{n}[i, j] = n_M - n[i, j]$. In the array of numbers $\{\bar{n}[i, j]\}$, the smaller numbers are the most probable, so they fit the favorable scenario of coding by recursive quadrisection procedure with adaptive thresholds, which is Algorithm 10.5. We give examples below to illustrate this approach.

Example 11.4 Example of coding a 4×4 array of thresholds.
Suppose that the initial bitplane thresholds are held in the following array:

$$
\begin{array}{cccc}
2 & 2 & 3 & 3 \\
3 & 2 & 2 & 3 \\
3 & 4 & 4 & 3 \\
3 & 3 & 2 & 3
\end{array}
$$

Subtracting each element from $n_M = 4$, the array to be coded is

$$
\begin{array}{cccc}
2 & 2 & 1 & 1 \\
1 & 2 & 2 & 1 \\
1 & 0 & 0 & 1 \\
1 & 1 & 2 & 1
\end{array}
$$

The bits to represent the block maximum of $n_M = 4$ are not counted here, because they are already sent as overhead anyway to assist entropy coding of the coefficients in the block. The overall difference array maximum $\bar{n}_{\max} = 2$. Splitting the array into four quadrants, the quadrant maxima are 2, 2, 1, and 2 in raster order. The 4-bit significance mask to be transmitted is therefore 1, 1, 0, 1. Splitting the first "1"-labeled quadrant gives four single elements with significance bits 1, 1, 0, 1. Likewise, splitting of the next two "1"-labeled quadrants gives significance values of 0, 0, 1, 0 and 0, 0, 1, 0, respectively. A "1" bit conveys that the element value is $\bar{n}_{\max} = 2$. The "0"-bits attached to the seven single elements mean that their vales are 0 or 1, so that they can each be encoded naturally with a single bit.

Returning to the original split, the insignificant third quadrant has its $\bar{n}_{\max} = 1$. The quad-split yields the significance pattern 1, 0, 1, 1, whose bits are exactly its values, which is always the case when the group maximum is 1.

The full array maximum $\bar{n}_{\max} = 2$ must also be encoded. Because it cannot exceed $n_M = 4$, 3 bits suffice to represent this value of 2. When split into four quadrants, only the maximum of the one "0"-labeled quadrant is unknown to the decoder. Therefore, its maximum of 1 is encoded with 1 bit, because it is known to be less than 2.

Counting the total number of code bits, we have 5 4-bit masks, 7 value bits, and 4 group maxima bits for a total of 31 bits. Compared to raw coding of the array of 16 thresholds, which consumes 3 bits for each, since we know the maximum is 4, we have saved 48-31=17 bits or about 35%.

Another way to represent the original threshold array is by successive differences in a linear or zigzag scan. For example, a zigzag scan from the top-left corner produces the zigzag and successive difference sequences below:

$$2 \quad 2 \quad 3 \quad 3 \quad 2 \quad 3 \quad 3 \quad 2 \quad 4 \quad 3 \quad 3 \quad 4 \quad 3 \quad 3 \quad 2 \quad 3$$
$$0 \quad -1 \quad 0 \quad 1 \quad -1 \quad 0 \quad 1 \quad -2 \quad 1 \quad 0 \quad -1 \quad 1 \quad 0 \quad 1 \quad -1$$

Probably the simplest and most effective way to code the successive difference sequence is by a unary (comma) code.[6] But now there are negative integers, so we must map these numbers to positive ones. For an integer k,

$$k \rightarrow 2k \qquad \text{if } k \geq 0$$
$$k \rightarrow 2|k| - 1 \quad \text{if } k < 0.$$

The successive difference sequence maps to

$$0 \quad 1 \quad 0 \quad 2 \quad 1 \quad 0 \quad 2 \quad 3 \quad 2 \quad 0 \quad 1 \quad 2 \quad 0 \quad 2 \quad 1$$

The unary code for this sequence consumes only 32 bits. We also need 3 more bits to represent the leading 2 for a total of 35 bits, 4 more than the recursive quadrature splitting method.

The recursive quadrisection method is especially efficient when all the thresholds in an array are equal. For example, suppose now that all the elements in the initial 4×4 initial threshold arrary equal 4. The difference from maximum array has all elements of 0. So the difference array maximum of 0 represents the entire array. It requires only 3 bits to represent this 0 maximum and hence the original array of 16 thresholds of 4.

The above recursive quadrisection procedure for coding the thresholds usually consumes fewer bits than methods using a linear scan, because it exploits better the two-dimensional statistical dependencies. It also is quite simple, in that it requires just comparisons and set splitting and does not use entropy coding. The masks when uncoded account for more than one-half of the code bits. Simple entropy coding, e.g., fixed Huffman coding with 15 symbols, might be employed beneficially to reduce the mask bit rate.

11.4.9 The SBHP method

Now that we have described dividing subbands into subblocks and the means to code their initial thresholds, we now explain two particular methods to code the subblocks: SBHP and JPEG2000. First, we consider the simpler of the two methods, called SBHP [1]. The coding method of SBHP is SPECK, initialized by an $\mathcal{S}$ set consisting of the 2×2 array of coefficients in the top-left corner of the subblock and with the $\mathcal{I}$ as the remainder of the subblock, as illustrated in Figure 11.14. When the threshold is lowered such that the $\mathcal{I}$ set becomes significant, the $\mathcal{I}$ set is split into three $\mathcal{S}$ sets adjoining the existing $\mathcal{S}$ and another $\mathcal{I}$ set for the remainder. Recall that the $\mathcal{S}$ sets are encoded by recursive quadrature splitting according to significance tests with fixed decreasing power of 2 thresholds. For each subblock, the method uses a LIS list and a LSP list to store coordinates of insignificant sets (including singleton sets) and significant coefficients, respectively. However, the subblock is fully encoded from the top threshold down to the threshold associated either with some large bit rate for lossy coding or through the 2^0 bitplane for lossless coding. The lists are then cleared and re-initialized for the next subblock to be coded. All the subblocks in every subband are independently coded to the same lossy bit rate or to its lossless bit rate in the same way. We pass through the subbands in the progressive resolution order and through subblocks within a subband in raster order. We obtain a sequence of codestreams, one for each subblock, where each codestream is bit-embedded, but when assembled together into one composite codestream is not embedded in any sense. We may reorganize this composite codestream by interleaving the subcodestreams at the same bitplane level to obtain an embedded codestream. The effect is the same as holding the threshold in the current pass across the subblocks and subbands, before lowering the threshold for the next pass. We can exercise rate control by cutting the reorganized codestream at the desired file size.

Another way to control the rate is the method to be described later in more detail in Section 11.6. By this method, a series of increasing slope parameters $\lambda_1, \lambda_2, \ldots, \lambda_J$ are specified and every subcodestream is cut, so that the magnitude of its distortion-rate slope matches each parameter in turn until the desired total number of bits among the codestreams is reached. Because SPECK codes subblocks in progressive bitplane order, approximate, but fairly accurate calculations of distortion changes with rate can be made very fast and in-line by counting the numbers of newly significant bits at each threshold. Then the codestream can be reorganized for quality and/or resolution scalability.

11.4.9.1 Entropy coding in SBHP

The sign and refinement bits in SBHP are not entropy coded. However, the 4-bit masks are encoded with a simple, fixed Huffman code for each of three contexts. These contexts are defined in the following table.

For each context, a simple, fixed Huffman code is used for coding the 15 possible mask patterns. Recall that all-zero cannot occur. The label of a pattern is the decimal number corresponding to the four binary symbols read in raster order, left-to-right and top-to-bottom. Table 11.2 contains Huffman codes for each context.

Table 11.1 Contexts for SBHP entropy coding

Context	Meaning
0	Sets with more than four pixels
1	Sets with four pixels revealed during a previous pass
2	Sets with four pixels revealed in the current pass

Table 11.2 Huffman codes for four-bit masks in SBHP

	Context					
	0		1		2	
Symbol	Length	Codeword	Length	Codeword	Length	Codeword
1	3	000	3	000	2	00
2	3	001	3	001	3	010
3	4	1000	4	1000	4	1010
4	3	010	3	010	3	011
5	4	1001	4	1001	4	1011
6	5	11010	4	1010	5	11100
7	5	11011	5	11010	6	111010
8	3	011	3	011	3	100
9	5	11100	5	11011	6	111011
10	4	1010	4	1011	4	1100
11	5	11101	5	11100	6	111100
12	4	1011	4	1100	4	1101
13	5	11110	5	11101	6	111101
14	5	11111	5	11110	6	111110
15	4	1100	5	11111	6	111111

This entropy code having just three simple contexts, requiring no calculation, and composed of non-adaptive codes of only 15 symbols for each context, adds very little complexity and is quite effective for natural, non-synthetic gray-level images and leads to very fast encoding and decoding.

11.4.10   JPEG2000 coding

The framework for the entropy coding engine of the JPEG2000 coding system is the same as that of SBHP. JPEG2000 encodes subblocks of subbands independently and visits the subbands and subblocks in the same order as previously described. The entropy coding engine, EBCOT, is the brainchild of David S. Taubman [5]. In the parlance of JPEG2000, the subblocks are called *code-blocks*. The method of coding is context-based, binary arithmetic coding (CABAC) of bitplanes from top (most significant) to bottom (least significant), as will be described.

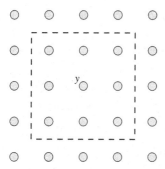

Figure 11.17 Eight pixel neighborhood of pixel y for context formation.

First of all, there is no sorting procedure, as in the set partitioning coders, to partially order the coefficients' quantizer values (indices) by their most significant (highest non-zero) bitplanes. Henceforth, we shall call the locations of a coefficient a *pixel* and its quantizer level (index) a *pixel value*. We start with the highest non-zero bitplane in the code-block, i.e., having index n_{max}. When encoding any given bitplane, we execute three passes:

1. the significance propagation pass (SPP)
2. the magnitude refinement pass (MRP)
3. the clean-up pass (CUP)

Each of these passes encodes a different set of bits and together these passes encode the bits of all pixels in the bitplane. That is why these passes are called *fractional bitplane coding*. The specific order above is chosen to code first those bits that are likely to give the greatest distortion reductions. The first pass, the SPP, visits only (previously) insignificant pixels with a "preferred" neighborhood, where at least one of its eight nearest neighbors is (previously) significant. An 8-pixel neighborhood of the current pixel y is depicted in Figure 11.17. If a "1" is encountered, changing the state from insignificant to significant, then its associated sign is encoded. The SPP pass is skipped for the first (n_{max}) bitplane, because nothing is yet significant, meaning that there is no preferred neighborhood.

The second pass, the MRP, is the same as the refinement pass for the set partitioning coders. It visits only (previously) significant pixels (those with their first "1" in a higher bitplane) and codes the associated bits in the current bitplane. This pass is also skipped for the n_{max} bitplane for the same reason.

The third pass (CUP) visits the locations not yet visited in the SPP and MRP. Clearly, these locations store insignificant pixels without preferred neighborhoods. Therefore, this pass is the only one for the n_{max} bitplane.

We explain next how we encode the bits encountered in these three passes. The short explanation is that each pass has associated with it sets of contexts that are functions of significance patterns for the 8-pixel neighborhood of the bit to be encoded. Probability estimates of the bit values (0 or 1) given each context are accumulated as encoding

Table 11.3 Contexts for code-blocks in LL and LH subbands

Context label	Sum H	Sum V	Sum D
8	2		
7	1	≥ 1	
6	1	0	≥ 1
5	1	0	0
4	0	2	
3	0	1	
2	0	0	≥ 2
1	0	0	1
0	0	0	0

Sum H – sum of significance states (0 or 1) of two horizontal neighbors
Sum V – sum of significance states (0 or 1) of two vertical neighbors
Sum D – sum of significance states (0 or 1) of four diagonal neighbors

proceeds and the bit is arithmetically encoded using these probability estimates. Of course, the devil is in the details, on which we shall elaborate.

11.4.10.1 Significance coding – normal mode

We now describe the normal mode of significance coding that is used exclusively in the SPP and partially in the CUP passes. As mentioned, the coding method is context-based, adaptive arithmetic coding of the binary symbols within a bitplane. The context template is the 8-pixel neighborhood of Figure 11.17, where the number of possible (binary) significance patterns is $2^8 = 256$. So-called context quantization or reduction is necessary for reasons of lowering computational complexity and collection of sufficient statistical data (to avoid so-called *context dilution*) for each pattern. A reduction to nine contexts is achieved by counting numbers of significant pixels in the horizontal, vertical, and diagonal neighbors. Table 11.3 lists the nine contexts for code-blocks in LL and LH subbands.

The context label descends in value from strongest to weakest horizontal neighbor significance. The HL subbands show strong dependence in the vertical direction, so their contexts, shown in Table 11.4, are numbered with the reversal of the Sum H and Sum V columns in Table 11.3.

The HH subbands show diagonal dependencies, so the contexts for code-blocks in these subbands are numbered by descending diagonal significance, accordingly in Table 11.5.

11.4.10.2 Sign coding

In the SPP or CUP of a bitplane, whenever a "1" is encountered, changing the state from insignificant (0) to significant (1), the pixel value's sign is encoded also. The context template for sign coding consists just of the two horizontal and two vertical neighbors

Table 11.4 Contexts for code-blocks in HL subbands

Context label	Sum H	Sum V	Sum D
8		2	
7	≥ 1	1	
6	0	1	≥ 1
5	0	1	0
4	2	0	
3	1	0	
2	0	0	≥ 2
1	0	0	1
0	0	0	0

Sum H – sum of significance states (0 or 1) of two
horizontal neighbors
Sum V – sum of significance states (0 or 1) of two
vertical neighbors
Sum D – sum of significance states (0 or 1) of four
diagonal neighbors

Table 11.5 Contexts for code-blocks in HH subbands

Context label	Sum H + Sum V	Sum D
8		≥ 3
7	≥ 1	2
6	0	2
5	≥ 2	1
4	1	1
3	0	1
2	≥ 2	0
1	1	0
0	0	0

Sum H – sum of significance states (0 or 1) of two
horizontal neighbors
Sum V – sum of significance states (0 or 1) of two
vertical neighbors
Sum D – sum of significance states (0 or 1) of four
diagonal neighbors

in Figure 11.17. An intermediate value characterizes the significance and sign pattern either of the horizontal or vertical neighbors. This intermediate value, denoted as $\bar{\chi}$, takes values as follows:

$\bar{\chi} = 1$ when one neighbor is insignificant and the other significant and positive, or both neighbors significant and positive;

$\bar{\chi} = -1$ when one neighbor is insignificant and the other significant and negative, or both neighbors significant and negative;

Table 11.6 Contexts for code-blocks for sign coding

Context label	$\bar{\chi}_H$	$\bar{\chi}_V$	χ^{flip}
14	1	1	1
13	1	0	1
12	1	−1	1
11	0	1	1
10	0	0	1
11	0	−1	−1
12	−1	1	−1
13	−1	0	−1
14	−1	−1	−1

$\bar{\chi}_H$ – intermediate value for horizontal neighbors
$\bar{\chi}_V$ – intermediate value for vertical neighbors
$\bar{\chi}_D$ – intermediate value for diagonal neighbors

$\bar{\chi} = 0$ when both neighbors are significant and opposite in sign, or both neighbors insignificant.

Since the horizontal and vertical neighbor pairs have three intermediate values each, there are $3^2 = 9$ possible patterns or context values. We expect the conditional distribution of the sign to be coded given a particular pattern to be identical to the negative of that sign given the sign complementary pattern. Therefore, we group a pattern and its sign complementary pattern into a single context, thereby reducing to five distinct sign contexts. A flipping factor $\chi^{flip} = 1$ or −1 is used to change the sign of the bit to be coded according to the pattern or its complement, respectively. The contexts, their labels, and flipping factors are exhibited in Table 11.6. Denoting the sign to be coded as χ, the coded binary symbol κ_{sign} for the given context is

$$\kappa_{sign} = 0 \quad \text{if} \quad \chi \cdot \chi^{flip} = 1$$
$$\kappa_{sign} = 1 \quad \text{if} \quad \chi \cdot \chi^{flip} = -1 \qquad (11.4)$$

11.4.10.3 Magnitude refinement coding

Only previously significant pixels are visited in the MRP through a given bitplane. The associated bits refine these pixel values. The coded bit equals the actual bit of the pixel in the bitplane. We define three contexts in the context template for magnitude refinement coding. The value of each context is determined by whether or not the bit to be coded is the first refinement bit and the sum of the significance states of the eight pixels in the template. We distinguish three contexts as follows:

- not the first refinement bit
- first refinement bit and sum of 8-pixel significance states is 0
- first refinement bit and sum of 8-pixel significance states is non-zero

We summarize the context definitions and labels in Table 11.7.

Table 11.7 Contexts for magnitude refinement coding

Context label	Pixel's first refinement bit?	Sum H + Sum V + Sum D
17	No	
16	Yes	> 0
15	Yes	0

Sum H – sum of significance states (0 or 1) of two horizontal neighbors
Sum V – sum of significance states (0 or 1) of two vertical neighbors
Sum D – sum of significance states (0 or 1) of four diagonal neighbors

When a refinement bit is 0, the value of the associated coefficient is in the lower half of the quantization interval. When it is 1, the value is in the upper half. The typical probability density function of a coefficient is peaked and steep near the origin for small magnitudes and much flatter and shallower for high magnitudes. Therefore, when the coefficient has small magnitude, the probability is higher for the lower part of the interval. When the coefficient has large magnitude, the probability is close to 1/2 for both the lower and upper half. That is the reason for distinguishing between the first refinement bit, when the magnitude is larger, and the subsequent refinement bits of a pixel when the magnitude is smaller. Because of statistical dependence among immediate neighbors, we distinguish whether or not there is at least one significant neighbor when the first refinement bit of a pixel is being coded.

11.4.10.4 Run mode coding

The set partitioning coders employ a sorting pass that locates significant pixels and insignificant sets of pixels for a given threshold or bitplane. The insignificant sets are encoded with a single 0 symbol. The EBCOT method in JPEG2000 has no such sorting pass and must visit and encode the bit of every pixel in the bitplane. The pixels visited in the CUP are all previously insignificant with insignificant neighborhoods. These pixels are likely to remain insignificant, especially for the higher bitplanes, so we employ a run-length coding mode in addition to the normal mode. This run mode is entered when four consecutive pixels are insignificant with insignificant neighborhoods. Therefore, the initial coding mode for the CUP is the run mode. However, when we encounter pixels with 1's in the scan of the bitplane, they become significant and make the neighborhoods of succeeding pixels "preferred," thereby triggering normal mode coding.

The preceding coding modes were not dependent on the scan pattern through the code-block. However, in order to describe the details of run mode coding, we need to specify this scan pattern. The code-block is divided into horizontal stripes of four rows per stripe. Within each stripe, the scan proceeds down the four-pixel columns from the leftmost to rightmost column. After the last pixel of the rightmost column is visited, the scan moves to the leftmost column of the next stripe below.[7] This scan pattern is illustrated in Figure 11.18.

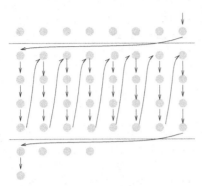

Figure 11.18 Column-wise scan pattern of stripes within a code-block.

The run mode is entered when all three of the following conditions are met.

1. Four consecutive pixels in the scan shown in Figure 11.18 must currently be insignificant.
2. All four of these pixels must have insignificant neighborhoods.
3. The four consecutive pixels must be aligned with a column of a stripe.

In run mode, a binary "run interruption" symbol is coded to indicate whether (0) or not (1) all four pixels remain insignificant in the current bitplane. This symbol is coded using its natural context of four consecutive pixels remaining insignificant or not. JPEG2000 labels this context of four consecutive pixels as number 9.[8] A pixel becomes significant when a "1" is read in the current bitplane. We count the number r of "0"s until a "1" is reached and encode this run number followed by the sign of the pixel associated with the "1" terminating the run. The run-length r ranges from 0 to 3, because it has already been signified that there is at least one "1." The run-length r is close to uniform in distribution, so its values are encoded with 2 bits, the most significant one sent first to the codestream.[9] The significances of the remaining samples of the four, if any, are encoded using the normal mode. Coding continues in normal mode until conditions to enter run mode are again encountered.

Example 11.5 Run mode coding example

Suppose that we are in run mode and the run interruption symbol "1" signifies that one or more of the four pixels in a column has become significant. Let the bitplane bits of the four pixels be 0010. Since $r = 2$, 1 and then 0 are sent to the codestream. Furthermore, the sign associated with the 1 following 00 is coded in the usual way and sent to the codestream. The next bit of 0 is then coded using the normal mode.

The operation of JPEG2000's arithmetic coder, which is the MQ-coder, has purposely not been described here. We refer the reader to JPEG2000 textbooks [6, 7] for such material. Clearly, JPEG2000 coding is heavily dependent on arithmetic coding, much more so than set partitioning coders, and for that reason is considered to be computationally complex. The payoff is excellent compression efficiency. Due to its bitplane coding from most to least significant, the resulting codestream of every code-block is bit-embedded. Also, the coding of subblocks of subbands independently allows resolution scalability and random access to spatial regions directly within the codestream. Despite its excellent performance and range of scalability options, some applications, especially those involving hardware implementation constrained in size and power consumption, cannot tolerate the high complexity. This complexity is one of the reasons that JPEG2000 has not seen such wide adoption thus far.

11.4.10.5 Scalable lossy coding in JPEG2000

The JPEG2000 method of coding of subblocks produces bit-embedded subcodestreams. Each subcodestream has a distortion versus rate characteristic, because it is associated with a different part of the wavelet transform. In order to achieve the best lossy reconstruction at a given rate, each subcodestream must be cut to its optimal size, so that the total of the sizes equals the size corresponding to the given rate as closely as possible. Sets of subcodestream sizes or cutting points corresponding to a number of rates can be calculated during encoding. The codestream is then considered to be built up in *layers*. A method to determine the optimal cutting points for a number of layers will be described later in Section 11.6. Briefly, according to this method, a series of increasing slope parameters $\lambda_1, \lambda_2, \ldots, \lambda_J$ are specified and every subcodestream is cut, so that the magnitude of its distortion-rate slope matches each parameter in turn until the desired total number of bits among the codestreams is reached.

11.4.11 The embedded zero-block coder (EZBC)

The arithmetic coding that is mandatory in JPEG2000 can be applied optionally in the embedded block coders, such as SPECK and SBHP, to encode sign and refinement bits. Furthermore, both SPECK and SBHP used simple, fixed Huffman coding of the 4-bit masks that signify the significance state of the four quadrants obtained in splitting a significant set. Adaptive versions of Huffman or arithmetic coding, either unconditioned or conditioned on context, can also be used for coding these masks. Another method called EZBC [3, 4] that we describe now does use these more complex forms of entropy coding.

The EZBC coder utilizes SPECK's quadrisection coding of full wavelet subbands and visits them in the zigzag order as shown in Figure 11.8. The transition from one subband to the next can be either at the same threshold or after passing through all thresholds from top to bottom. The former produces a bit-embedded composite codestream and the latter a resolution-progressive one with separate bit-embedded subband codestreams. What distinguishes EZBC from normal SPECK is the entropy coding of the significance map, sign bits, and refinement bits. The original SPECK used arithmetic coding of the 4-bit masks, but did no entropy coding of sign and refinement bits. EZBC entropy encodes the sign and refinement bits in the manner of JPEG2000. It also performs entropy coding of the significance decision bits in the masks adaptively

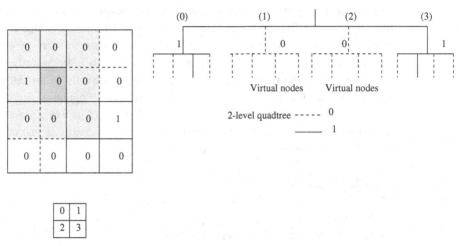

Figure 11.19   Two levels of quadrisection of 8 × 8 block and associated quadtree, including virtual nodes.

using a context template consisting of significance bits in neighboring masks and parent subbands. We shall describe next this coding in more detail.

The coding of bits in the masks is best illustrated by an example. Consider the case shown in Figure 10.8 of three levels of splitting of an 8 × 8 block of wavelet coefficients. Shown are the successively smaller quadrants created by the splittings and the associated quadtree and its code. Let us redraw this example for two levels of splitting in Figure 11.19. We have created virtual splits, marked by dashed lines, of the insignificant quadrants 1 and 2 resulting from the first level split. Likewise, in the quadtree, we have created corresponding virtual nodes. The four "0"s in these quadrants are associated with these virtual nodes and are not part of the significance map code, because they are redundant. Once a set is deemed insignificant, all its subsets are known to be insignificant. However, the inclusion of these virtual splits allow us to define a context template of the eight nearest neighbor quadtree nodes. The context is the significance bit pattern of these eight nodes. We show in Figure 11.19 a node or significance bit to be coded in solid shading of its set and its eight-neighbor context in lighter pattern shading. Every non-virtual node at this second level of the quadtree is coded in this manner. In fact, the (non-virtual) nodes in each level of the quadtree, corresponding to the significance decision bits in each level of quadrisection, are coded in this manner. Nodes associated with sets near edges of the block will have some nearest neighbors ouside the subband. Such external neighbor nodes are considered always to be "0." Subband sizes are usually much larger than 8 × 8, so nodes associated with edges will be relatively few in most levels of the quadtree.

Actually, the context just described applies only to the bottom or last level of the quadtree, where the nodes correspond to individual pixels. It was chosen to account for strong dependence among neighboring quadtree nodes (sets) within a subband. For the other levels of the quadtree, we add to this context the parent in the SOT, in order to exploit dependence across scales of a wavelet transform. It is mainly beneficial at low bit rates where relatively few coefficients and hence nodes have become significant.

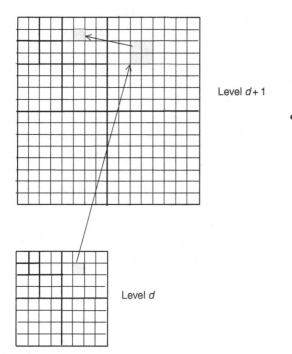

Level $d+1$

Level d

Figure 11.20 Two successive levels of quadtree with nodes located in wavelet transform and dependency relationships.

However, this parent is associated with a quadtree node at the next higher splitting level. The reason is that the current node to be coded is a fusion of the four nodes in its subband that are the children of this particular parent node in the SOT. These associations are depicted in Figure 11.20.

Tables similar to those for JPEG2000 define the context labels. For those readers interested, these tables may be found in Hsiang's doctoral thesis [4]. The number of labels is about the same as that for JPEG2000. Probability estimates are gathered for each context as coding proceeds in order for the arithmetic coder to compute the length and bits of the codeword that specifies a corresponding location in the [0, 1) unit interval.

11.4.11.1 Conditional dequantization

EZBC also utilizes conditional dequantization to reduce the reconstruction error. The probabilistic centroid of the quantization interval minimizes the mean squared error of a quantizer. When you decode one of these wavelet transform coders, you obtain a sequence of quantization bin indices that identify the quantization intervals. The so-called dequantizer sets the reconstruction value within the decoded quantization interval. The midpoint or the 38% point is often used arbitrarily, because the actual probability distribution is unknown. The 38% point is based on a probability model assumption that skews more of the probability toward the lower end of the interval.

An estimate of the true reconstructed value can be obtained by realizing that the refinement bit from the next lower bitplane tells whether or not the value is in the lower or upper half of the current quantization interval $[\tau, 2\tau)$, where τ is the current threshold. Specifically, a refinement "0" from the next lower bitplane indicates that the value is in the first half of the interval. Therefore, except for decoding the bottom bitplane, we can count the number of next lower "0" refinement bits from previously decoded coefficients to estimate the probability $\hat{p}_0$ that the true value is in the interval $[\tau, 1.5\tau)$. The probability density function is assumed to be step-wise uniform, with step heights $\hat{p}_0/(\tau/2)$ for the first half and $(1 - \hat{p}_0)/(\tau/2)$ for the second half of the interval. Using this estimated probability density function in the centroid definition, a formula for the centroid can be easily derived in terms of $\hat{p}_0$ and τ. These quantities are substituted into this formula to determine the reconstruction value. The centroid y_c of a probability density function $p_0(y)$ in the interval is given by

$$y_c = \int_{\tau}^{2\tau} y p_0(y) dy \tag{11.5}$$

Substituting the step-wise uniform density function

$$p_0(y) = \begin{cases} \frac{2\hat{p}_0}{\tau}, & \tau \le y < 1.5\tau \\ \frac{2(1-\hat{p}_0)}{\tau}, & 1.5\tau \le y < 2\tau \\ 0, & \text{elsewhere} \end{cases} \tag{11.6}$$

produces the reconstruction value

$$y_c = \left(\frac{7}{4} - \frac{1}{2}\hat{p}_0 \right) \tau \tag{11.7}$$

The performance of EZBC surpasses that of JPEG2000, usually by 0.3 to 0.6 dB in PSNR, depending on the image and the bit rate. Presumably, the use of conditional dequantization and the parent context and the overhead in JPEG2000 to enable random access and other codestream attributes account for this advantage. Neither the conditional dequantization nor the parent context adds much computation. The JPEG2000 Working Group deliberately chose not to use parent context, so that each subblock would be coded independently. Such a choice preserves the features of random access, ROI enhancement coding, and transmission error recovery. The conditional dequantization utilizes the counts of "0" bits already accumulated for arithmetic coding of the refinement bits, and substitutes into the reconstruction formula (11.7), so involves very little in extra computation. EZBC is less complex computationally than JPEG2000, because it codes fewer symbols. JPEG2000 must pass through the bitplanes of every coefficient, while EZBC passes only through bitplanes of coefficients that have become significant. Furthermore, EZBC passes through each bitplane only once versus three times for JPEG2000. Also, in its bit-embedded mode, there is no post-compression Lagrangian optimization for rate control.

Any bitplane coder can utilize this method of conditional dequantization. For the images tested in trials of EZBC, PSNR gains of 0 to 0.4 dB are claimed. Presumably these gains are relative to midpoint dequantization. Most likely, there will be smaller gains with respect to SPIHT and JPEG2000, because they use the usually more accurate 38% (or 3/8) point for reconstruction.

11.5 Tree-based wavelet transform coding systems

We have described several block-based wavelet transform coding systems, so now we turn to those which are tree-based. Already in the last chapter, we have presented the SPIHT and EZW algorithms operating on trees in the transform domain, because these algorithms require the energy compaction properties of transforms to be efficient. These algorithms were described in their breadth-first search or bit-embedded coding modes and without entropy coding of their output bits. Here, we shall explain how to realize the attributes of resolution scalability and random access of the block-based systems. Furthermore, we shall describe the entropy coding normally utilized in these tree-based systems. To do so without being redundant, SPIHT will be the test case for the methods that follow. The same methods work for EZW with obvious modification.

11.5.1 Fully scalable SPIHT

The SPIHT algorithm is described in detail in Chapter 10. Here, we shall review the basic idea of locating individual coefficients and sets of coefficients whose magnitudes fall below a given threshold. The transform space is partitioned into trees descending from three of the four pixels in every 2×2 group plus the remaining upper left corner pixels in the lowest frequency subband, denoted by $\mathcal{H}$. A partition generated from a 2×2 group is called a tree-block and is depicted in Figure 11.21. The members of this partition or tree-block are three spatial orientation trees (SOTs), branching vertically, diagonally, and horizontally, and the corner pixel. The constituent sets of an SOT are all descendants from a root node, called a $\mathcal{D}$ set, the set of four offspring from a node, called an $\mathcal{O}$ set, and the grand-descendant set, denoted by $\mathcal{L}$, consisting of all descendents of members of the offspring set, $\mathcal{O}$. These sets are illustrated in Figure 11.21 for the particular case where the root node is in the lowest frequency subband.

11.5.1.1 The sorting pass

First, the largest integer power of 2 threshold, $\tau_{max} = 2^{n_{max}}$, that does not exceed the largest magnitude coefficient in the entire transform is found. The algorithm maintains three control lists: the LIP, LIS, and LSP. The LIP is initialized with all coordinates in $\mathcal{H}$, the LIS with all coordinates in $\mathcal{H}$ with descendants, and LSP as empty. The initial entries in the LIS are marked as Type A to indicate that the set includes all descendants of their roots. The algorithm then proceeds to test members of the LIP for significance at the current threshold. A binary symbol of 0 for "insignificant" or 1 for "significant" is sent to the codestream for each test. When a pixel is significant, its sign bit is also sent to the codestream and its coordinates are moved to the LSP. Then the entries in the LIS are visited to ascertain whether or not their associated trees are significant. A tree is said to be significant if any member pixel is significant. Again, the binary result

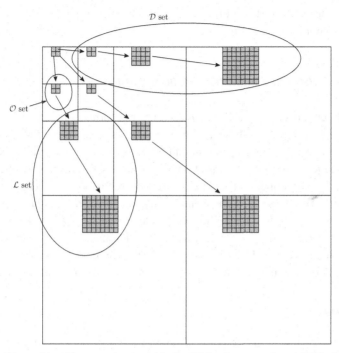

Figure 11.21   Illustration of set types in a tree-block with its three constituent SOTs descending from a 2 × 2 group in the lowest frequency subband in a 3-level, two-dimensional wavelet transform. A full descendant $\mathcal{D}$ set, an offspring $\mathcal{O}$ set, and a grand-descendant $\mathcal{L}$ set are encircled in the diagram. All pixels (coefficients) in gray, including the upper-left corner pixel in the 2 × 2 group, belong to the tree-block.

of the significance test is sent to the codestream. If insignificant (0), then the algorithm moves to the next LIS member. If significant, the tree is split into its child nodes (pixels) and the set of all descendants of these children, the so-called *grand-descendant set*. The children are then individually tested for significance. As before, if a child node is significant, a 1 and its sign are written to the codestream and its coordinates are moved and appended to the end of the LSP. If insignificant, a 0 is written and its coordinates are moved and appended to the end of the LIP. The grand-descendant set is placed on the LIS and is designated by the coordinates of its tree root and a Type B tag to indicate that it is a grand-descendant set. That distinguishes it from LIS sets of Type A. The grand-descendant set (if non-empty) is tested for significance and, as usual, the binary result is written immediately to the codestream. If significant, the set is divided into the four subtrees emanating from the children nodes. The coordinates of the roots of these subtrees, which are the children nodes, are appended to the LIS and tagged as Type A. They will be tested at the current threshold after the starting LIS entries of this sorting pass. The grand-parent node of these subtrees is now removed from the LIS. If the grand-descendant set is insignificant, it stays on the LIS as a Type B set.

During the course of the algorithm, new entries are being added to the end of the LIS, through the splitting of significant grand-descendant sets into four subtrees, so these entries will be tested for significance at the same threshold under which they were added. Since at least one of these subtree sets is significant, that will engender a sequence of analogous splitting of successively smaller sets at later generations of the SOT until the significant single elements are located. Later generations of the SOT are equivalent to higher resolution levels. Therefore, we note that pixels and sets belonging to different resolutions are intermingled in the lists. We also note that the LIS significance search is breadth first, because LIS entries in the starting list of the pass are tested before those created at a later generation of the tree corresponding to the next higher level of resolution.

When all entries on the LIS have been processed, the sorting pass at the current threshold is complete. The threshold is then lowered by a factor of 2 to start the next sorting pass. The algorithm visits the LIP left from the last pass to test it against the new lower threshold and to code its pixels as described above. It then enters the top of the LIS remaining from the previous pass to process its entries for the new threshold.

11.5.1.2 The refinement pass

The refinement pass is executed immediately after every sorting pass, except for the first one at threshold $2^{n_{\max}}$, before lowering the threshold. The refinement pass at a threshold 2^n consists of outputting the n-th bit of the magnitude of pixels in the LSP that became significant at a higher threshold. This corresponds to collecting and writing the bits through bitplane n of all coefficients previously found to be significant.

11.5.1.3 The decoder

The conveyance of the outcomes of the significance tests allows the decoder to recreate the actions of the encoder. The decoder initializes the same lists and populates them according to the significance test bits received from the codestream. Once all of the codestream has been read, the coefficients belonging to the LSP coordinate entries can be decoded and reconstructed, and all other coefficients are set to 0. The inverse wavelet transform then produces the reconstruction of the source image.

11.5.2 Resolution scalable SPIHT

The original SPIHT algorithm just described above is scalable in quality at the bit level. Each additional bit written to the bitstream reduces the distortion no more than any of the preceding bits. However, it can not be simultaneously scalable in resolution. Scalable in resolution means that all bits of coefficients from a lower resolution are written to the bitstream before those from a higher resolution. We see in the original quality scalable SPIHT that resolution boundaries are not distinguished, so that bits from higher bitplanes in high resolution subbands might be written to the codestream before lower bitplane bits belonging to lower resolutions. We therefore have to change the priority of the coding path to encode all bitplanes of lower resolution subbands before higher resolution ones.

Referring to Figure 11.8, the correspondence of subbands to resolution level r is described in Table 11.8. The missing LL_r subbands in this table are the union of all

Table 11.8 Correspondences of resolution levels to subbands

Resolution level r	Subbands
0	LL (LL_3)
1	HL_3, LH_3, HH_3
2	HL_2, LH_2, HH_2
3	HL_1, LH_1, HH_1

the subbands at resolutions lower than r. To reconstruct an image at a given resolution $r > 0$, all subbands at the current and lower resolution levels are needed. For example, we need subbands $HL_2, LH_2, HH_2, HL_3, LH_3, HH_3, LL_3$ in order to reconstruct the image at resolution $r = 2$.

Let us assume that there are m levels of resolution numbered $r = 0, 1, \ldots, m - 1$ ($m - 1$ is the number of wavelet decompositions.) The way to distinguish resolution levels and avoid their intermingling on the lists is to maintain separate LIP, LIS, and LSP lists for each resolution. Let LIP_r, LIS_r, and LSP_r be the control lists for resolution r, $r = 0, 1, \ldots, m - 1$. That means that coordinates located in subbands belonging only to resolution r reside in LIP_r, LIS_r, and LSP_r. As usual, whenever a pixel or set is tested, the binary result of 0 or 1 is immediately written to the codestream. For each resolution, starting with $r = 0$, we proceed with the usual SPIHT search for significance of pixels and sets at all thresholds from highest to lowest. This process creates entries to the lists at the current and next higher resolution. To see how this works, suppose that we enter the process for some resolution r. Consider testing LIP_r at some threshold $\tau = 2^n$ and if significant, we move its coordinates to LSP_r. If not significant, its coordinates stay on LIP_r. After exiting LIP_r, LIS_r is tested at the same threshold. If an entry in LIS_r is insignificant, it stays on this list. However, if significant, it is removed from this list and the tree descending from the associated location is split into its four offspring and its grand-descendant set. The four offspring belong to the next higher resolution $r + 1$, so are placed on LIP_{r+1}. The grand-descendant set is designated by the coordinates of its grandparent, so is then appended to LIS_r as Type B. The algorithm tests this LIS_r Type B entry immediately. If not significant, it stays as is. If significant, its four subtrees descending from offspring are placed on LIS_{r+1} as Type A entries designated by their associated offspring coordinates.

There are complications that arise with this procedure. Since we pass through all thresholds at resolution r before entering the lists of resolution $r + 1$, we must indicate the threshold at which single pixels are placed on LIP_{r+1}. These pixels are the offspring created from the splitting of a significant grand-descendant set. Instead of just putting these offspring immediately on LIP_{r+1}, they are first tested for significance and placed on LIP_{r+1} if insignificant or on LSP_{r+1} if significant, along with tags indicating the threshold at which they were tested. When testing reaches resolution $r + 1$, new entries to LIP_{r+1} and LSP_{r+1} should be inserted into the lists in positions that preserve the order of decreasing thresholds. These threshold tags, just like Type A or B, do not have to be

ALGORITHM 11.5 _____

Resolution scalable SPIHT
Notation:

- $c_{i,j}$ is value of wavelet coefficient at coordinates (i, j).
- Significance function for bitplane n (threshold 2^n) $\Gamma_n(\mathcal{S}) = 1$, if entity (coefficient or set) $\mathcal{S}$ significant; otherwise 0.

Initialization step:

- Initialize n to the number of bitplanes n_{max}.
- $LSP_0 = \emptyset$
- LIP_0 : all the coefficients without any parents (the eight root coefficients of the block)
- LIS_0 : all coefficients from the LIP_0 with descendants (three coefficients as only one has no descendant)
- For $r \neq 0$, $LSP_r = LIP_r = LIS_r = \emptyset$

For each r from 0 to maximum resolution $m - 1$
For each n from n_{max} to 0 (bitplanes)
Sorting pass :
For each entry (i, j) of the LIP_r which had been added at a threshold strictly greater to the current n

- Output $\Gamma_n(i, j)$
- If $\Gamma_n(i, j) = 1$, move (i, j) to LSP_r and output the sign of $c_{i,j}$ **(1)**

For each entry (i, j) of the LIS_r **which had been added at a bitplane greater than or equal to the current n**

- If the entry is type A
  – Output $\Gamma_n(\mathcal{D}(i, j))$
  – If $\Gamma_n(\mathcal{D}(i, j)) = 1$ then
    - For all $(k, \ell) \in \mathcal{O}(i, j)$: output $\Gamma_n(k, \ell)$; If $\Gamma_n(k, \ell) = 1$, add (k, ℓ) to the LSP_{r+1} and output the sign of $c_{k,\ell}$ else, add (k, ℓ) to the end of the LIP_{r+1} **(2)**
    - If $\mathcal{L}(i, j) \neq \emptyset$, move (i, j) to the LIS_r as a type B entry
    - Else, remove (i, j) from the LIS_r
- If the entry is type B
  – Output $\Gamma_n(\mathcal{L}(i, j))$
  – If $\Gamma_n(\mathcal{L}(i, j)) = 1$
    - Add all the $(k, \ell) \in \mathcal{O}(i, j)$ to the LIS_{r+1} as a type A entry
    - Remove (i, j) from the LIS_r

Refinement pass:

- For all entries (i, j) of the LSP_r **which had been added at a bitplane strictly greater than the current n** : Output the n-th most significant bit of $c_{i,j}$

_____

sent, because the decoder can recreate them when it populates its lists. The procedure is stated in its entirety in Algorithm 11.5.

This resolution scalable algorithm puts out the same bits as the original quality scalable algorithm, but in different order. In order to distinguish the resolutions, counts of bits coded at each resolution are written into a codestream header. Then we can decode

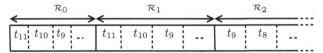

Figure 11.22 Resolution scalable bitstream structure. $\mathcal{R}_0, \mathcal{R}_1, \ldots$ denote the segments with different resolutions, and $t_{11}, t_{10}, \ldots$ the different thresholds ($t_n = 2^n$.) Note that in $\mathcal{R}_2$, t_{11} and t_{10} are empty.

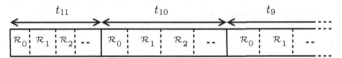

Figure 11.23 Quality scalable bitstream structure. $\mathcal{R}_0, \mathcal{R}_1, \ldots$ denote the segments of different resolutions, and $t_{11}, t_{10}, \ldots$ the different thresholds ($t_n = 2^n$). At higher thresholds, some of the finer resolutions may be empty.

up to a resolution level less than the full resolution with confidence that we will get the same reconstruction if we had encoded the source image reduced to the same resolution. However, we have lost control of the rate in this algorithm. The steps in Algorithm 11.5 describe coding through all the bitplanes, where the final file size or bit rate is undetermined. In order to enable rate control, we require knowledge of the bitplane boundaries in every resolution level. It is simple enough when encoding to count the number of bits coded at every threshold in every resolution level and put these counts into a file header. We depict the resolution scalable codestream in Figure 11.22 with resolution and threshold boundaries indicated by these bit counts. Then, to reconstruct a desired resolution level r_d, we discard the bits coded from higher resolution levels and reorganize the codestream to the quality scalable strucure, such as that shown in Figure 11.23. This structure is obtained by interleaving the code bits belonging to the remaining resolutions $r = 0, 1, 2, \ldots, r_d$ at the same thresholds. One follows the zigzag scan of the subbands through resolution level r_d to extract first the bits coded at maximum threshold $2^{n\max}$, then rescans to extract bits coded at threshold $2^{n\max-1}$, and so forth. Then the codestream can be truncated at the point corresponding to the target bit rate. To decode, one must use the quality scalable version of SPIHT or reverse the reorganization and use the resolution scalable version.

Actually, the algorithm for quality scalable SPIHT can be described just by interchanging the order of the two "For" loops in Algorithm 11.5. That accomplishes passing through all the resolution levels at every threshold, instead of vice versa. Clearly, this interchange is exactly what we need to achieve quality scalability.

11.5.3   Block-oriented SPIHT coding

The resolution scalable SPIHT algorithm can be further modified to provide *random access* decoding. Random access decoding means that a designated portion of the

ALGORITHM 11.6

Block-oriented SPIHT coding

- Form tree-blocks $\mathcal{B}_1, \mathcal{B}_2, \cdots, \mathcal{B}_K$ that partition the wavelet transform.
- for $k = 1, 2, \ldots, K$, code $\mathcal{B}_k$ either with Algorithm 10.4 (neither quality nor resolution scalable) or Algorithm 11.5 (quality and resolution scalable).

image can be reconstructed by extracting and decoding only an associated segment of the codestream. Recall that a block-tree of a wavelet transform, illustrated in Figure 11.21, corresponds to a region of the image geometrically similar in size and position of its 2×2 root group within the LL subband. Using the example of this figure, the dimensions of the LL subband and image are 8×8 and 64×64, respectively. Therefore, the second 16×16 block at the top of the image is similar to the 2×2 group root in the LL subband. In reality, this 2×2 wavelet coefficient group does not reconstruct the corresponding 16×16 image block exactly, because surrounding coefficients within the filter span contribute to values of pixels near the borders on the image block. However, the correct image block is reproduced, although not perfectly. The tree-blocks are mutually exclusive and their union forms the full wavelet transform.

Random access decoding can be realized, therefore, by coding all the tree-blocks separately. In fact, we can use either the quality or resolution scalable SPIHT to encode each tree-block in turn. We just initialize the lists of the SPIHT algorithm in the usual way, but only with the coordinates of the 2×2 group in the LL subband that is the root of the tree-block to be coded. The following steps of the quality or resolution scalable SPIHT algorithm are exactly as before. When the algorithm finishes a tree-block, it moves to the next one until all the tree-blocks are coded. These steps are delineated in Algorithm 11.6. Within every tree-block, the codestream is either quality or resolution scalable, depending on which SPIHT algorithm is used.

Example 11.6 Using block-oriented SPIHT to decode region of interest. We demonstrate an example of using block-oriented SPIHT coding (Algorithm 11.6) to reconstruct a region of interest from a portion of the codestream. Again, as in Example 11.1, the Goldhill source image's wavelet transform is quantized with step size 1/0.31, and the bin indices are losslessly coded to a codestream size corresponding to 1.329 bpp. The lossless coding method follows Algorithm 10.4, which is not quality scalable. The coding algorithm is block-oriented, so a segment of the codestream containing the coded blocks corresponding to the desired region of interest is extracted and decoded. Figure 11.24 shows the reconstruction of the region of interest beside that of the fully coded image.

Figure 11.24   Reconstructions from same codestream of 512×512 Goldhill, quantizer step size $= 1/0.31$, coded to rate 1.329 bpp, and of 70×128 region at coordinates (343, 239).

11.6   Rate control for embedded block coders

In order to exercise rate control when encoding the wavelet transform in separate blocks, such as we have in JPEG2000, SBHP, and block-oriented SPIHT, we need a method to determine the rate in each block that minimizes the distortion for a given average bit rate target. When the blocks are bit-embedded, their codestreams can simply be truncated to the sizes corresponding to the bit rates determined by such an optimization method. We assume that blocks have been coded to the bottom bitplane or threshold and are bit-embedded either naturally or by reorganization.

Given the target number of codestream bits B_T, the task is to assign rates (measured in bits) $b_1, b_2, \ldots, b_K$ to the blocks $\mathcal{B}_1, \mathcal{B}_2, \ldots, \mathcal{B}_K$, respectively, so that the average distortion $D(B_T)$ is a minimum. Stated more formally, we seek the solution of rates $b_1, b_2, \ldots, b_K$ that minimizes

$$D(B_T) = \sum_{k=1}^{K} d_{\mathcal{B}_k}(b_k) \quad \text{subject to} \tag{11.8}$$

$$B_T = \sum_{k=1}^{K} b_k$$

where $d_{\mathcal{B}_k}(b_k)$ denotes the distortion in block $\mathcal{B}_k$ for rate b_k. We assume that this function is convex and monotonically non-increasing. Since the distortion measure is squared error, the total distortion is the sum of those of the blocks. The minimization of the objective function,

ALGORITHM 11.7 _____

Optimal bit allocation to independent blocks

Notation:
- i: index of points increasing with rate
- i-th rate in bits to block $\mathcal{B}_k$: b_k^i
- Distortion $d_{\mathcal{B}_k}^i$ for rate b_k^i

1. *Initialization.*
   (a) Obtain sequence of (rate, distortion) points $\{(b_k^i, d_{\mathcal{B}_k}^i)\}$ for every block $\mathcal{B}_k$, $k = 1, 2, \ldots, K$.
   (b) Set a bit budget target B_T to allocate.
   (c) Choose $\lambda > 0$.

2. *Main:*
   For $k = 1, 2, \ldots, K$ (for all blocks)
   (a) set initial i large for large rate
   (b) for steadily decreasing i calculate
   $$\frac{\delta d_k^i}{\delta b_k^i} = \frac{d_{\mathcal{B}_k}^i - d_{\mathcal{B}_k}^{i+1}}{b_k^{i+1} - b_k^i}$$
   only until $\geq \lambda_j$ and stop[10]. Let $i = i_k^*$ be the stopping index at smaller of the two rates.
   (c) Denote rate solution for k-th block $b_k(\lambda) = b_k^{i_k^*}$

3. Total number of bits is $B(\lambda) = \sum_{k=1}^{K} b_k^{i_k*}$.

_____

$$J(\lambda) = \sum_{k=1}^{K}(d_{\mathcal{B}_k}(b_k) + \lambda b_k), \tag{11.9}$$

with Lagrange multiplier $\lambda > 0$, yields the solution that at the optimal rates, all the individual distortion-rate slopes must be equal to $-\lambda$. Since we do not have formulas for the block distortion-rate functions, we offer the following practical procedure to approximate this solution.

In the course of coding any block, say $\mathcal{B}_k$, the algorithm can calculate and record a set of distortion-rate points $(d_{\mathcal{B}_k}^i, b_k^i)$. The index i of these points increases from low to high rates. According to the convex assumption, the distortion decreases with increasing rate and index i. For a given λ, we start calculating distortion-rate slope magnitudes starting from high rates to lower rates. These magnitudes increase with decreasing rate. We stop when we first reach

$$\frac{d_{\mathcal{B}_k}^i - d_{\mathcal{B}_k}^{i+1}}{b_k^{i+1} - b_k^i} \geq \lambda. \tag{11.10}$$

and declare the smaller of the rates b_k^i as the rate solution for block $\mathcal{B}_k$. We repeat for all $k = 1, 2, \ldots, K$. Algorithm 11.7 details this method.

When we choose λ, the total rate, $B(\lambda) = \sum_k b_k$, is revealed only after the optimization procedure. We do know that λ is the magnitude of the common slope at the rate solution points. Therefore, we start with small λ corresponding to a high rate and steadily increase λ to test slopes at smaller rates. In this way, we can store a table of correspondences of λ to $B(\lambda)$ and then know the value of λ needed for the target rate B_T.

ALGORITHM 11.8 _____

Bisection procedure for optimal bit allocation

1. Set target rate $B_T = \sum_{k=1}^{K} b_k(\lambda^*)$. To find λ^*. Set precision parameter $\epsilon > 0$.
2. Find slope parameters $\lambda_1 > 0$ and $\lambda_2 > \lambda_1$ such that $B(\lambda_2) < B_T < B(\lambda_1)$.
3. Let $\lambda_3 = (\lambda_2 + \lambda_1)/2$.
4. Set $i = 3$.
5. If $B(\lambda_i) < B_T$, $\lambda_{i+1} = (\lambda_i + \lambda_{j*})/2$, $j^* = \arg\min_{j<i} B(\lambda_j) > B_T$.
    If $B(\lambda_i) > B_T$, $\lambda_{i+1} = (\lambda_i + \lambda_{j*})/2$, $j^* = \arg\max_{j<i} B(\lambda_j) < B_T$.
    (Recursively determines half interval containing target rate.)
6. If $|B(\lambda_{i+1}) - B_T| < \epsilon$, stop. Let $\lambda_{i+1} = \lambda^*$.
    Else, set $i = i + 1$ and return to Step 5.

One of the problems with this procedure is that a rather odd set of total rates usually results. More often, one would like to find the cutting points for a given set of rates, such as $0.25, 0.50, 0.75, 1.0$, and 2.0 bpp perhaps. Therefore, one needs to determine the unique slope parameter λ that produces each one of these rates. The bisection method starts with an interval of slope magnitudes that contains the solution for the prescribed rate and determines in which half of this interval lies the solution. Then the solution point is narrowed to one of the halves of this half-interval and so forth until the interval containing the solution is sufficiently small. The bisection procedure for a single rate is detailed in Algorithm 11.8.

In calculating these rates for the various values of λ, we needed to calculate the distortion-rate slopes as in Step 2(b) in Algorithm 11.7. When you need to find the optimal parameters λ^* for a series of rates, it is particularly easy to set initial intervals once the first λ^* for the largest rate is found. Taking the rate targets in decreasing order, the distortion-rate slopes are increasing, so one can always use the last solution as the the lower end point of the next slope interval.

11.7   Conclusion

In this chapter we have attempted to present the principles and practice of image or two-dimensional wavelet compression systems. Besides compression efficiency as an objective, we have shown how to imbue such systems with features of resolution and quality scalability and random access to source regions within the codestream. The simultaneous achievement of more than one of these features is a unique aspect of the systems described here. We have tried to present the material at a level accessible to students and useful to professionals in the field.

The coding methods featured in this chapter are the basic ones. The purpose of this chapter was to explain the fundamental methods, so as to provide the student and professional with the tools to extend or enhance these methods according to their own purposes. Many enhancements and hybrids appear in the literature. We shall mention just a few of them to suggest further readings. For example, only the basic JPEG2000 (Part 1) coding algorithm was described along with its scalability features. There is a

Part 2 standard with many extensions. Even in Part 1, built-in features of error resilience and region-of-interest encoding (max-shift method) have not been presented. Interested readers may consult the JPEG2000 standardization documents [2, 8] and the book by Taubman and Marcellin [6].

There have been published many extensions and enhancements of the SPIHT and SPECK methods. SPIHT especially has inspired thousands of attempts to improve upon the original work [9]. One way was to dispense with the control lists, especially the LSP, because they grow long, requiring large memory, for higher rates and large images. Articles by Wheeler and Pearlman [10] and Shively *et al.* [11] report the use of fixed memory arrays to store the states of pixels and sets, thereby alleviating the memory problem. A problem of SPIHT that detracts from its efficiency, especially at low bit rates, is the identification of an insignificant pixel at a high threshold. For each such pixel, 1 bit has to be encoded for every lower threshold. Therefore, there are attempts in the literature to cluster these insignificant pixels, so that a single "0" indicates their common insignificance. One such successful attempt is the block-tree approach of Moinuddin and Khan [12]. Another approach by Khan and Ghanbari [13] clustered insignificant offspring together by creating virtual root nodes on top of the real root nodes of the LL subband.

There have been some notable attempts to combine SPIHT or EZW and vector quantization. The idea is that the significance search can sort vectors according to their locations between shells in higher dimensional spaces. Among such efforts are those by da Silva *et al.* [14] and Cosman *et al.* [15, 16], who developed vector EZW coders, and Mukherjee *et al.* [17], who developed a vector SPIHT coder.

Notes

1. Wavelet transforms and their implementation are explained in Chapter 7.
2. For non-square images, the splitting is into quadrants as nearly equal in size as possible. At the last splitting stages, when a single row or column with more than one element is reached, binary splitting is employed.
3. Indicators are markers or counts of bits inserted into the codestream header.
4. This phenomenon holds also for full-size subband blocks, because of the successive upsampling, filtering, and combining needed to synthesize an image from its subbands.
5. Often, instead of markers, the counts of code bytes representing the subblocks are written to the codestream header.
6. The unary code for a non-negative integer is the same number of 0's followed by a 1, which acts as a comma. For example, $0 \rightarrow 1$, $1 \rightarrow 01$, $2 \rightarrow 001$, and so forth.
7. We are assuming that code-block dimensions are multiples of four. For non-square images, some code-blocks may not have such dimensions, leaving stripes with fewer than four rows. Run mode is not invoked for stripes with fewer than four rows.

8. Some textbooks count this context as two states, significant and insignificant four consecutive pixels.

9. In the standard, the MQ-coder is set to a non-adaptive mode with a fixed uniform probability distribution to code the run lengths.

10. In practice the convexity assumption may fail and $\frac{\delta d_k^i}{\delta b_k^i} < \frac{\delta d_k^i}{\delta b_k^i}$. In that case, the point $(b_k^i, d_{B_k}^i)$ and both previous slope calculations involving this point must be discarded and replaced with $\frac{\delta d_k^i}{\delta b_k^i} = \frac{d_{B_k}^{i-1} - d_{B_k}^{i+1}}{b_k^{i+1} - b_k^{i-1}}$.

References

1. C. Chrysafis, A. Said, A. Drukarev, A. Islam, and W. A. Pearlman, "SBHP – a low complexity wavelet coder," in *Proceedings of the IEEE International Conference on Acoustics Speech Signal Process.*, vol. 4, Istanbul, Turkey, June 2000, pp. 2035–2038.

2. *Information Technology – JPEG2000 Image Coding System, Part 1: Core Coding System*, ISO/IEC Int. Standard 15444-1, Geneva, Switzerland, 2000.

3. S.-T. Hsiang and J. W. Woods, "Embedded image coding using zeroblocks of sub-band/wavelet coefficients and context modeling," in *Proceedings of the IEEE International Symposium Circuits Systems*, vol. 3, Geneva, Switzerland, May 2000, pp. 662–665.

4. S.-T. Hsiang, "Highly scalable subband/wavelet image and video coding," Ph.D. thesis, Electrical, Computer and Systems Engineering Dept., Rensselaer Polytechnic Institute, Troy, NY, USA, 2002.

5. D. S. Taubman, "High performance scalable image compression with EBCOT," *IEEE Trans. Image Process.*, vol. 9, no. 7, pp. 1158–1170, July 2000.

6. D. S. Taubman and M. W. Marcellin, *JPEG2000: Image Compression Fundamentals, Standards, and Practice*. Norwell, MA: Kluwer Academic Publishers, 2002.

7. T. Acharya and P.-S. Tsai, *JPEG2000 Standard for Image Compression: Concepts, Algorithms and VLSI Architectures*. Hoboken, NJ: Wiley-Interscience, Oct. 2004.

8. *Information Technology – JPEG2000 Extensions, Part 2: Core Coding System*, ISO/IEC Int. Standard 15444-2, Geneva, Switzerland, 2001.

9. A. Said and W. A. Pearlman, "A new, fast, and efficient image codec based on set partitioning in hierarchical trees," *IEEE Trans. Circuits Syst. Video Technol.*, vol. 6, no. 3, pp. 243–250, June 1996.

10. F. W. Wheeler and W. A. Pearlman, "SPIHT image compression without lists," in *Proceedings of the IEEE International Conference on Acoustics Speech Signal Process.*, vol. 4, Istanbul, Turkey, June 2000, pp. 2047–2050.

11. R. R. Shively, E. Ammicht, and P. D. Davis, "Generalizing SPIHT: a family of efficient image compression algorithms," in *Proceedings of the IEEE International Conference on Acoustics Speech Signal Processing*, vol. 4, Istanbul, Turkey, June 2000, pp. 2059–2062.

12. A. A. Moinuddin and E. Khan, "Wavelet based embedded image coding using unified zeroblock-zero-tree approach," in *Proceedings of the IEEE International Conference on Acoustics Speech Signal Processing*, vol. 2, Toulouse, France, May 2006, pp. 453–456.

13. E. Khan and M. Ghanbari, "Very low bit rate video coding using virtual SPIHT," *IEE Electron. Lett.*, vol. 37, pp. 40–42, Jan. 2001.

14. E. A. B. da Silva, D. G. Sampson, and M. Ghanbari, "A successive approximation vector quantizer for wavelet transform image coding," *IEEE Trans. Image Process.*, vol. 5, no. 2, pp. 299–310, Feb. 1996.

15. S. M. Perlmutter, K. O. Perlmutter, and P. C. Cosman, "Vector quantization with zerotree significance map for wavelet image coding," in *Rec. Twenty-Ninth Asilomar Conference on Signals Systems and Computers*, vol. 2, Pacific Grove, CA, Oct. 1995, pp. 1419–1423.

16. P. C. Cosman, S. M. Perlmutter, and K. O. Perlmutter, "Tree-structured vector quantization with significance map for wavelet image coding," in *IEEE Data Compression Conf.*, Snowbird, UT, Mar. 1995, pp. 33–41.

17. D. Mukherjee and S. K. Mitra, "Successive refinement lattice vector quantization," *IEEE Trans. Image Process.*, vol. 11, no. 12, pp. 1337–1348, Dec. 2002.

Further reading

E. Christophe and W. A. Pearlman, "Three-dimensional SPIHT coding of volume images with random access and resolution scalability," *EURASIP J. Image Video Process.*, vol. 2008, no. Article ID 248905, 2008, 13 pages.

A. Islam and W. A. Pearlman, "Embedded and efficient low-complexity hierarchical image coder," in *Proceedings of the SPIE Vol. 3653: Visual Communication Image Processing*, San Jose, CA, Jan. 1999, pp. 294–305.

W. A. Pearlman, A. Islam, N. Nagaraj, and A. Said, "Efficient, low-complexity image coding with a set-partitioning embedded block coder," *IEEE Trans. Circuits Syst. Video Technol.*, vol. 14, no. 11, pp. 1219–1235, Nov. 2004.

A. Said and W. A. Pearlman, "An image multiresolution representation for lossless and lossy compression," *IEEE Trans. Image Process.*, vol. 5, no. 9, pp. 1303–1310, Sept. 1996.

J. M. Shapiro, "Embedded image coding using zerotrees of wavelet coefficients," *IEEE Trans. Signal Process.*, vol. 41, no. 12, pp. 3445–3462, Dec. 1993.

12 Methods for lossless compression of images

12.1 Introduction

In many circumstances, data are collected that must be preserved perfectly. Data that are especially expensive to collect, require substantial computation to analyze, or involve legal liability consequences for imprecise representation should be stored and retrieved without any loss of accuracy. Medical data, such as images acquired from X-ray, CT (computed tomography), and MRI (magnetic resonance imaging) machines, are the most common examples where perfect representation is required in almost all circumstances, regardless of whether it is really necessary to preserve the integrity of the diagnostic task. The inaccuracies resulting from the acquisition and digitization processes are ignored in this requirement of perfection. It is only in the subsequent compression that the digitized data must be perfectly preserved. Physicists and materials scientists conduct experiments that produce data written as long streams or large arrays of samples in floating point format. These experiments are very expensive to set up, so there is often insistence that, if compressed, the decompressed data must be identical to the original.[1]

Nowadays, storage and transmission systems are overwhelmed with huge quantities of data. Although storage technology has made enormous strides in increasing density and reducing cost, it seems that whatever progress is made is not enough. The users and producers of data continue to adapt to these advances almost instantaneously and fuel demand for even more storage at less cost. Even when huge quantities of data can be accommodated, retrieval and transmission delays remain serious issues. Retrieval is cumbersome and slow in large databases and transmission time for an urgently needed large dataset may be too slow.

As we have seen with lossless data compression in earlier chapters, potential reductions in file size seem to average about 2 to 1 with the degree highly dependent on the data characteristics. The latter property of data dependence holds true for images as well, but reductions in file size can be larger than those seen with one-dimensional data using analogous methods. It is also useful and perhaps even necessary for a large losslessly compressed file to have the capability to be decoded progressively in resolution and quality. In Chapter 11, we have described subband/wavelet coding systems, which, when implemented with the proper choice of filters, can write a compressed lossless file that can be decoded progressively in resolution and quality.

At the outset, it is worth repeating that only integers can be represented in compressed form without loss. The acquisition apparatus of digital images writes 8, 16, or 32 bit integer values to a buffer, so there is no problem in compressing these data without loss. If the data are non-integers and discrete, then they can be put into one-to-one correspondence to integers before lossless compression. For example, when the data samples are written in a floating point format with six decimal places, multiplication by 10^6 converts them to integers. If no such integer exceeds the maximum possible that can be handled by the computer (usually $2^{63} - 1$), then the original data can be represented in compressed form without loss.

Because of the integer requirement for lossless compression, one must be careful that mathematical operations on data prior to compression use fixed point arithmetic. Otherwise, there may occur loss of precision due to rounding in the computer, so if one decides to use a transform compression technique, the transform coefficients cannot be floating point values. The transform must produce integer values and be perfectly reversible.

For the above reasons and for the sake of speed and simplicity, lossless image compression started with direct operations on the integer values. More recently, integer transforms have been developed that can be computed quickly and that convert the data to forms amenable to efficient compression and progressive decompression.

12.2　Lossless predictive coding

Among non-transform methods, predictive coding seems to be the one of choice, since it takes into account the correlation among neighboring samples. The idea is to predict the value of the (integer) sample to be encoded and to encode the difference between the actual and predicted values. Unlike lossy DPCM coding, there is no quantization here, so that we need not recover in the encoder these differences via a feedback mechanism in order to predict the next encoded sample.

The paradigm of lossless predictive coding pertains to any source, but we shall describe it in the framework of images. Figure 12.1 portrays the model of lossless predictive coding for an image. Referring to the figure, each image sample (called a *pixel* for picture element) is predicted from previously seen samples (pixels), and is subtracted from the actual pixel. This difference, called the *prediction residual* is then encoded losslessly by an entropy encoder and written to the compressed bitstream. The decoder reproduces the same prediction from past residuals and adds the current residual to produce the original pixel value. The encoder must be initialized with actual values of the pixels used in the first prediction.

12.2.1　Old JPEG standard for lossless image compression

As a prelude to formulating the general principles involved in a lossless predictive coding scheme, we shall describe the old JPEG standard for lossless image compression [1], which incorporates the simpler aspects of these principles. The three nearest

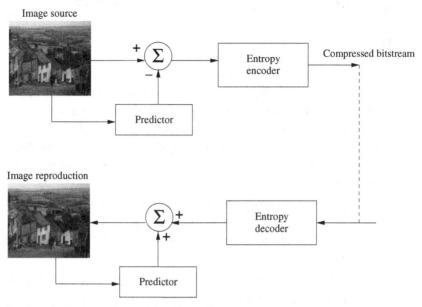

Figure 12.1 Lossless predictive coding and decoding of image sources.

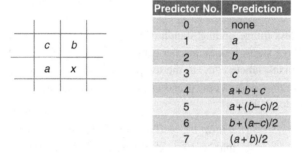

Predictor No.	Prediction
0	none
1	a
2	b
3	c
4	$a + b + c$
5	$a + (b-c)/2$
6	$b + (a-c)/2$
7	$(a + b)/2$

Figure 12.2 Neighborhood and prediction modes for lossless JPEG.

neighbors already visited in a raster scan form the prediction of the current pixel. These neighbors and the current pixel are labeled "a," "b," "c" and "x," respectively in Figure 12.2. There is one of eight possible predictions that are formed from these neighbors and subtracted from the actual value of "x." These predictions are also displayed in Figure 12.2. The divisions by two in Predictors 5, 6, and 7 are integer divisions realized by a single bit shift to the right. The particular predictor is often predetermined from the class of images to be encoded. However, the combination of image size and encoding method often results in fast enough encoding that all the predictors can be tried in turn to find the one that yields the smallest file size. Then the 3-bit identifier of the winning predictor and the entropy coded residuals are written to the compressed bitstream. The encoding of the residuals is usually done with a simple adaptive arithmetic or Huffman code.

12.3   State-of-the-art lossless image coding and JPEG-LS

The old JPEG standard does not achieve state-of-the-art compression efficiencies. The reason is that it does not make judicious use of the past samples available to the prediction and coding steps. The neighborhood of past values, called the *causal template*, is too restricted and the residual coding is not conditioned on the *context* in which it occurs. The context, like the causal template, is a function of past values used to calculate an empirical conditional probability of the values of the current pixel x. These conditional probabilities are used to determine the codewords and their lengths in the adaptive entropy coder, so the steps to follow for successful lossless image compression are as follows:

1. Form prediction $\hat{x}$ of x based on a defined *causal template*.
2. Define a *context* and determine in which realization of this context x occurs.
3. Code the prediction residual $e = x - \hat{x}$ using a probability model conditioned on the context of x.

Based on these steps as guiding principles, Hewlett-Packard researchers developed the (low complexity lossless compression for images) (LOCO-I) algorithm [2] that was chosen for the new JPEG-LS standard.

12.3.1   The predictor

The new JPEG lossless image coding standard (part of JPEG2000), called JPEG-LS, incorporates these three steps. It uses a richer causal template, depicted in Figure 12.3, that adds one more pixel, labeled "d," to the template of the old standard. This additional pixel is used in the coding context and not in the prediction to follow. There is a single prediction mode, defined as

$$\hat{x} = \begin{cases} \min(a, b), & \text{if } c \geq \max(a, b) \\ \max(a, b), & \text{if } c \leq \min(a, b) \\ a + b - c, & \text{otherwise} \end{cases} \tag{12.1}$$

The motivation for this prediction formula derives from the characteristics of images. This prediction tries to detect an edge by the first two conditions, and failing that,

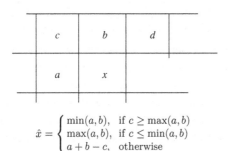

Figure 12.3   Causal template and predictor for JPEG-LS.

uses the average value in the planar triangle formed by the heights of a, b, and c. For example, suppose that $a \leq b$, the predictor picks $\hat{x} = a$, sensing a horizontal edge above x because c exceeds both a and b. Likewise, the predictor chooses $\hat{x} = b$ when it senses a vertical edge to the left of x by the occurrence of both a and b exceeding c.

12.3.2 The context

The context involves all four pixels in the causal template. Three gradients are calculated as follows:

$$g_1 = d - b$$
$$g_2 = b - c$$
$$g_3 = c - a. \tag{12.2}$$

One of the basic tenets of context definition is keep the number of realizations small. For images with gray levels of 0 to 255, the possible realizations of the triplet (g_1, g_2, g_3) are about 511^3. Empirical calculations of conditional probabilities of every one of the 256 possible values of x are not only impractical, but would yield poor estimates, due to insufficient data. This phenomenon is called *context dilution*. To avoid context dilution, each gradient is quantized to nine levels each, resulting in a reduction to $9^3 = 279$ context realizations.

The quantization rule is the same for each gradient. For images with maximum value of 255, the rule to map each g_i to its quantized value q_i for $i = 1, 2, 3$ is specified in Table 12.1.

Except for the zero bin, these quantization bins are intended to be approximately equal in probability for a Laplacian or exponentially decaying probability model. When q_1 is negative, a sign flag is set to -1 and (q_1, q_2, q_3) is mapped to $(-q_1, -q_2, -q_3)$. Therefore, using this sign flag, the number of realizations is reduced to 365. The standard stipulates that every triplet (q_1, q_2, q_3) corresponds uniquely to an integer q in the range 0 to 364.

Table 12.1 Quantizer bins and indices for context in JPEG-LS for 8-bit grayscale images.

Quantizer index q_i	Quantizer bin contents (Set of g_i's)
-4	$\{\leq -21\}$
-3	$\{-20, -19, \ldots, -7\}$
-2	$\{-6, -5, -4, -3\}$
-1	$\{-2, -1\}$
0	$\{0\}$
1	$\{1, 2\}$
2	$\{3, 4, 5, 6\}$
3	$\{7, 8, \ldots, 20\}$
4	$\{\geq 21\}$

12.3.3 Golomb–Rice coding

The prediction residuals are encoded with a sequential entropy coder that accumulates counts of occurrences in each context and issues codewords based on these counts. A Golomb–Rice entropy coder has been chosen for JPEG-LS, due to its low complexity. The Golomb codes, including the Golomb–Rice special case, have been described in detail in Chapter 4. Here, we present just a brief summary.

Golomb codes are optimal for exponentially decaying (geometric) probability distributions of non-negative integers. Given a positive integer parameter m, the Golomb code G_m encodes an integer n in two parts: a unary representation of the quotient $n_q = \lfloor n/m \rfloor$, and a binary representation of the remainder $n_r = n \pmod{m}$. When m is an integer power of two, say $m = 2^k$, the code consists of n_q 1 bits and a 0 bit,[2] followed by the k least significant bits in the binary representation of n. For example, suppose that $n = 19$ and $m = 2^3 = 8$. Then, in binary $n = 10011$ and $G_3 = 110|011$.

The Golomb code for an integer number n therefore depends only on the parameter m. For a geometric probability distribution $P(n) = (1 - \theta)\theta^n$, for $n \geq 0$, the optimal selection of m is given by

$$m = \lceil \log(1 + \theta)/\log(\theta^{-1}) \rceil. \tag{12.3}$$

In other words, θ^m is as close to 1/2 as possible without being strictly less than 1/2. The average or expected value of n, $\overline{n} = \theta(1 - \theta)^{-1}$, so in practice, θ is estimated through calculation of $\overline{n}$.

The prediction residuals can be negative or positive, so do not fit without change into the non-negative integer domain of the Golomb codes. Therefore, the negative values are interleaved between the positive ones and numbered according to their positions in the sequence. This results in the mapping of a residual e to a non-negative integer e' as follows:

$$e' = \begin{cases} 2e, & e \geq 0 \\ 2|e| - 1, & e < 0. \end{cases} \tag{12.4}$$

The determination of the best m or, equivalently k, depends on finding an estimate of θ in Equation (12.3). An estimate of θ is related to the average of residual (error) magnitudes. Thus, the creators of JPEG-LS develop the following simple, approximate relationship between the average error magnitudes, denoted $\overline{|e|}$ and k:

$$k \approx \lceil \log(\overline{|e|}) \rceil. \tag{12.5}$$

For every context q, the encoder records a running count of occurrences (N) and an accumulation of error magnitudes $|e|$ (A), which are used to estimate the average error magnitude $\overline{|e|} \approx A/N$ and hence, the nearly optimal k for each image sample.

12.3.4 Bias cancellation

In context-based coding, there may occur biases of the prediction error away from zero, which may accumulate as coding proceeds. Furthermore, the success of Golomb–Rice

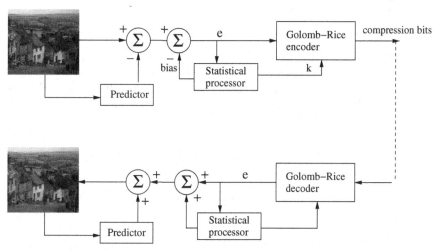

Figure 12.4 Coding and decoding system in JPEG-LS.

coding depends heavily on the assumption of an even-symmetric exponentially decaying probability model. The mapping of negative integers to positive ones can also contribute to these biases. A clever idea of cancelling biases during coding originated in the lossless image coder called CALIC [3, 4] and is utilized in JPEG-LS. For every context q, an additional variable (B) of the running sum of prediction errors e is recorded. When the estimated average error $\bar{e} \approx B/N$ exceeds 0.5, the current prediction error is decremented by 1. Similarly, when $\bar{e}$ is less than -0.5, the current prediction error is incremented by 1. The accumulated error B sums the errors after incrementing or decrementing. In this way, the average error is kept between -0.5 and +0.5. The decoder can reproduce these actions of prediction and bias cancellation, because they are based on values of past decoded samples. The full system for encoding and decoding is depicted in Figure 12.4.

12.3.5 Run mode

The above Golomb–Rice coding takes place only when one or more of the gradients are non-zero. When all the gradients are zero, $q_1 = q_2 = q_3 = 0$, and the encoder enters a run-length coding mode, where it expects a run of values equal to that of a. The encoder in this mode continues its scan to determine the length ℓ of the run, which can be 0 up to the end of the line. Interestingly, the run-lengths themselves are not encoded, but their ranks r_ℓ based on decreasing frequency of occurrence. In other words, the most frequent ranks 0, the next most frequent ranks 1, and so forth. The correspondence of rank r_ℓ to run-length ℓ can be reconstructed at the decoder. These ranks are Golomb–Rice encoded by looking up in a table of associations of k ($m = 2^k$) to ranks to form the codewords consisting of the unary code of the quotient of $\lfloor r_\ell/m \rfloor$ and the k least significant bits of r_ℓ. A run mode is terminated either when the end of a line is reached or when x does not match a. In the latter case, a special "end-of-run" state is entered and the non-zero $e = x - a$ is encoded.

12.3.6   Near-lossless mode

JPEG-LS can also operate in what is called a near-lossless mode. In this mode, the reconstruction of every pixel is guaranteed to have error no larger than an integer δ in magnitude. The principle behind this operation is to quantize every image pixel x with quantization error no larger than δ and then encode this quantized image losslessly. The following mapping $g(x)$ produces the quantizer levels for this quantization. Given integers x and δ,

$$g(x) = \begin{cases} \left\lfloor \frac{x+\delta}{2\delta+1} \right\rfloor & x > 0 \\ \left\lceil \frac{x-\delta}{2\delta+1} \right\rceil & x < 0. \end{cases} \tag{12.6}$$

The reconstruction $\hat{x}$ from its quantization level $g(x)$ is obtained by

$$\hat{x} = (2\delta + 1)g(x). \tag{12.7}$$

In order to avoid passing through the image twice, this quantization is enacted on the prediction error e of every pixel. All the calculations and encoding use the quantizer levels of the prediction errors. The decoder reconstructs the quantized errors and adds to the prediction to produce every $\hat{x}$ within δ units of its original value x.

12.3.7   Remarks

We described the JPEG-LS image compression method to illustrate the steps that a lossless coder must take to achieve state-of-the-art results. However, JPEG-LS, although an international standard, is not the most efficient method available. Probably the most efficient known method is context-based adaptive lossless image coding (CALIC) developed by Wu and Memon [3, 4]. Both JPEG-LS and CALIC were presented before the JPEG Working Group charged to create a standard. Presumably, JPEG-LS won because it was considered less complex and suffered only a small loss in efficiency compared to CALIC. JPEG-LS borrowed some novel ideas from CALIC, most notably the feedback of accumulated error to cancel bias. CALIC uses a larger neighborhood template and computes six gradients, but combines them into two sums, one for horizontal and one for vertical. The prediction depends on several cases of the difference between these gradient sums. In coding, CALIC uses 576 contexts versus 365 for JPEG-LS. Anyway, we refer the reader to the cited literature for the details of CALIC.

12.4   Multi-resolution methods

In many circumstances, it is desirable to transmit and receive an image with progressively finer quality and resolution. We have already elaborated on these aspects in Chapter 11 in connection with wavelet transform coding systems. In these systems, lossless reconstruction can be achieved with integer transforms, meaning transforms resulting from integer or fixed-point mathematical operations. We have described some of the integer transforms used in image transform coding in Chapter 7. The transforms are almost always actualized through successive stages of half-band filtering. The filters are generally not orthonormal and require $\sqrt{2}$ factors to make them orthonormal. The idea is to use an even number of stages along the paths of the filtering tree, so that the product of these factors results in an integer multiple of 2. Then the multiplication is simply the operation of bit shifting. Then any of the subband/wavelet coding methods described in Chapter 11 can be applied to the transform with integer coefficients. In a lossless application, there would be no quantization, explicit or implicit.[3] These methods are generally not quite as efficient in lossless compression as the context-based, adaptive predictive coding methods of JPEG-LS and CALIC. The penalty in lossless compression for using transform methods, such as SPIHT or JPEG2000, is usually less than 10% expansion in file size and sometimes non-existent. The benefit is the progressive decoding capability in both quality and resolution.

One can obtain progressive resolution in lossless coding without resorting to wavelet transforms. One such method is named hierarchical interpolation (HINT) [1]. The idea is that the pixels of a subsampled image at a low resolution are used to predict the pixels of the next higher resolution iteratively until full resolution is reached. Coding is performed on the prediction residuals in every stage. This method brings successive build up of resolution, but is really not competitive in efficiency with any of the methods previously mentioned. The reason is that the predictions often produce large residuals, because they are not formed from close neighbors that are highly correlated.

12.5 Concluding remarks

The objective of this chapter is to explain the principles of achieving efficient lossless image coding and to present examples of systems that operate in accordance with these principles. These principles and methods can also be applied to samples of data associated with a two-dimensional grid when their values correspond uniquely to integers not exceeding the limit of the computer. We repeat the pinciples for lossless predictive coding:

1. Form prediction $\hat{x}$ of the target sample x based on a defined *causal template*.
2. Define a *context* and determine in which realization of this context x occurs.
3. Code the prediction residual $e = x - \hat{x}$ using a probability model conditioned on the context of x.

The prediction serves to exploit the dependence between neighboring samples, so as to reduce substantially this dependence. The block or subband transform prior to coding serves this same purpose. For lossless coding, the transform must be perfectly reversible and the coefficients must be integers. Predictive coding that does not cross subbands can follow such a transform. However, predictive coding does not produce embedded code, so cannot support progressive lossy to lossless coding. When such capability is required, an embedded coder, such as SPIHT, SPECK, or JPEG2000, is a clear choice.

Problems

12.1 Consider once again the image block

$$B = \begin{bmatrix} 90 & 91 & 65 & 42 \\ 86 & 78 & 46 & 41 \\ 81 & 58 & 42 & 42 \\ 82 & 51 & 38 & 43 \end{bmatrix}.$$

This time we wish to encode it losslessly, using a method similar to that of the JPEG or JPEG-LS lossless standard.

First, we select Predictor 4 $(a + b - c)$ in the lossless JPEG standard. The pixels in the first row and the first column are not surrounded with the full causal template. Except for the top-left pixel, use only the preceding pixel in the first row and the preceding pixel in the first column as the prediction. The top-left pixel must be encoded as its full binary value with 8 bits.

(a) Generate the sequence of predictions and the sequence of prediction residuals.

(b) Show that the sequence of prediction residuals plus the top-left pixel value reconstruct the original block without error.

(c) Encode the sequence of prediction residuals with a fixed Golomb–Rice code using the k given by Equation (12.5), where the average magnitude of the prediction residuals is calculated from the sequence in the previous part. What is the code rate in bits per pixel?

(d) Suppose you reverse a bit in the code sequence to simulate a channel error. Show that the decoding error propagates through the sequence. Do not put the bit error in the top-left pixel.

12.2 Repeat part (c) of the previous problem using an adaptive Golomb–Rice code. Use k calculated from the past accumulation of residual magnitudes in encoding the current residual. Initially use the fixed value of k found in the previous problem. What is the code rate in bits per pixel?

Decode the codestream to show that the original block is reconstructed without error.

12.3 We want to perform near-lossless coding in the framework of Problem 12.1.

(a) Generate the sequence of predictions and the sequence of prediction residuals.

(b) Quantize the prediction residuals with the step $\delta = 2$ and encode the bin indices with the same fixed Golomb–Rice code. What is the code rate in bits per pixel?

(c) Calculate the MSE in the reconstructed block.

Notes

1. More often than not, the data are written with a number of decimal places exceeding the precision of the measurement.
2. The unary code for n_q could also be n_q 0 bits followed by a 1 bit.
3. By *implicit quantization* is meant that the embedded bitstream is truncated before it reaches its full size for lossless recovery.

References

1. M. Rabbani and P. W. Jones, *Digital Image Compression Techniques*, ser. Tutorial Texts. Bellingham, WA: SPIE Optical Engineering Press, 1991, vol. TT7.
2. M. J. Weinberger, G. Seroussi, and G. Sapiro, "The LOCO-I lossless image compression algorithm: principles and standardization into JPEG-LS," *IEEE Trans. Image Processing*, vol. 9, no. 8, pp. 1309–1324, Aug. 2000.
3. X. Wu and N. Memon, "Context-based, adaptive, lossless image coding," *IEEE Trans. Commun*, vol. 45, no. 4, pp. 437–444, Apr. 1997.
4. X. Wu, N. Memon, and K. Sayood, "A context-based, adaptive, lossless/nearly-lossless coding scheme for continuous tone images," ISO/JPEG Working Group, Tech. Rep. ISO/IEC SC29/WG1N256, 1995.

13 Color and multi-component image and video coding

13.1 Introduction

Due to advances in the technology of metrology and storage density, acquisition of images is now increasingly being practiced in more than two dimensions and in ever finer resolutions and in numerous scientific fields. Scientific data are being acquired, stored, and analyzed as images. Volume or three-dimensional images are generated in clinical medicine by CT or MRI scans of regions of the human body. Such an image can be regarded as a sequence of two-dimensional slice images of a bodily section. Tomographic methods in electron microscopy, for example, produce images in slices through the material being surveyed. The material can be a biological specimen or a materials microstructure. In remote sensing, a surface is illuminated with broad-spectrum radiation and the reflectance spectrum of each point on the surface is measured and recorded. In this way, a reflectance "image" of the surface is generated for a number of spectral bands. One uses the term multi-spectral imaging when the number of bands is relatively small, say less than 20, and the term hyper-spectral imaging when the number of bands is larger, usually hundreds. In either case, one can view the data either as a sequence of images or a single image of vector pixels. One particular kind of multi-spectral image is a color image, where the spectrum is in the range of visible wavelengths. Because of the properties of the human visual system, only three particular color images are needed to generate a visual response in the human viewer. Therefore, color images for viewing require special treatment, despite being a subset of the multi-spectral image class. The default assumption for color images is that they are generated for human viewing. In this respect, the psychovisual properties of the human visual system in connection with color perception should be important considerations when processing or coding images. We make no attempt here to review the vast body of literature on the psychophysics of color vision in this regard and present only some conclusions that are germane to coding. Interested readers may consult references [1–4].

Video consists of a sequence of monochrome or color image frames in which illumination changes and objects move between successive frames. The unique aspect of video is motion, so that adds another dimension to the coding procedure. Because video has a substantial impact on everybody on a daily basis, coding of video remains the focus of assiduous efforts toward improvement and international standardization. It is not our intent here to present the various standards that have evolved to our present time. Our intent in this book is to present the principles of efficient coding and illustrate their application in certain systems in order to teach how to use coding effectively. There really are no new coding principles to be described in this chapter, but there will be described some non-obvious and clever ways to apply the learned coding principles to multi-component images.

13.2   Color image representation

A color image for human viewing is a two-dimensional array of vectors, each having three intensity components. The three components are the so-called tri-stimulus values (primaries) of red (R), green (G), and blue (B), or cyan (C), yellow (Y), and magenta (M). An additive mixture (linear combination) of either R, G, and B or C, Y, and M matches any visible color. They are said to be components in the RGB or CYM color space. It is customary to refer to the tristimulus system as RGB, although CYM could be used equally well. The color image pixel at coordinates (m, n) is denoted therefore as $i(m, n) = (R(m, n), G(m, n), B(m, n))$, where $R(m, n)$, $G(m, n)$, and $B(m, n)$ are the red, green, and blue intensities, respectively. These intensities do not evoke equivalent visual response for the same values and are highly correlated, because they are weighted averages of spectral responses in overlapping wavelength ranges. The visual system is more sensitive to green than the other two and least sensitive to blue. It is also convenient to represent color by a set of three values that are compatible with a monochrome display device. It transpires that such a representation saves transmission bandwidth and digital storage space, because it almost decorrelates the three components, thereby enabling a more compact encoding.

We shall assume that the color components $(R(m, n), G(m, n), B(m, n))$ are the gamma-corrected outputs of the camera at the coordinates (m, n).[1] We shall now explain some color transformations that are relevant for color image and video coding. The tranforms are the same for each coordinate, so we shall drop the explicit coordinate notation. The monochrome-compatible color transformation in the PAL television system (used in Europe, Australia, and in some Asian countries) is

$$Y = 0.299R + 0.587G + 0.114B$$
$$U = 0.492(B - Y)$$
$$V = 0.877(R - Y) \tag{13.1}$$

where Y is called the *luminance* component, and U and V the *chrominance* components. In matrix-vector form, the transformation is

$$\begin{bmatrix} Y \\ U \\ V \end{bmatrix} = \begin{bmatrix} 0.299 & 0.587 & 0.114 \\ -0.147 & -0.289 & 0.436 \\ 0.615 & -0.515 & -0.100 \end{bmatrix} \tag{13.2}$$

The form of the transformation in Equation (13.1) is the more easily computable form and reveals the inverse rather directly. Both U and V can be negative, so a constant level must be added to each for display purposes. However, the values in the above equations are those that are coded.

For NTSC television (standard in the United States and Japan), the chrominance (U, V) vector undergoes a $33°$ rotation to (I, Q) to reduce the bandwidth of Q. Thus,

$$I = -U \sin 33° + V \cos 33°$$
$$Q = U \cos 33° + V \sin 33° \qquad (13.3)$$

The components I and Q are so named, because they comprise the in-phase and quadrature components modulating the carrier of the composite TV signal.

In matrix-vector form, the transformation is

$$\begin{bmatrix} Y \\ I \\ Q \end{bmatrix} = \begin{bmatrix} 0.299 & 0.587 & 0.114 \\ 0.596 & -0.275 & -0.321 \\ 0.212 & -0.523 & 0.311 \end{bmatrix} \qquad (13.4)$$

For the JPEG and JPEG2000 image compression standards and the MPEGx video standards, the color space is called YC_bC_r or YCC and is a slight modification of YUV in (13.1), where Y is identical, and the chrominance components, C_b and C_r, take different scale factors to result in the following transformation (ITU 601 Recommendation):

$$Y = 0.299R + 0.587G + 0.114B$$
$$C_b = 0.564(B - Y)$$
$$C_r = 0.713((R - Y) \qquad (13.5)$$

The matrix-vector form of the YC_bC_r transformation becomes

$$\begin{bmatrix} Y \\ C_b \\ C_r \end{bmatrix} = \begin{bmatrix} 0.299 & 0.587 & 0.114 \\ -0.169 & -0.331 & 0.500 \\ 0.500 & -0.418 & -0.081 \end{bmatrix} \qquad (13.6)$$

Once again, this is the form used for encoding. The minimum and maximum of the two chrominance components are -0.5×255 and 0.5×255, respectively, for 8-bit input values. Therefore, the level of 128 will be added to C_b and to C_r when one wishes to display them.

The floating point values of the elements in these matrices and the necessity to round the color components to integers make the color transforms only approximately reversible and therefore suffer a small loss of accuracy. Normally, this small loss is imperceptible to viewing, but in some processing applications, it is necessary to reproduce the color-coded components perfectly without loss. For these purposes, JPEG2000 has adopted the optional use of a lossless or reversible color transform (RCT), that can be realized with fixed-point arithmetic. This transform is similar to the irreversible YC_bC_r tansform and takes the following form:[2]

$$Y' = \left\lfloor \frac{R + 2G + B}{4} \right\rfloor$$
$$D_b = B - G$$
$$D_r = R - G \tag{13.7}$$

It may not be so obvious that this transform is invertible with perfect precision unless one notices that

$$\left\lfloor \frac{D_b + D_r}{4} \right\rfloor = \left\lfloor \frac{R + B + 2G}{4} - G \right\rfloor$$
$$= Y' - G.$$

Because G is an integer, any fractional part truncated by the floor function exists only in the Y' term. Therefore, from the expression directly above and D_b and D_r in (13.7), the inverse solution of R, G, and B follows as shown below.

$$G = Y' - \left\lfloor \frac{D_b + D_r}{4} \right\rfloor$$
$$B = D_b - G$$
$$R = D_r - G \tag{13.8}$$

13.2.1   Chrominance subsampling

The luminance and chrominance components have different visual properties, in which case it is useful to regard a color image as consisting of three planes, one luminance and two chrominance planes. Vision is known to be less sensitive to spatial and intensity changes in chrominance components of a color transform than the same changes in the luminance component. In order to take advantage of this relative insensitivity to spatial changes, the chrominance planes are downsampled either by 2:1 horizontally (4:2:2 format) or by 2:1 both horizontally and vertically (4:2:0 format) and filtered. In the 4:2:2 format, the reduction in the number of data samples is a 3:2 ratio, while in the 4:2:0 format, the reduction is 2:1. The later standard definition video formats are YC_bC_r or YUV 4:2:0, depending on the country. Downsampling of chrominance planes does not seem to harm the perception of color to any noticeable degree for standard definition formats. For high-definition television and digital cinema, chrominance planes are not subsampled (4:4:4 format). In digital cinema, the pixel resolution in all three planes is 10 or 12 bits as opposed to 8 bits for the other video and image formats, because small brightness and color changes can be discerned in a large-screen display. The discussions here will be limited to the usual 8-bit resolution in each of the three color planes. No loss of generality is implied by this limitation, since extensions to higher pixel resolutions will be obvious.

13.2.2 Principal component space

These luminance-chrominance color spaces, although motivated by compatibility with an achromatic (grayscale) device, possess suitable charcteristics for reasonably efficient coding. For normal, natural color images, the luminance (Y) component contains more than 90% of the total energy and is nearly uncorrelated with its chrominance components. Theoretically, the best space for color coding would be the Karhunen-Loeve (KL) or principal component space, which gives uncorrelated components with maximum energy compaction. However, experimental studies have concluded that computed K-L spaces are only slightly more efficient for coding natural color images than these YUV, YIQ, or YCC spaces [5]. Their disadvantages are that the transform is image dependent and requires extensive calculation. A transform matrix must be written to the bitstream as overhead for every image. For non-visual images of data having a greater number of components, such as multi-spectral or hyper-spectral images, the K-L space seems to provide even greater efficiencies than these luminance-chrominance color spaces, but the amount of computation of the transform might be a formidable burden.

The KLT is explained in Chapter 7. Briefly, the KLT of RGB space involves calculating a 3×3 covariance matrix, finding the eigenvectors and forming a matrix consisting of these eigenvectors as columns, and calculating the inverse of this matrix. The covariances are averages of pairs of pixel values of the same axial separation over the common spatial extent of the image plane. Mathematically stated, let the "color" image with three image components be expressed by $\mathbf{X} = (X_1(m, n), X_2(m, n), X_3(m, n))$, $m = 1, 2, \ldots, M$, and $n = 1, 2, \ldots, N$. Then an estimate of the correlation $\hat{R}_{ij}, i, j = 0, 1, 2$ is calculated as

$$\hat{R}_{ij} = \frac{1}{MN} \sum_{m=1}^{M} \sum_{n=1}^{N} X_i(m, n) X_j(m, n)$$
$$i, j = 0, 1, 2 \tag{13.9}$$

Then the estimate of the covariance matrix $\hat{\Phi}_X$ is formed with elements

$$\phi_{ij} = \hat{R}_{ij} - <X_i><X_j> \quad m = 1, 2, \ldots, M, \tag{13.10}$$

where $<X_i> = (\sum_{m,n} X_i(m, n))/MN$ is the mean in the X_i plane and similarly for X_j. The estimated 3×3 covariance matrix of $\mathbf{X}$ is thus

$$\hat{\Phi}_X = \begin{bmatrix} \phi_{00} & \phi_{01} & \phi_{02} \\ \phi_{10} & \phi_{11} & \phi_{12} \\ \phi_{20} & \phi_{21} & \phi_{22} \end{bmatrix} \tag{13.11}$$

The matrix of normalized eigenvectors P is now calculated and its inverse P^{-1} applied to the input vector $\mathbf{X}$ gives the KL transformed three-component vector. (See Section 7.2.)

Although the KLT can be defined for a non-stationary source, its estimation here depends on the source being at least wide-sense stationary (stationary for first- and second-order statistical averages). Clearly, the statistics vary over the extent of a normal image, so that the KLT implemented as indicated here, will not produce uncorrelated transform planes. But, as mentioned before, this method usually produces a better space for compression than the fixed color transforms. One expects that statistics would vary less over smaller regions of an image, so that dividing it into rectangular blocks and KL transforming each block separately would yield better decorrelation and energy compaction overall. One must take care that the blocks not be too small in size and too large in number, in order to get reliable estimates and not increase substantially the overhead of transmission of the block transform matrices.

13.3   Color image coding

13.3.1   Transform coding and JPEG

In transform coding of color images, the two-dimensional DCT operates blockwise on each of the three color transform planes separately and encodes them separately. The standard color space is YC_bC_r, but the coding procedure applies to any of the color spaces described. Often, the chrominance components, C_b and C_r, are downsampled by 2:1 (4:2:2 format) prior to transformation and coding. In any event, each plane is treated as if it were a monochromatic image. The only difference is that the chrominance planes are quantized more coarsely and will consume less bit rate than the luminance plane, since they contain a small percentage of the image energy.

13.3.1.1   Generic color transform coding

Suppose the objective is to minimize the MSE distortion for a given bit rate in coding the block DCTs of the color transform images Y, C_b, and C_r. One method is to determine the optimal distribution of the target rate R_T among all the blocks of the three component DCTs. In other words, we wish to determine block rates $\{R_y(i)\}$, $\{R_{cb}(j)\}$, and $\{R_{cr}(k)\}$, such that

$$\sum_i R_y(i) + \sum_j R_{cb}(j) + \sum R_{cr}(k) \le R_T,$$

and the distortion D is a minimum. The solution to this problem has already been derived in Chapter 8 for the distortion versus rate model of an optimum quantizer. This solution appears in Equations (8.15) and (8.16). Following this solution and denoting the variances of the blocks by $\{\sigma_y^2(i)\}$, $\{\sigma_{cb}^2(j)\}$, and $\{\sigma_{cr}^2(k)\}$, the rate solution is expressed in terms of a parameter θ as

$$R_y(i) = \min\{0, \frac{1}{2} \log \frac{\sigma_y^2(i)}{\theta}\},$$

$$R_{cb}(j) = \min\{0, \frac{1}{2} \log \frac{\sigma_{cb}^2(j)}{\theta}\},$$

$$R_{cr}(k) = \min\{0, \frac{1}{2} \log \frac{\sigma_{cr}^2(k)}{\theta}\}, \qquad (13.12)$$

where the block indices i, j, and k range from 1 to N_y, N_{cb}, and N_{cr}, the number of blocks in Y, C_b, and C_r, respectively. The total rate per sample R_T is the numerical average of the above rates, i.e.,

$$R_T = \frac{1}{N_B}\left[\sum_i R_y(i) + \sum_j R_{cb}(j) + \sum_k R_{cr}(k)\right], \qquad (13.13)$$

where $N_B = N_y + N_{cb} + N_{cr}$, the total number of blocks.

The accompanying minimum distortion D is an average of the block distortions given by

$$N_B D = \sum_{i:\sigma_y^2(i)>\theta} \theta + \sum_{i:\sigma_y^2(i)\le\theta} \sigma_y^2(i) + \qquad (13.14)$$

$$\sum_{j:\sigma_{cb}^2(j)>\theta} \theta + \sum_{j:\sigma_{cb}^2(j)\le\theta} \sigma_{cb}^2(j) + \qquad (13.15)$$

$$\sum_{k:\sigma_{cr}^2(k)>\theta} \theta + \sum_{k:\sigma_{cr}^2(k)\le\theta} \sigma_{cr}^2(k), \qquad (13.16)$$

In words, wherever the block variance exceeds θ, the rate is 1/2 the base-2 logarithm of the ratio of the variance to θ and zero otherwise. The distortion contributed by a block whose variance exceeds θ is θ and otherwise it is equal to the variance, since it receives zero rate.

A block variance is calculated as the sum of the squares of the AC coefficients divided by the number of coefficients in the block. The DC coefficient is proportional or equal to the mean, so it is excluded in the sum of squares and is encoded separately, usually with its natural 8 bits, so as not to create artificial discontinuities between blocks.

If one wishes to assign rate to account for the lower visual sensitivity of chrominance relative to luminance, one weights the chrominance variances in the formulas above by fractional values (less than one). The effect is to increase the bit rate sent to the luminance plane.

Once the rate is assigned to every block, the coefficients are again encoded following the optimal solution in (8.15) and (8.16). The DC coefficients are omitted, since they are coded separately. Among quantizers, the uniform quantizer is nearly optimum, so one uses the step sizes that produce the designated block rates. Minimal distortion in a block is obtained when the distortions of every coefficient that receives positive rate are equal. Therefore, the same step size is applied to every such coefficient. The step size will vary among the blocks, because they have different variances.

13.3.1.2 Baseline JPEG

As one example of a non-generic method, consider the JPEG standard. The coding of the luminance plane (image) is exactly that of a monochromatic image, as described in Chapter 9. Both chrominance images are DCT transformed in 8×8 blocks. Within each plane, every block is quantized using the same quantization table and entropy coded by

Table 13.1 JPEG quantization table for chrominance images

17	18	24	47	99	99	99	99
18	21	26	66	99	99	99	99
24	26	56	99	99	99	99	99
47	66	99	99	99	99	99	99
99	99	99	99	99	99	99	99
99	99	99	99	99	99	99	99
99	99	99	99	99	99	99	99
99	99	99	99	99	99	99	99

the same Huffman code. The suggested quantization table for the chrominance images is shown in Table 13.1.

Scaling of the entries in this table by the same factor gives the step sizes for uniform quantization of the DCT coefficients. Different scale factors may be chosen for luminance and chrominance to further exploit the relative perceptual importance of luminance. The quantizer bin indices are coded in the same manner, as the luminance quantizer indices, except with different Huffman codes. The reason is that the chrominance indices have a different probability distribution than the luminance indices. Recall that differences of DC indices from block to block are coded separately from the AC coefficients in any given block. The coding procedure is explained in detail in Section 9.3.

13.3.2   Wavelet transform systems

Henceforth, we treat only luminance-compatible color transforms, where it does not matter what are the exact chrominance components. We shall designate the color components as Y, U, and V, keeping in mind that UV stands also for IQ and C_bC_r. For wavelet transform coding of color images, a two-dimensional wavelet transform acts upon the Y, U, and V planes in turn. In most systems, including JPEG2000, these planes are coded as monochrome images. Rate is distributed optimally among the three planes, so as to minimize a distortion criterion, usually plain or perceptually weighted squared error.

The manner of the distribution of rate optimally among the three planes distinguishes the different coding methods. JPEG2000, as described in Chapter 11, uses post-compression truncation of the code-block bitstreams to reach the same distortion-rate slope, which is the solution for an arbitrary set of quantizers given in Equation (8.23). The algorithm truncates the high-rate bitstream of every code-block among the three wavelet transformed color transform planes to the same distortion-rate slope. The reason that simple truncation can be employed in this manner is that the bitstream of any code-block is completely embedded and rate-scalable, i.e., the bits are ordered from highest to lowest significance.

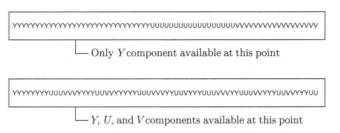

Figure 13.1 Compressed color bitstreams: conventional (top) and embedded (bottom).

For any compression algorithm that is embedded and rate-scalable, one can use post-compression truncation of the bitstreams of the coding units at their equal slope points to achieve optimal rate distribution. The coding units are the typical 32×32 or 64×64 subband subblocks, such as in JPEG2000 and SBHP, tree-blocks that are coded with SPIHT, or full subbands, such as in SPECK, EZBC, or other algorithms with these properties that code whole subbands independently. See Chapters 10 and 11 for descriptions of these coding algorithms.

13.3.2.1 Progressive bitplane coders

Algorithms such as SPIHT and SPECK that use progressive bitplane coding offer the advantage of creating a color-embedded bitstream directly without the necessity of a post-compression optimized truncation. The previously considered algorithms produced the Y, U, and V compressed bitstreams (codestreams) sequentially, so must be interleaved to obtain lower rate full color images embedded in any higher rate codestream. Figure 13.1 depicts both non-embedded and embedded color codestreams. In both SPIHT and SPECK, a search ensues for sets of coefficients that pass or fail a series of threshold tests. The sets in SPIHT are tree-structured, whereas those in SPECK are block structured. The thresholds decrease at every new stage, beginning with the threshold corresponding to the most significant (highest) bitplane of the highest magnitude coefficient ($2^{n_{max}}$) and proceeding to the next threshold of the next lower bitplane ($2^{n_{max}-1}$) and so forth.

Instead of coding from highest to lowest bitplane in one image plane at a time, the search at a particular threshold can cross through all three planes. We maintain the control lists LIP and LIS for each plane separately, but maintain only a single LSP list. When a set tests as significant, it is partitioned only within the plane to which it belongs. When a significant pixel is found, its coordinate is placed on the common LSP. In this way, the significant points of the three planes are intermingled in the common LSP. The LSP entries are eventually encoded into the bitstream in the order in which they are put onto the list. Therefore, an embedded color codestream results and coding can stop when the desired rate is reached (bit budget is exhausted).

Let us see how this works for color SPECK. Consider the color space YUV (4:2:0 format) where the chrominance U and V planes are one-quarter the size of the luminance Y-plane. Each plane is separately wavelet transformed and partitioned into sets $\mathcal{S}$ and $\mathcal{I}$ as shown in Figure 13.2. In this figure we show just a few levels of wavelet

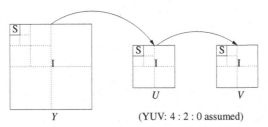

(YUV: 4 : 2 : 0 assumed)

Figure 13.2   Set partitioning and color plane traversal for CSPECK. Coding proceeds in order of Y, U, and V at the same significance threshold.

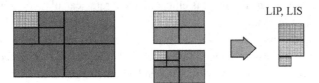

LIP, LIS

Figure 13.3   Initial control lists of color SPIHT for the Y, U, and V transform planes in 4:2:0 subsampling format.

decomposition for ease of illustration. An LIS is maintained for each of the three transform planes,[3] each one initialized with the corner coordinates of its top level $\mathcal{S}$. There is just one LSP list. The coding proceeds as in the normal SPECK algorithm starting with Y, but then crosses to U and then V at the same significance threshold, as depicted in Figure 13.2. Starting with the maximal significance level n among the color planes (almost always the maximum in Y), SPECK's first sorting pass proceeds for the whole Y plane. Then at the same n, this sorting pass is enacted in turn on the full U and V planes. Significant points among the three planes are mixed in the single LSP. Then, the significance level is lowered to $n - 1$, and the LIS sets of Y, U, and V are visited in turn on the three lists for the SPECK sorting passes. After completion of these passes at level $n - 1$, the refinement pass takes place on the single LSP by sending the n-level bits of the binary expansion of the magnitude of points found significant in previous passes. As before, the procedure repeats at lower significance levels until the bit budget is exhausted or all bits have been sent in the lowest bitplane.

SPIHT follows a similar procedure. The LIP and LIS lists of each plane are accessed separately in turn, but they are usually stored contiguously. Figure 13.3 illustrates the initial coordinates of the LIP and LIS lists of the three color planes joined together (but not interleaved) in storage. When a (tree-structured) set is found to be significant, it is partitioned within the color plane to which it belongs. When a pixel is found to be significant, its coordinates are moved to the common LSP list. As before, the significant coordinates that are moved to the LSP may belong to different planes. The associated significance bits belong to different color planes and are therefore intermingled when deposited in the codestream. They are ordered according to decreasing levels of significance, so that the code is embedded in rate and color simultaneously. Notice that the wavelet decomposition level may vary among the planes, but it is better to decompose the two chrominance planes identically.

13.4 Multi-component image coding

Other images with multiple components, besides three-component color images, are often acquired through CT or electron microscopy. Volume medical images, which contain a sequence of cross-sectional slices through a part of the human or animal body, and materials microstructures, which are comprised of high-resolution measurements of slices through a small specimen of material, are two examples. Another category of multi-component imagery is hyperspectral imagery, acquired by remote sensing. A hyperspectral image consists of intensity measurements of reflected or radiant transmission from a planar surface at a series of closely spaced wavelengths. Every sample in the measured surface can be viewed as a vector pixel whose components are intensity values at the different wavelengths. Equivalently, the hyperspectral image can be viewed as a sequence of monochrome images, each image corresponding to a different wavelength. All these multi-component images are still images, so we do not consider video or moving images in this context. We shall explore some of the unique aspects of coding video in a later section of this chapter.

An image with multiple components will be treated as a sequence of monochrome images. Adjacent images are correlated, so coding efficiencies can be gained by applying a mathematical transform across the components of the sequence. The optimal transform is the KL or principal components transform, which we visited for color images. The KLT was not too impractical for color images, having only three components, but would be for images having a large number of components. Usually, volume medical and hyperspectral images have more than a hundred components. A multispectral image is similar to a hyperspectral image, except that it contains typically fewer than ten or twenty components. So the KLT would be generally unfeasible for multi-component images, except perhaps for some multi-spectral images. Instead, a fixed transform, either wavelet or discrete cosine transform, is applied across the components to reduce the intercomponent correlation. Often, the components are transformed and coded in fixed-length subgroups, so as to reduce the memory requirement and prevent the loss of the entire image or large portions thereof through error propagation.

13.4.1 JPEG2000

The JPEG2000 standard for multi-component images allows a choice of the KL or wavelet transform across the components. Wavelet transformation of the two-dimensional $(x-y)$ slices precedes the transformation across the slices (z-direction). No matter which z-direction transform is chosen, the encoder operates on the two-dimensional spatial $(x-y)$ slices that have been transformed in the three directions. The encoding method for these slices is the baseline JPEG2000, as described in detail in Chapter 11. Post-compression optimized truncation of the code-blocks among all the slices, as described for color image coding, may be chosen for rate control, but might be prohibitively burdensome in computation and time delay, especially for real-time applications. Often, this optimized truncation is done only within one slice or a small group of slices given the same rate target for each slice or group of slices. An exception is the coding system for distribution of digital cinema, which applies no transform across the frames. For this application, in which the amount and time of computation are virtually unlimited, optimized truncation is performed over all the code-blocks in the entire sequence. A two-hour movie contains approximately 43 000 frames of 8×10^6 three-color pixels per frame and therefore approximately 2.5×10^9 64×64 code-blocks. Despite the time to perform the encoding and optimized truncation over this enormous number of blocks, the process is done off-line without time constraints. No such operations take place in the decoder, so that decoding and display are virtually instantaneous in the dedicated hardware of the cinema projector.

JPEG2000 Part 10, called JP3D, has issued a draft recommendation for volume image coding. But prior to this issuance, several researchers had already published articles that used the basic method [6, 7]. This three-dimensional extension of the Part 1 standard, codified in JPEG2000, Part 10, prescribes coding of three-dimensional code-blocks instead of square ones. Otherwise, everything else is basically the same in accommodating to this modification. The scanning of the code-blocks becomes three-dimensional, as do the contexts for arithmetic coding. The details will not be exposed here. Interested readers may consult the JP3D document [8]. A general overview of the features of JP3D may be found in Schelkens *et al* [9].

13.4.2   Three-dimensional wavelet transform coding

13.4.2.1   Three-dimensional wavelet transforms

In the methods that use three-dimensional wavelet transform, the order in which the directional axes are decomposed affects the overall compression performance. We have so far described full spatial $(x–y)$ decomposition followed by axial (z) decomposition using separable wavelets. Such a transform is said to be *asymmetric, anisotropic* or *non-dyadic* and is called a *wavelet packet transform*. The subbands of of a wavelet packet transform are shown in Figure 13.4. The figure depicts three levels of decomposition spatially and two axially. The number of levels in the axial direction are not tied to that in the spatial direction. Within the spatial planes, decomposing (filtering and downsampling) in x and y dimensions must alternate, so x and y must reach the same number of levels. The solid and dashed lines mark the boundaries of the subbands. Notice that the boundaries in the axial direction extend across the full extent of the spatial slices. That is not so with the dyadic or symmetric wavelet transform in all three dimensions. For the dyadic transform, all three directions alternate in their decompositions to produce the kind of subband structure shown in Figure 13.5. This figure shows two levels of alternating dyadic decomposition that produce 15 subbands. The packet transform having three levels spatially and two axially has 40 subbands, because each of the 10 spatial subbands is decomposed into four axial subbands. When the packet transform has two levels both spatially and axially, it comprises 28 subbands. For the same number of decomposition levels spatially and axially, the packet transform contains more subbands than the dyadic one, which may account for its superior performance when used with most coding algorithms.

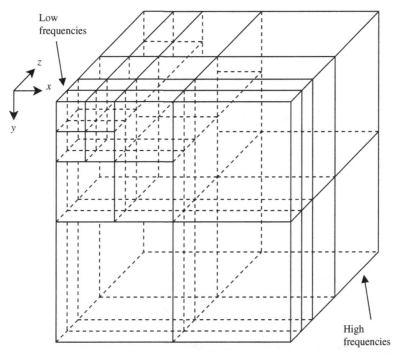

Figure 13.4   Subbands of a wavelet packet transform with three spatial levels and two axial levels.

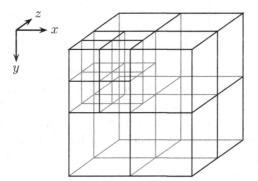

Figure 13.5   Subbands of a dyadic wavelet transform with alternating decomposition to two levels in the three directions.

13.4.2.2   Three-dimensional SPIHT

The SPIHT algorithm is insensitive to the dimensionality of the wavelet transform that it encodes. In fact, in its description in Algorithm 10.8, all one has to do is change the coordinate entries (i, j) to (i, j, k) for three dimensions. SPIHT partitions tree-structured sets in its search for significance, so we must now specify the structure of these sets in three dimensions. The obvious generalization of the tree structure for the dyadic transform is for each intermediate node to branch to a $2 \times 2 \times 2$ block of coordinates in the same relative position in next higher frequency subband. By this rule, except

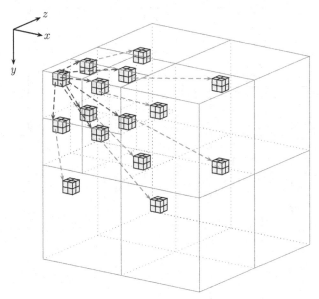

Figure 13.6   Symmetric three-dimensional tree structure for SPIHT coding.

for the lowest and highest frequency subbands, the offspring of a coordinate (i, j, k) are located on the vertices of a cube with upper-left corner point at $(2i, 2j, 2k)$. So the eight offspring coordinates are

$$(2i, 2j, 2k), (2i + 1, 2j, 2k), (2i, 2j + 1, 2k), (2i, 2j, 2k + 1), (2i + 1, 2j + 1, 2k),$$
$$(2i + 1, 2j, 2k + 1), (2i, 2j + 1, 2k + 1), (2i + 1, 2j + 1, 2k + 1). \quad (13.17)$$

Following the analogy to two dimensions, the upper-left corner point of a $2 \times 2 \times 2$ cube in the lowest frequency subband has no offspring, but the other vertices of the cube have eight offspring at the same orientation in one of the seven subbands of the same size in the same resolution. Their coordinates follow the same rule as in (13.17) above: all combinations of double the coordinate index and its unit increment in the three dimensions. Figure 13.6 illustrates the resulting symmetric tree in a two-level, three-dimensional wavelet transform. The coordinates in the highest frequency subbands terminate the tree, so they have no offspring. The descendant sets $\mathcal{D}$, the offspring sest $\mathcal{O}$, and the grand-descendant sets $\mathcal{L}$ are defined as before. $\mathcal{D}(i, j, k)$ is the set of all descendants of (i, j, k); $\mathcal{O}(i, j, k)$ is the set of all offspring of (i, j, k); and $\mathcal{L}(i, j, k) = \mathcal{D}(i, j, k) - \mathcal{O}(i, j, k)$. Once a $\mathcal{D}$ (type A) set in the LIS tests significant, it is divided into its $\mathcal{O}$ and $\mathcal{L}$ sets. Once an $\mathcal{L}$ (type B) set in the LIS tests significant, it is divided into its eight descendant sets from the offspring in its associated $\mathcal{O}$ set.

   In the wavelet packet transform, the subbands are no longer the same shape in different scales, so the symmetric tree structure is no longer applicable. There are various ways to construct an asymmetic tree, but the one that has proved to be the best has the structure shown in Figure 13.7. The packet transform in this figure consists of two levels of spatial decomposition followed by two levels of axial decomposition for ease of

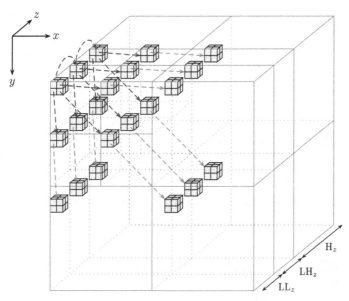

Figure 13.7 Asymmetric three-dimensional tree structure for SPIHT coding.

illustration. Nonetheless, the figure illustrates the general rules for constructing the tree as follows.

1. For each $2 \times 2 \times 2$ group of pixels in the lowest spatial frequency (top-left corner) subband within an axial subband, the top-left corner pixel branches to the $2 \times 2 \times 2$ group in the corresponding location in the next high frequency subband in the axial (z) direction.
2. Each of the other seven pixels in this group in the lowest frequency subband branches to a $2 \times 2 \times 2$ group in the next spatial subband within the same axial subband.
3. In other spatial subbands (except the highest frequency ones) contained within the same axial subband, all pixels in a $2 \times 2 \times 2$ group branch to its corresponding $2 \times 2 \times 2$ group in the next high spatial frequency subband at the same spatial orientation within the same axial subband.

These branching rules hold true for any packet transform, regardless of the number of spatial and axial decompositions. This asymmetric tree structure exhibits longer paths to its terminal nodes than does the symmetric tree structure. That may account for its superior performance in SPIHT coding.

13.4.2.3 Block coding

Algorithms that code two-dimensional transform blocks are easily adaptable to three-dimensional blocks. For set-partitioning algorithms, one can generalize splitting in two dimensions to three dimensions. In SPECK and EZBC, which splits and codes whole subbands, one replaces splitting rectangles into four equal parts with splitting rectangular parallelopipeds[4] into eight equal parts. This process is depicted in

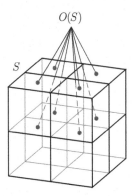

Figure 13.8   Splitting a significant block S into its eight offspring blocks $O(S)$.

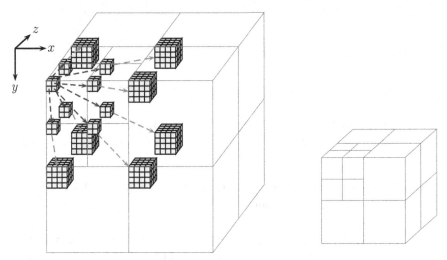

Figure 13.9   The block tree (right) of a $2 \times 2 \times 2$ root group in the lowest frequency subband (left, top-left corner).

Figure 13.8. The splitting is executed by dividing the range by two in each of the three dimensions when the *block* tests as significant. For SBHP, which splits subblocks of subbands, the same *cubic block* splitting occurs. When the three-dimensional transform is dyadic, octave band partitioning may be used to determine the order in which the subbands are visited for coding. Otherwise, the algorithm can visit subbands in order of increasing resolution from lowest to highest. The same generalization obtains for block-tree coding in the case of the dyadic transform. The blocks containing the nodes of a symmetric tree can be assembled into a three-dimensional tree-block, analogous to the tree-blocks in two dimensions. A portrayal of this assemblage appears in Figure 13.9 for a single $2 \times 2 \times 2$ root group in the lowest frequency subband. The structure and arrangement of subbands of the block tree are geometrically similar to the full wavelet transform. Contiguous $2 \times 2 \times 2$ groups in the lowest frequency subband can also serve as the root block of a block tree. Referring to the case of a single such

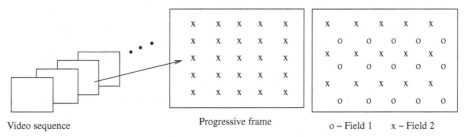

Video sequence   Progressive frame   o – Field 1   x – Field 2

Figure 13.10 A video sequence (left) and the progressive (middle) and interlaced (right) scan formats.

group, the tree-block consists of eight low-frequency subbands of size $2 \times 2 \times 2$ and seven high-frequency subbands of size $4 \times 4 \times 4$. The three-dimensional tree-block can now be coded with the SPECK or EZBC algorithm, using either octave band partitioning or increasing resolution to determine the order of coding the subbands. All the tree-blocks that comprise the transform can be encoded independently in this manner. Rate allocation can be accomplished either by optimized truncation, as described previously, or by coding through the end of the same bitplane in every tree-block.

13.4.3   Video coding

Sampling a moving scene at regular time intervals produces an image sequence called a *video sequence*, or simply *video*, when reproduced for television viewing. The sampling is usually done with a stationary camera that takes a series of snapshots with rectangularly shaped field of view. Nowadays, there seems to be activity toward sampling three-dimensional, time-varying scenes for digital television and digital cinema. Here we shall treat encoding of two-dimensional video, i.e., sequences of moving images for television viewing. The images in a video sequence are in color and in formats already described previously in the section on color images earlier in this chapter. Therefore, we shall narrow our considerations to monochrome video. The extension to color will be obvious from our treatment of color images. Figure 13.10 depicts a video sequence and the pixels in a typical frame. The pictured scanning or sampling pattern is called *progressive*. Progressive scan is the standard for digital television. Analog television uses an interlaced pattern synthesized from two fields interleaved with a half sample shift to form a frame, as shown also in Figure 13.10. The digital standards use the progressive mode, so only that mode will be treated here.

The dimensions of the frames and the temporal sampling rate depend on the particular standard and the country. Table 13.2 summarizes some of this information. Dimensions and sampling rate differences are not relevant to the operation of a coding method, so they remain arbitrary in our treatment.

The standard methods encode the video sequence in groups of frames, called GOFs (for groups of frames) or GOPs (for groups of pictures). The number of frames in a GOP is typically 16 and sometimes less. Encoding the sequence separately in GOPs is necessary for transmission over lossy channels in order to resynchronize the decoder when

Table 13.2 Digital video formats

Frame sizes (width x height)	Format name
1280×720, 1920×1080	720p HD, 1080p HD
720×480 (US), 720×576 (Europe)	CCIR 601
352×288	CIF (common intermediate format)
352×240	SIF (standard interchange format)
176×144	QCIF
Luminance-chrominance format	YUV (PAL), YCbCr (NTSC)
Sampling rate (frames per sec.)	30 (US), 25 (Europe)

bit errors or packet losses occur. The first frame in every GOP is encoded separately, that is, intraframe encoded, and is called an intra-coded frame or simply an *intra* frame. The remainder of the frames are encoded differentially with compensation for movement of objects between frames. These frames are called interframe coded frames or simply *inter* frames. The intra-coded frame serves as a stable, high quality reference for the interframe coding. We shall now explain the method used for interframe coding in a GOP.

13.4.3.1 Motion compensated predictive coding (MCPC)

The implied assumption in a video sequence is that objects in a scene move and illumination changes in the time between successive frames. Deviations from this assumption, such as object deformation and occlusion, complicate matters greatly, but do not change the underlying principles, so will not be treated here. The idea is to estimate to where a pixel in the current frame moves in the next frame and predict its intensity. Then the difference between the current and predicted intensities is quantized and coded. The standard methods for coding video use this form of differential PCM (DPCM) that incorporates compensation for motion between frames. A block diagram of such a system appears in Figure 13.11. This system is called a *hybrid* system, because it combines both predictive coding and motion estimation of interframe pixel displacements. The motion information is derived from the last source frame, since this information is sent to the decoder. The predictive coding is derived from the last reconstructed frame, because only that frame, not its source frame, is known at the decoder. In order to understand the workings of this system, its components need to be explained, starting with the motion estimation and compensation.

Motion compensation

Let a pixel value at location (x, y) in a frame at time t be denoted by $I_t(x, y)$. Suppose we can determine that, in the previous frame at time $t - 1$, this pixel was located at coordinates $(x - \delta_x, y - \delta_y)$. In other words, the motion or displacement vector of this pixel in one unit of time is (δ_x, δ_y). The objective is to encode the difference,

$$\Delta I_t(x, y) = I_t(x, y) - I_{t-1}(x - \delta_x, y - \delta_y), \tag{13.18}$$

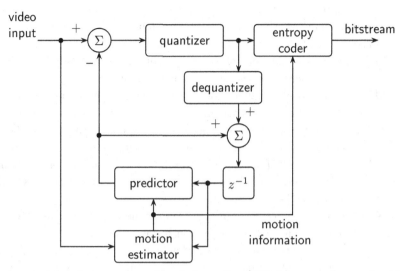

Figure 13.11 System for motion-compensated predictive coding (MCPC) between frames of a video sequence.

which for all (x, y) in the current frame is called the *displaced frame difference* (DFD) or *motion-compensated frame difference* (MCFD). If the determination of the motion vectors is reasonably accurate, the DFD values will be smaller than those of the straight or uncompensated frame difference and will require smaller coding rate. However, the motion vectors are unknown to the decoder, so have to be sent as side information. This side information has the potential to consume any rate savings from the motion compensation. So one has to be careful to send as few bits as possible to represent the motion vectors.

The usual strategy to reduce the motion vector overhead is to estimate a single motion vector for a block of pixels. In many of the standard video coders, the size of the block for motion estimation is 16×16. This block is called a *macroblock*, because the frames are coded typically by 8×8 blocks, called *code-blocks*. So a macroblock comprises four code-blocks that share a single motion vector in common. If it takes 16 bits to represent a motion vector (8 bits per component), the side information rate now is reduced to $16/256 = 1/16$ bits per pixel and even further when the motion vectors are entropy coded.

The estimation of motion vectors for blocks is called *block matching,* because a search ensues for the block in frame I_{t-1} that most closely matches a given block in the current frame I_t. The range of the search is limited to save computation and time. Typical search ranges are -7 to +7 grid points for δ_z and δ_y. The criterion for a match is usually the sum of absolute differences. The motion vector that results in the best matching block in frame I_{t-1} to the block B in frame I_t is therefore the solution

$$(\hat{\delta}_x, \hat{\delta}_y) = \arg \min_{(\delta_x, \delta_y)} \sum_{(x,y) \in B} |I_t(x, y) - I_{t-1}(x - \delta_x, y - \delta_y)|. \qquad (13.19)$$

The blocks can be overlapping or non-overlapping. The later video standards use overlapping blocks and get better coding results, despite the extra overhead. The search for the best match can be exhaustive or selective. There are methods for fast, selective search that do almost as well as exhaustive search.

Since pixels are located on a regular rectangular grid, the displacements searched in the above match equation are integer grid units. That limits the solution to integer pixel (or pel) resolution. However, the actual motion vector may actually fall between grid points, so it may be better to admit solutions of higher resolution, so-called fractional pixel resolution. As the standards evolved, 1/2, 1/4, and even 1/8 pel resolutions were adopted. Fractional resolution pixels are interpolated from surrounding grid point pixels. Bilinear interpolation is the most popular, although Lagrange interpolation has also been used. Clearly, fractional pixel resolution increases the computation quite substantially. In fact, the most computationally intensive part of the latest video coding standards is the block-matching algorithm.

Wavelet transforms are sometimes used to encode video. The availability of different scales (resolutions) in spatial wavelet transforms makes convenient *hierarchical block matching*, by which motion vectors yielding matches at lower scales are scaled by two to initialize the search for motion vectors that yield matches at the next higher scale. Consider three scales of the past and current frames, as shown in Figure 13.12. The block dimensions of the lowest scale are 1/4 those of the full scale, so both dimensions of the search range for a match are reduced by a factor of 4 from those of a full scale search. The motion vector producing the match is then scaled by 2 in each component and used as the initial motion vector in the search for a match at 1/2 scale. This initial motion vector produced a lower resolution match, so it should be reasonably close to the matching one at this scale. Therefore, the search is limited to a small range. When the match is found, the resulting motion vector is again scaled as before to initialize the search at full scale. Hierarchical block matching finds equally accurate motion vectors with savings in computation over its full scale counterpart.

Quantization and coding

The box in Figure 13.11 labeled "Quantizer" is often a cascade of a two-dimensional DCT and quantization of a code-block. The output of the quantizer is an array of bin indices. The bin indices are entropy coded outside the loop, just before transmission

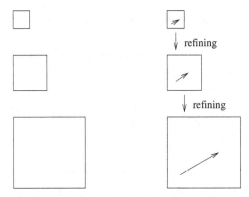

Figure 13.12   Hierarchical motion estimation in three scales.

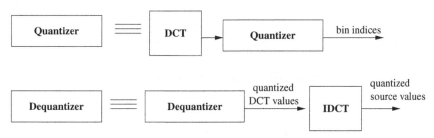

Figure 13.13 Quantizer and dequantizer equivalences to transform types in MCPC loop.

or storage. The box labeled "Dequantizer" is a dequantizer, a mapping from a bin index to a reconstruction point inside the associated bin, followed by an inverse two-dimensional DCT. The functional expansions of the quantizer and dequantizer diagram blocks are shown in Figure 13.13. Inside the loop, the bin indices are reconstructed as displaced frame differences and added to values of the current frame predicted from the previous frame to obtain the reconstruction of the current frame. The "Predictor" uses a neighborhood of $(x - \hat{\delta}_x, y - \hat{\delta}_y)$, which is (x, y) in the current frame displaced by the estimated motion vector $(\hat{\delta}_x, \hat{\delta}_y)$. The motion vector estimator uses the actual current frame block and the corresponding block of the decoded previous frame containing the displaced pixel $(x - \delta_x, y - \delta_y)$. The Predictor needs the estimate $(\hat{\delta}_x, \hat{\delta}_y)$, so it is inserted directly into the Predictor, as shown in Figure 13.11. The bin indices at the output of the Quantizer and the motion vector output at the output of the Motion estimator are both fed directly to the Entropy coder, which entropy encodes them and emits the compressed bitstream.

Backward motion estimation

The MCPC system just described uses a motion estimation method called *forward estimation*, because it estimates motion forward in time to a frame that has not been decoded. Estimation of current motion from past frames that have already been decoded is called *backward estimation*. This method envisions estimating the motion vectors from the two most recent decoded frames, one at time $t - 1$ and the other at time $t - 2$. Its advantage is that the estimated motion vectors do not need to be sent as side information, because the receiver can perform the same estimation from the same decoded frames. Its disadvantages are that motion estimation has to be performed at the receiver and that the estimate may be inaccurate, because current motion may differ too much from the recent past motion. Some of the video standards use weighted averages of backward and forward estimation to determine the current motion vector. We choose not to reveal here the details or depict the corresponding expansion of the block diagram in Figure 13.11 and point the interested reader to the references at the end of this chapter.

13.4.3.2 Wavelet transform coding

Video can be regarded as a three-dimensional image, so could be coded as described in the previous section on multi-component image coding. In certain cases, motion

1   8

Figure 13.14   Side view of a motion thread through an 8-frame GOP.

Figure 13.15   Temporal subbands of a 2-level wavelet transform of an 8-frame GOP: LL = low-low; LH = low-high; H = high.

estimation and compensation may result in little to no improvement and add only to the burdens of computation and time delay. Those cases occur when there is very slow motion between frames or when the motion is too fast and complicated to be captured by the estimation technique. Straight three-dimensional wavelet coding in groups (GOPs) of eight or sixteen frames has proved to work quite well in such cases.

In the intermediate cases of motion, which covers more than 90% of all video, it is advisable to use motion compensation within the three-dimensional wavelet transform framework. One method that has yielded good results is motion-compensated temporal filtering (MCTF). The idea is to track the motion of pixels in a frame from the beginning to end of a GOP. Assume that this can be done for every pixel in the beginning frame, so that we obtain a motion trajectory or thread for every such pixel. An example of a motion thread through an eight-frame GOP is shown in Figure 13.14. Then we take a one-dimensional wavelet transform of all the pixels along every motion thread. Thus, we end up with a motion-compensated temporal wavelet transform. The subbands of a two-level transform are depicted in Figure 13.15. Then the spatial planes of the temporal transform are individually (two dimension) wavelet transformed to complete the transformation of the GOP in three dimensions. Then coding may be done using any of the aforementioned wavelet coders, such as SPIHT [10, 11], SPECK, EZBC [12], or JPEG2000, using either two- or three-dimensional coding units. The early reference software for the H.264/AVC standard under development was EZBC applied to two-dimensional frames that had undergone MCTF and spatial wavelet transformation.

One may encounter certain pitfalls in the motion compensation of MCTF. Pixels may disappear and reappear when occluded and then uncovered as objects move. Pixels belonging to objects that newly enter a scene within a GOP must be tracked and filtered. Special procedures have been invented to cover these occurrences. Woods [13] discusses these procedures, along with the hierarchical block matching motion estimation, and the lifting filters for the temporal transform.

Some of the advantages of three-dimensional wavelet transform coding are scalability in spatial and temporal resolution, as well as in rate and region of interest. In the case of temporal scalability, using the example in Figure 13.15, one may extract from the bitstream the LL and LH subband bits to decode the video at one-half the frame rate or the LL subband bits to decode the video at one-fourth of the frame rate. Such capability can be inherent only in wavelet transform coding methods.

13.5   Concluding remarks

The intention of this chapter was to present the essence of the coding operations on color images, multi-component or volume images, and video. Commensurate with this narrowness of scope, there was deliberate omission of certain subjects, such as the psychophysics of color and color science, selective search in block matching, other forms of motion estimation, details differentiating the various video standards, and fidelity layers in video standards. There are several modern textbooks, such as those by Sayood [14], Shi and Sun [15], and Woods [13], that treat one or more of these subjects in depth. Our approach in this chapter differs, as it presents just the framework to which coding is applied to these kinds of imagery. To our knowledge, no other textbooks have explained generic approaches toward coding multi-dimensional imagery in this manner.

Notes

1. The display device is a power law (power γ) function of the camera sensor's output voltage that is corrected by $1/\gamma$ in the camera electronics so that the display will show the true color.
2. Two bit shifts to the right actuates division by 4 and the floor function ($\lfloor \ \rfloor$).
3. A LIP list is not used in SPECK. Instead, single insignificant points are placed on top of the LIS.
4. They are usually called *cubic blocks* or simply *blocks* when the dimension is understood.

References

1. J. O. Limb, C. R. Rubinstein, and J. E. Thompson, "Digital coding of color video signals – a review," *IEEE Trans. Commun*, vol. COM-25, no. 11, pp. 1349–1384, Nov. 1977.
2. D. L. MacAdam, "Visual sensitivities to color differences in daylight," *J. Opt. Soc. Am.*, vol. 32, no. 5, pp. 247–274, May 1942.
3. W. T. Wintringham, "Color television and colorimetry," *Proc. IRE*, vol. 39, no. 10, pp. 1135–1172, Oct. 1951.
4. A. N. Netravali and B. G. Haskell, *Digital Pictures*. New York: Plenum Press, 1988.
5. W. K. Pratt, "Spatial transform coding of color images," *IEEE Trans. Commun. Technol.*, vol. COM-19, no. 6, pp. 980–992, Dec. 1971.
6. J. Xu, Z. Xiong, S. Li, and Y.-Q. Zhang, "3-D embedded subband coding with optimal truncation (3-D ESCOT)," *Appl. Comput. Harmon. Anal.: Special Issue on Wavelet Applications in Engineering*, vol. 10, no. 5, pp. 290–315, May 2001.
7. P. Schelkens, A. Munteanu, J. Barbarien, M. Galca, X. G. i Nieto, and J. Cornelis, "Wavelet coding of volumetric medical datasets," *IEEE Trans. Med Imaging*, vol. 22, no. 3, pp. 441–458, Mar. 2003.
8. *Information Technology JPEG 2000 Image Coding System: Extensions for Three-dimensional Data: ITU-T Recommendation T.809*. Geneva, Switzerland: International Telecommunications Union, Aug. 2007.
9. P. Schelkens, C. Brislawn, J. Barbarien, A. Munteanu, and J. Cornelis, "Jpeg2000 – part 10:volumetric imaging," in *Proceedings of SPIE Annual Meeting, Applications of Digital Image Processing XXVI*, San Diego CA, July 2003, pp. 296–305.
10. B.-J. Kim, Z. Xiong, and W. A. Pearlman, "Low bit-rate scalable video coding with 3D set partitioning in hierarchical trees (3D SPIHT)," *IEEE Trans. Circuits Syst. Video Technol.*, vol. 10, pp. 1374–1387, Dec. 2000.
11. W. A. Pearlman, B.-J. Kim, and Z. Xiong, "Embedded video subband coding with 3D SPIHT," in *Wavelet Image and Video Compression*, ed. P. N. Topiwala Boston/Dordrecht/London: Kluwer Academic Publishers, 1998, ch. 24, pp. 397–434.
12. S.-T. Hsiang and J. W. Woods, "Embedded image coding using invertible motion compensated 3-d subband/wavelet filter bank," *Signal Proc. Image Commun.*, vol. 16, no. 5, pp. 705–724, May 2001.
13. J. W. Woods, *Multidimensional Signal, Image and Video Processing and Coding*. Amsterdam: Academic Press (Elsevier), 2006.

14. K. Sayood, *Introduction to Data Compression*, 3rd edn. San Francisco, CA: Morgan Kaufmann Publishers (Elsevier), 2006.
15. Y. Q. Shi and H. Sun, *Image and Video Compression for Multimedia Engineering*. Boca Raton: CRC Press, Taylor & Francis Group, 2008.

14   Distributed source coding

In this chapter, we introduce the concept that correlated sources need not be encoded jointly to achieve greater efficiency than encoding them independently. In fact, if they are encoded independently and decoded jointly, it is theoretically possible under certain conditions to achieve the same efficiency as when encoded jointly. Such a method for coding correlated sources is called *distributed source coding (DSC)*. Figure 14.1 depicts the paradigm of DSC with independent encoding and joint decoding. In certain applications, such as sensor networks and mobile communications, circuit complexity and power drain are too burdensome to be tolerated at the transmission side. DSC shifts complexity and power consumption from the transmission side to the receiver side, where it can be more easily handled and tolerated. The content of this chapter presents the conditions under which DSC is ideally efficient and discusses some practical schemes that attempt to realize rate savings in the DSC paradigm. There has been a plethora of recent work on this subject, so an encyclopedic account is impractical and ill-advised in a textbook. The goal here is to explain the principles clearly and elucidate them with a few examples.

14.1   Slepian–Wolf coding for lossless compression

Consider two correlated, discrete scalar sources X and Y. Theoretically, these sources can be encoded independently without loss using $H(X)$ and $H(Y)$ bits, respectively, where $H(X)$ and $H(Y)$ are the entropies of these sources. However, if encoded jointly, both these sources can be reconstructed perfectly using only $H(X, Y)$ bits, the joint entropy of these sources.

From information theory, we know that

$$H(X, Y) = H(X) + H(Y/X)$$
$$\leq H(X) + H(Y), \tag{14.1}$$

with equality if and only if X and Y are statistically independent, so we can save rate when the two sources are dependent and are encoded together. However, this saving comes at a cost in complexity of storage and computation. For example, in the usual arithmetic or Huffman entropy coding, it means maintenance of a probability table of

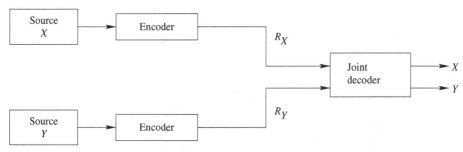

Figure 14.1   Distributed source coding: independent encoding and joint decoding of correlated sources.

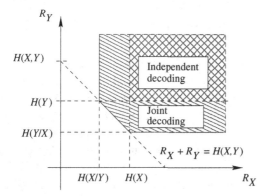

Figure 14.2   Rate regions for achieving lossless reconstruction for joint and independent decoding.

size of the product of the sizes of the alphabets of X and Y ($|\mathcal{X}| \times |\mathcal{Y}|$), instead of their sum $|\mathcal{X}| + |\mathcal{Y}|$, as in independent encoding.

Let us now state the Slepian–Wolf theorem [1] that says that joint decoding can reconstruct both sources without loss, whether or not they are both known to each other at the encoder.

THEOREM 14.1 *When scalar sources X and Y with joint probability distribution $p(x, y) \neq p(x)p(y)$ are encoded independently with rates R_X and R_Y, respectively, they can be decoded exactly if and only if these rates obey the following conditions:*

$$R_X \geq H(X/Y), \tag{14.2}$$

$$R_Y \geq H(Y/X), \tag{14.3}$$

$$R_X + R_Y \geq H(X, Y). \tag{14.4}$$

The set of rate pairs (R_X, R_Y) that satisfy these conditions can be pictured as a region in two dimensions, as shown in Figure 14.2. This region is called the *achievable region* for lossless reconstruction.

We omit here the formal details of the proof, which can be found in the original paper [1] or the textbook by Cover and Thomas [2]. However, we shall explain the logic of the proof, because it illuminates the path to a practical method. We must

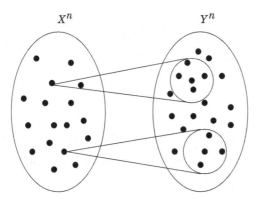

Figure 14.3　Fans of points in $\mathcal{Y}^n$ typically linked to points **x** in $\mathcal{X}^n$.

take independent variates from these sources X and Y to form the i.i.d. sequences **x** and **y** of long length n. (Since X and Y are correlated, so are **x** and **y**.) Consider the point $R_X = H(X)$ and $R_Y = H(Y/X)$ in the achievable region. For any given **x** and n sufficiently large, there are $2^{nH(Y/X)}$ sequences **y** that are highly probable to occur (probability close to 1) out of the $2^{nH(Y)}$ total that are highly probable. We say that there are $2^{nH(Y/X)}$ typical sequences **y** linked to **x** out of the total of $2^{nH(Y)}$ typical sequences for **y**. We picture it as a fanout containing $2^{nH(Y/X)}$ points from the point **x** in the n-dimensional space $\mathcal{Y}^n$. A two-dimensional rendition is illustrated in Figure 14.3. The X encoder encodes the sequence **x** with $R_X \geq H(X)$ bits per dimension, so the decoder reconstructs **x** exactly. So how can we convey **y** using only $H(Y/X)$ bits per dimension? If the Y encoder knows **x**, then it can send the index of **y** within the fan using $nH(Y/X)$ bits. However, it does not know **x**, so it must resort to another strategem. This encoder randomly colors all sequences **y** in $\mathcal{Y}^n$ with 2^{nR_Y} colors. Then it sends the color of the particular **y** with $nH(Y/X)$ bits. Since **x** is known at the decoder, the decoder looks for the received color of **y** within the fanout around **x**. If the number of colors is sufficiently high, then with probability close to one, all the colors within the fanout about **x** will be different and the color of the **y** sequence within this fan will uniquely identify **y**. The number of colors is sufficiently high if $R_Y > H(Y/X)$, so that 2^{nR_Y} is exponentially larger than $2^{nH(Y/X)}$. In the limit of large n, the probability of error approaches zero and the rates converge to the lower limits in the Slepian–Wolf conditions.

14.1.1　Practical Slepian–Wolf coding

This method of random coloring of long sequences is impractical, because it requires long coding delay and large storage. For fixed and finite length sequences, we use a different method, first suggested by Wyner [3]. What is needed is a division of $\mathcal{Y}^n$ space into regions where the operative **y** sequence can be uniquely identified. A channel code on $\mathcal{Y}^n$ divides this space into regions wherein their sequences are mapped to codewords. The criterion for this mapping is choosing the most probable transmitted codeword for the given received sequence. The region containing a codeword is called the *correctable* region for that codeword. Let us consider a code $\mathcal{C}$ consisting of the $M = 2^k$ codewords, $\mathbf{c}_0, \mathbf{c}_1, \ldots, \mathbf{c}_{M-1}$, in $\mathcal{Y}^n$. In other words, there are M codewords, each with n dimensions, separated one from another by at least distance d. Such a code is denoted as an (n, k, d) code. The correctable regions may be thought of as n-dimensional spheres of radius $d/2$ surrounding the codewords, i.e., $\|\mathbf{y} - \mathbf{c}_i\| < d/2$ for $i = 0, 1, \ldots, M - 1$, $M = 2^k$. We can now view the sequences $\mathbf{y}$ as belonging to cosets of the code. Cosets of the code $\mathcal{C} = \{\mathbf{c}_0, \mathbf{c}_1, \ldots, \mathbf{c}_{M-1}\}$ are the set of sequences $\mathcal{C} + \mathbf{e} = \{\mathbf{c}_0 + \mathbf{e}, \mathbf{c}_1 + \mathbf{e}, \ldots, \mathbf{c}_{M-1} + \mathbf{e}\}$ for a given error sequence $\mathbf{e}$. Just as a sequence $\mathbf{y}$ was randomly associated with a certain color, a sequence $\mathbf{y}$ is now labeled with the index of the coset, equivalent to coloring the coset, that identifies its error sequence from some codeword. However, such labeling does not identify the particular codeword. We illustrate this situation in Figure 14.4, where we show the correctable regions as clusters of colored balls representing the points. Like-colored balls belong to the same coset. Say that the decoder is informed that $\mathbf{y}$ belongs to the "red" coset, i.e., the color of $\mathbf{y}$ is "red." When the decoder receives $\mathbf{x}$, it searches for the most probable "red" ball and declares it to be $\mathbf{y}$. The most probable "red" ball is illustrated as the one closest to $\mathbf{x}$ in Figure 14.4.

Let us consider first the point $(H(X), H(Y/X))$ on the graph in Figure 14.2. Let the alphabets $\mathcal{X}$ and $\mathcal{Y}$ of X and Y be binary, so that $\mathbf{y}$ and $\mathbf{x}$ are binary sequences. The most probable sequence $\mathbf{y}$ for a given $\mathbf{x}$ is the one closest in Hamming distance, which is the one with fewest bit differences. Suppose that $\mathbf{x}$ is sent with $R_X = H(X)$ bits per dimension, so that it is known exactly at the decoder. Suppose, however, that only the coset of $\mathbf{y}$ is known at the decoder. Members of a coset have the same distance properties as the original code. So we can view the coset members as a new equivalent code for the decoder. If the code is designed properly according to the known correlation of $\mathbf{x}$ and $\mathbf{y}$, then $\mathbf{x}$ should lie in the correctable region of the true $\mathbf{y}$. The point $(H(X/Y), H(Y))$ is obtained just by exchanging the roles of X and Y in the argument above. Time sharing of these two strategies achieves the intermediate points on the $R_X + R_Y = H(X, Y)$ line.

The question arises as to how many bits are needed to index the cosets. To provide an index for all the cosets would require n bits, as many as indexing $\mathbf{y}$ itself, so nothing is gained. However, we need only index the cosets associated with error sequences in the correctable regions. Since there are $M = 2^k$ codewords spread out in

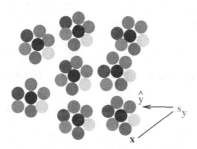

Figure 14.4 Illustration of correctable regions as clusters of balls and cosets as like-colored balls. Only the red color of $\mathbf{y}$ is known, so is indicated by the red color of its syndrome $\mathbf{s}_y$. Therefore, $\mathbf{y}$ is declared to be the red ball closest to $\mathbf{x}$ and is marked as $\hat{\mathbf{y}}$. (The blank spaces are there just to exaggerate the display of the clusters and are not to be interpreted as distances.)

a space of 2^n sequences, there are no more than 2^{n-k} error sequences in each correctable region. Therefore, we need $n - k$ bits to index these cosets associated with correctable regions. For i.i.d. binary n-sequences, $\mathbf{y} = \mathbf{x} + \mathbf{w}$, and $\mathbf{w}$ independent of $\mathbf{x}$, $H(\mathbf{x}, \mathbf{y}) = H(\mathbf{y}/\mathbf{x}) + H(\mathbf{x})$, with $H(\mathbf{x}) = n$ and $H(\mathbf{y}/\mathbf{x}) = H(\mathbf{w}) = n - k$, so the minimal rates of the Slepian–Wolf theorem are achieved.

For a linear block code, indexing of the cosets belonging to correctable regions may be done through calculation of the syndrome of a sequence. Recall that associated with a linear block code is a $(n - k) \times n$ parity-check matrix H, such that its codewords $\mathbf{c}_i, i = 0, 1, \ldots, M - 1$ satisfy

$$\mathbf{c}_i H^T = \mathbf{0}, \text{ for all } \mathbf{c}_i \in \mathcal{C}. \tag{14.5}$$

As is customary in the communications literature, we use row vectors. The superscript T denotes matrix transpose. Since $\mathbf{y}$ is not necessarily a codeword, $\mathbf{y} = \mathbf{c}_i + \mathbf{e}_y$, for an unknown i, so

$$\begin{aligned} \mathbf{y} H^T &= (\mathbf{c}_i + \mathbf{e}_y) H^T \\ &= \mathbf{c}_i H^T + \mathbf{e}_y H^T \\ &= \mathbf{0} + \mathbf{e}_y H^T \end{aligned} \tag{14.6}$$

The $(n - k)$-dimensional vector $\mathbf{s} \equiv \mathbf{y} H^T$ is called the *syndrome* of $\mathbf{y}$. In view of Equation (14.6), the non-zero syndrome is caused by an error vector $\mathbf{e}_y$ of n dimensions. Since there is more than one solution for any syndrome, we associate the most probable error vector with the given syndrome. The most probable error vector is the one with the least Hamming weight (least number of "1"s). Since there are 2^{n-k} possible syndromes (including the $\mathbf{0}$ syndrome), they comprise all the correctable error vectors. We can use the $n - k$ syndrome bits as the index of the coset of $\mathbf{y}$ or, for a systematic code,[1] we can use the $n - k$ parity bits as this index.

We now summarize the steps for encoding and decoding of the binary n-sequences $\mathbf{x}$ and $\mathbf{y}$ modeled as above.

Encoder operations:

1. Design a linear block code $\mathcal{C} = \{\mathbf{c}_0, \mathbf{c}_1, \ldots, \mathbf{c}_{M-1}\}$ on the space $\mathcal{Y}^n$ with $(n - k) \times n$ parity check matrix H.
2. Source encode $\mathbf{x}$ with $nH(X)$ bits. (Leave $\mathbf{x}$ uncoded; $\mathbf{x}$ sequences equiprobable.)
3. Compute syndrome $\mathbf{s}_y = \mathbf{y} H^T$ and send its $n - k$ bits to the decoder.

Decoder operations:

1. Coset decode $\mathbf{s}_y$: associate $\mathbf{s}_y$ to its most probable error vector $\mathbf{e}_y$ to form decoder's code $\mathcal{C} + \mathbf{e}_y$.
2. Decode $\mathbf{x}$ in decoder code; determine nearest (most probable) codeword $\hat{\mathbf{y}}$ in $\mathcal{C} + \mathbf{e}_y$.
3. Set $\hat{\mathbf{y}}$ as reconstruction of $\mathbf{y}$.

Because of the unique properties of linear binary codes, the decoder operations can be carried out using the original parity-check matrix H. In the decoder we receive $\mathbf{s}_y = \mathbf{y} H^T$. We compute then $\mathbf{s}_x = \mathbf{x} H^T$ upon receiving $\mathbf{x}$. Then

$$\mathbf{s}_x + \mathbf{s}_y = (\mathbf{e}_x + \mathbf{e}_y)H^T = (\mathbf{e}_{xy})H^T.$$

where we determine the minimum weight (most probable) $\mathbf{e}_{xy}$, which is the error vector between $\mathbf{x}$ and $\mathbf{y}$. Therefore, for a linear block code, the decoder operations may become:

1. Receive $\mathbf{x}$ and compute $\mathbf{s}_x = \mathbf{x}H^T$.
2. Receive $\mathbf{s}_y$ and compute $\mathbf{s}_x + \mathbf{s}_y = \mathbf{e}_{xy}H^T$.
3. Determine minimum weight $\mathbf{e}_{xy}$ (the one with fewest "1"s); equivalently associate syndrome sum with its correctable error vector $\mathbf{e}_{xy}$.
4. Let $\hat{\mathbf{y}} = \mathbf{x} + \mathbf{e}_{xy}$.

Actually, the code need not be a block code, because any linear code, such as a convolutional code, has an associated matrix H whose codewords obey Equation (14.5).

The success of the procedure can only be guaranteed with high probability. In other words, even with proper design, there will remain a small probability that the reconstruction $\mathbf{y}$ will be wrong. This probability of failure is the probability that $\mathbf{x}$ and $\mathbf{y}$ are separated by $d/2$ or more. Let us illustrate how we might follow Step 1 to design a proper code.

Since the decoder has incomplete knowledge of $\mathbf{y}$, it must have a statistical model of its dependence on $\mathbf{x}$. We postulate that $\mathbf{y} = \mathbf{x} + \mathbf{v}$, where $\mathbf{v}$ is a random binary sequence independent of $\mathbf{x}$. (Addition is modulo 2.) We assume that its components are independent with probability of "1" p. A "1" in a component of $\mathbf{v}$ signifies a bit difference between the corresponding components of $\mathbf{y}$ and $\mathbf{x}$. The number of "1"s K_v in $\mathbf{v}$ is therefore a binomial random variable with probability distribution

$$P_v(\ell) = \binom{n}{\ell} p^\ell (1 - p)^{n-\ell}, \quad \ell = 0, 1, \ldots, n. \tag{14.7}$$

We shall attempt to compute the probability that $\mathbf{x}$ and $\mathbf{y}$ are less than $d/2$ apart in Hamming distance or, in other words, the error vector between them has fewer than $d/2$ "1"s. The number of "1"s in a binary sequence is called the Hamming weight and denoted by the function $w(\mathbf{e}_{xy})$ to count the number of "1"s in $\mathbf{e}_{xy}$, for example. So we seek the probability that $w(\mathbf{e}_{xy}) < d/2$, conditioned on knowledge of $\mathbf{x}$. Now $\mathbf{e}_y = \mathbf{e}_x + \mathbf{v}$, so we wish to compute

$$Pr\{w(\mathbf{e}_{xy}) < d/2|\mathbf{x}\}.$$

According to our model, $\mathbf{e}_{xy} = \mathbf{x} + \mathbf{y} = \mathbf{v}$, so that the probability of successful reconstruction P_S is

$$\begin{aligned}
P_S &\equiv Pr\{w(\mathbf{e}_{xy}) < d/2|\mathbf{x}\} \\
&= Pr\{w(\mathbf{v}) < d/2\} \\
&= Pr\{K_v < d/2\}.
\end{aligned}$$

Table 14.1 Probability of perfect reconstruction , using Golay (23, 12, 7) 3-error correcting channel code.

p	0.10	0.08	0.06	0.04	0.02	0.01
P_S	0.8073	0.8929	0.9540	0.9877	0.9989	0.9999

Therefore, the probability of successful reconstruction becomes

$$P_S = \sum_{\ell=0}^{k_c} \binom{n}{\ell} p^\ell (1-p)^{n-\ell}, \quad \ell = 0, 1, \ldots, n, \tag{14.8}$$

where we define $k_c \equiv \lceil d/2 - 1 \rceil$, which is the largest integer smaller than $d/2$. Designing the channel code so that the probability of successful reconstruction P_S in Equation (14.8) is greater than $1 - \epsilon$ for small $\epsilon > 0$ means that the probability of failure in perfect reconstruction is less than ϵ.

Example 14.1 Numerical example of probability of successful reconstruction.
We shall take as an example the triple-error correcting binary Golay (23, 12, 7) channel code. This code has exactly $2^{11} = 2048$ binary vectors within distance $k_c = 3$ in every one of the $2^{12} = 4192$ correctable regions of a codeword.[2] Substituting $n = 23$, $k_c = 3$, and some values of p into the formula for P_S in Equation 14.8, we obtain the results shown in Table 14.1.

As expected, as p decreases, the probability of successful reconstruction increases, reaching 0.9999 for $p = 0.01$.

14.2   Wyner–Ziv coding for lossy compression

The counterpart of the Slepian–Wolf theorem for lossy compression is the Wyner–Ziv theorem [4]. When the source values are continuous in amplitude, as they are in many situations, they cannot be encoded with finite rate without loss. Therefore, the goal of distributed source coding in this case is to represent the source with a finite rate and reconstruct it within a given distortion at the decoder with the benefit of side information that is correlated with the source. The model for this problem is shown in Figure 14.5. Note that the side information is shown to be resident only at the decoder, because it is considered to be lossless and cannot be transmitted without loss, if continuous in amplitude. The minimum source rate required to achieve the distortion target, as expressed by the rate-distortion function, does not turn out to be the same, whether or not the side information is available at the encoder, except in the special case of the source and side information being jointly Gaussian.

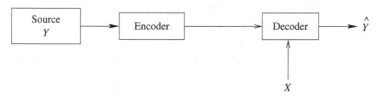

Figure 14.5 Source decoding with side information at the decoder.

We shall now state the rate-distortion function for the case of side information available only to the decoder. The rate-distortion function is the minimum achievable rate for a given distortion limit. Let $\mathcal{Y}$, $\hat{\mathcal{Y}}$, and $\mathcal{X}$ denote the alphabets of the source Y, its reconstruction $\hat{Y}$, and the side information X, respectively. Let $\rho(y, \hat{y})$ measure the distortion between source and reconstruction values y and $\hat{y}$. The rate-distortion function $R^*(d)$ for this system is expressed as

$$R^*(d) = \min_{p(w/y)} \min_{f} (I(Y; W) - I(X; W)), \qquad (14.9)$$

where the minimizations of the difference between $I(Y; W)$, the average mutual information between Y and W, and $I(X; W)$, the average mutual information between X and W, are over all functions $f : \mathcal{X} \times \mathcal{W} \rightarrow \hat{\mathcal{Y}}$ and conditional probability distributions $p(w/y)$, such that the average distortion between source and reconstruction does not exceed d, i.e.,

$$E[\rho(Y, \hat{Y})] \leq d.$$

The random variables X, Y, and W are assumed to form a Markov chain, i.e., $p(x, y, w) = p(x)p(y/x)p(w/y)$ for all x, y, w. W has the role of the encoded version of Y. The function f takes realizations of W and X to produce $\hat{Y}$, its best estimate of Y.

Wyner and Ziv prove that $R^*(d)$ is achievable in the limit of long sequences of the correlated variates X and Y drawn independently from the probability distribution $p(x, y)$. The steps of this proof are described in the textbook by Cover and Thomas [2] and its details are delineated in the original article [4].

The logic of the proof has much in common with that of the Slepian–Wolf theorem for the lossless case. We shall omit the logic here and state only the practical counterpart. Starting at the source, the space of source sequences $\mathcal{Y}^n$ must be partitioned into a finite number of regions. The centroids of these regions are the set $\hat{Y}^n$ of the reconstruction sequences $\hat{y}^n$. For a source rate of R_s bits per sample, the number of these regions is 2^{nR_s}, each of which is indexed with nR_s bits. So now our continuous-amplitude source sequences have been transformed into sequences of these indices, so W above is an alphabet of 2^{nR_s} integers and $\mathcal{W}^n$ is the space of these integer sequences $\mathbf{w}$, which are commonly called the active codewords. Now we consider that there is a channel code that partitions this $\mathcal{W}^n$ space into 2^{nR_c} regions, the so-called correctable regions, where R_c is the channel code bit rate per dimension. Assuming that these regions have the same volume, the number of elements in each region is $2^{n(R_s - R_c)}$, with the channel codeword coinciding with an active codeword. The set C of channel codewords and the set $\mathcal{C} + \mathbf{e}$ for every correctable $\mathbf{e}$ (the sequence $\mathbf{c} + \mathbf{e}$ is in the correctable region for $\mathbf{c}$) form the cosets. As before, we send the coset label $\mathbf{s}_y$ corresponding to $\mathbf{y}$, but now use $R = R_s - R_c$ bits per dimension. $R_s - R_c$ corresponds to $I(Y; W) - I(X; W)$ in the rate-distortion function in Equation 14.9. The decoder observes the side information sequence $\mathbf{x}$ (in X^n) and then deduces which member of the coset indicated by $\mathbf{s}_y$ is most probable for the given $\mathbf{x}$ and declares it to be the quantization bin number $\mathbf{w}_y$ of the source $\mathbf{y}$. It then sets the reconstruction $\hat{\mathbf{y}}$ of $\mathbf{y}$ to be the centroid of quantization bin corresponding to $\mathbf{w}_y$. With proper design of the source and channel codes, the probability is close to one that the correct quantization bin and hence correct reconstruction $\hat{\mathbf{y}}$ result.

14.2.1   Scalar Wyner–Ziv coding

In order to make the procedure above more concrete, we develop it for a hypothetical case of $n = 1$, which is scalar quantization of the source Y. For a rate of R_s bits per source sample, the quantization alphabet $\hat{\mathcal{Y}}$ consists of 2^{R_s} points, which impose a partition of the continuous source alphabet $\mathcal{X}$ into 2^{R_s} regions, called decision regions or quantization bins. The reconstruction points are the centroids of these bins. In Figure 14.6 is shown the case of $R_s = 3$ with eight quantization bins. We label each of these bins with a unique index W in $\mathcal{W} = \{0, 1, 2, \ldots, 2^{R_s} - 1\}$. Now we partition $\mathcal{W}$ into 2^{R_c} sets of the same size, so that there are $2^{R_s - R_c}$ elements in each set. For this scalar case, we can form $2^{R_s - R_c}$ groups of successively labeled 2^{R_c} bins. For the example in Figure 14.6, $R_s = 3$ and $R_c = 1$, so we group the eight bins into two groups of four elements each, $\{0,1,2,3\}$ and $\{4,5,6,7\}$. The cosets are $\{0,4\}, \{1,5\}, \{2,6\}, \{3,7\}$. For the general case, the cosets, each containing 2^{R_c} elements are

$$S_\ell = \{\ell, 2^{R_s - R_c} + \ell, 2 \cdot 2^{R_s - R_c} + \ell, \ldots, (2^{R_c} - 1) \cdot 2^{R_s - R_c} + \ell\},$$
$$\ell = 0, 1, 2, \ldots, 2^{R_s - R_c} - 1.$$

We use $R_s - R_c$ bits to convey the coset S_ℓ corresponding to Y. The decoder must then choose the most probable element in S_ℓ for the side information $\mathbf{x}$. The reconstruction $\hat{Y}$ is the centroid of the bin indicated by the chosen element in S_ℓ.

Continuing with the example in Figure 14.6, suppose that the decoder receives coset $S_1 = \{1, 5\}$ for Y and side information $X = 0.6$. Let us assume that the most probable element of S_1 corresponds to the bin interval at minimum distance from X. The decoder

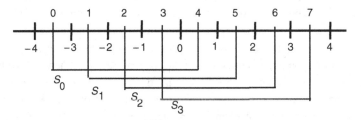

Figure 14.6   Quantization bins and cosets in scalar Wyner–Ziv uniform quantization ($R_s = 3$, $R_c = 1$).

chooses element 5 from S_1, because it corresponds to the interval $[1, 2)$, which is the closest to X. The reconstruction $\hat{Y}$ is the centroid of the interval $[1, 2)$.[3]

14.2.2 Probability of successful reconstruction

14.2.2.1 The joint Gaussian model

This procedure succeeds if X truly points to the most probable quantization bin for Y. Otherwise it fails to produce the proper reconstruction $\hat{Y}$. High probability of success depends on the number of quantization points, 2^{R_s}, and the number of coset elements, 2^{R_c}. Setting the rates R_s and R_c appropriately according to the statistical dependence between X and Y assures a high probability of successful reconstruction. We shall examine these relationships by deriving an expression for the probability of success P_S. First, we postulate the statistical model that X and Y are zero mean and jointly Gaussian with probability density function

$$f_{XY}(x, y) = \frac{1}{2\pi\sigma_x\sigma_y\sqrt{1 - r^2}} \exp$$

$$-\frac{1}{2(1 - r^2)} \left[\left(\frac{x}{\sigma_x}\right)^2 - 2r\left(\frac{x}{\sigma_x}\right)\left(\frac{y}{\sigma_y}\right) + \left(\frac{y}{\sigma_y}\right)^2 \right], \quad (14.10)$$

where σ_x^2 and σ_y^2 are the variances of X and Y, respectively, and r is the correlation coefficient. Also pertinent is the conditional probability density function of Y given X,

$$f_{Y/X}(y/x) = \frac{1}{\sqrt{2\pi\sigma_y^2(1 - r^2)}} \exp -\frac{1}{2(1 - r^2)} \left[\left(\frac{y}{\sigma_y}\right) - r\left(\frac{x}{\sigma_x}\right) \right]^2. \quad (14.11)$$

Let the quantization intervals, $I_j = [z_j, z_{j+1})$, have uniform width $|z_{j+1} - z_j| = h$, for all $j = 0, 1, \ldots, 2^{R_s} - 1$. For a given $X = x$, we wish to choose the interval I_j that maximizes the probability

$$Pr\{Y \in I_j/X\} = \int_{z_j}^{z_{j+1}} f_{Y/X}(y/x)dy.$$

Without explaining the derivation, maximizing this probability is equivalent to the criterion:

$$\text{Choose } I_j \text{ that minimizes } |z_j - r\left(\frac{\sigma_y}{\sigma_x}\right)x|. \quad (14.12)$$

Therefore, the reconstruction is successful if the scaled x is closer to the interval associated with the true coset element in S_ℓ than any other. Suppose that the separation of the coset elements between the bottom ends of their associated intervals is d. Then the reconstruction succeeds if

$$|z_j - r\left(\frac{\sigma_y}{\sigma_x}\right)x| < d/2,$$

when Y is in I_j, so the probability of successful reconstruction $P_S(I_j)$ for Y in I_j may be expressed by

$$P_S(I_j) = Pr\{|z_j - aX| < d/2 \,|\, Y \in I_j\},$$

where $a = r\left(\frac{\sigma_y}{\sigma_x}\right)$. The expression may also be recast as

$$P_S(I_j) = Pr\{(z_j - d/2)/a < X < (z_j + d/2)/a \,|\, Y \in I_j\}.$$

In terms of $f_{XY}(x, y)$ in Equation 14.10,

$$P_S(I_j) = \frac{\int_{x=(z_j-d/2)/a}^{(z_j+d/2)/a} \int_{y=z_j}^{z_{j+1}} f_{XY}(x, y)\,dy\,dx}{\int_{z_j}^{z_{j+1}} \frac{1}{\sqrt{2\pi\sigma_y^2}} e^{-u^2/2\sigma_y^2}\,du}. \tag{14.13}$$

Now, because Y is not uniformly distributed, we must average over the intervals I_j to obtain the overall probability of successful reconstruction P_S.

$$P_S = E[P_S(I_j)]$$

$$= \sum_{j=0}^{2^{R_s}-1} \int_{x=(z_j-d/2)/a}^{(z_j+d/2)/a} \int_{y=z_j}^{z_{j+1}} f_{XY}(x, y)\,dy\,dx. \tag{14.14}$$

This formula for P_S also works for a general joint density function $f_{XY}(x, y)$ using the distance minimization criterion in Equation (14.12). Substituting the joint Gaussian density in (14.10) and exercising several algebraic manipulations and changes of variables, we obtain the following result:

$$P_S = \sum_{j=0}^{2^{R_s}-1} \int_{(z_j-d/2)/\sigma_x}^{(z_j+d/2)/\sigma_x} \frac{1}{\sqrt{2\pi}} e^{-u^2/2} \left(\int_{(z_j-r\sigma_y u)/(\sigma_y\sqrt{1-r^2})}^{(z_{j+1}-r\sigma_y u)/(\sigma_y\sqrt{1-r^2})} \frac{1}{\sqrt{2\pi}} e^{-v^2/2}\,dv \right) du. \tag{14.15}$$

One popular model for Gaussian statistical dependence between X and Y is $Y = X + V$, where X and V are both zero-mean Gaussian with variances σ_x^2 and σ_v^2, respectively. With this model, the formula for P_S in (14.15) is applicable. Evaluating $r = \sigma_x/\sigma_y$ and $\sigma_y^2 = \sigma_x^2 + \sigma_v^2$ and substituting into (14.15), the formula for P_S becomes

$$P_S = \sum_{j=0}^{2^{R_s}-1} \int_{(z_j-d/2)/\sigma_x}^{(z_j+d/2)/\sigma_x} \frac{1}{\sqrt{2\pi}} e^{-u^2/2} \left(\int_{(z_j-\sigma_x u)/\sigma_v}^{(z_{j+1}-\sigma_x u)/\sigma_v} \frac{1}{\sqrt{2\pi}} e^{-v^2/2}\,dv \right) du. \tag{14.16}$$

Example 14.2 Continuation of example for calculation of P_S.

Assume $R_s = 3$ results from uniform scalar quantization with eight levels. Again we use $R_c = 1$ and form four cosets $S_0 = \{0, 4\}$, $S_1 = \{1, 5\}$, $S_2 = \{2, 6\}$, and $S_3 = \{3, 7\}$. Let the decision levels $\{z_0, z_1, \ldots, z_8\}$ be $\{(-4, -3, -2, -1, 0, 1, 2, 3, 4)\sigma_x\}$. The

Table 14.2 Probability of successful reconstruction for additive Gaussian model, $R_S = 3$ and $R_C = 1$.

β	0.8	1	1.8	2	3	4
P_S	0.85089	0.91548	0.99167	0.99544	0.99973	0.99989

inter-element distance in a coset is $d = 4\sigma_x$. Plugging into the expression for P_S in Equation (14.16),

$$P_S = \sum_{j=0}^{7} \int_{-6+j}^{-2+j} \frac{1}{\sqrt{2\pi}} e^{-u^2/2} \left(\int_{(-4+j-u)\beta}^{(-3+j-u)\beta} \frac{1}{\sqrt{2\pi}} e^{-v^2/2} dv \right) du,$$

where $\beta \equiv \sigma_x/\sigma_v$ is a signal-to-noise ratio. It is related to the correlation coefficient by

$$r = \frac{\beta}{\sqrt{\beta^2 + 1}}.$$

Therefore, for large β, r approaches 1, and for small β, $r \approx \beta \ll 1$.

We now tabulate the results of calculating P_S for several values of β and present them in Table 14.2. Better than 99% success rate is reached for $\beta = 1.8$, which corresponds to correlation coefficient $r = 0.874$. So reasonably tight correlation between X and Y is required to achieve a high success rate for this particular code. Many real-life sources, such as natural images, normally have adjacent pixels with correlation coefficients 0.9 or even higher. A combination of larger dimension n, lower R_c, and larger $R_s - R_c$ to lead to a larger minimum distance d are needed in general to achieve high probabilities of successful reconstruction with looser statistical dependence. In any event, a non-zero probability of reconstruction failure cannot be avoided in distributed source coding.

14.2.2.2 General additive noise models

Although the joint Gaussian model handles the addition of Gaussian sources, it does not apply to the addition of non-Gaussian sources. For instance, when $Y = X + V$ and V is independent of X and is Laplacian, the formula for P_S in (14.15) is not applicable. Therefore, we shall derive a formula that applies to the general additive noise case.

When the probability density function of the additive noise V is an exponential function of the magnitude of its argument v to some power, such as in a generalized Gaussian density, then the criterion for selecting the coset member of Y is the one closest in distance to the given $X = x$. Therefore, given Y in interval I_j, we shall derive an expression for the probability that $|x - z_j|$ is less than $d/2$. We start with the probability

$$Pr\{|X - z_j| < d/2, Y \in I_j\} = Pr\{z_j - d/2 < X < z_j + d/2, z_j < X + V < z_{j+1}\}. \tag{14.17}$$

If we view this probability as an integration over $f_{XV}(x, v)$, the joint probability density of X and V over a region in the $x - v$ plane, we obtain

$$Pr\{|X - z_j| < d/2, Y \in I_j\} = \int_{x=z_j-d/2}^{z_j+d/2} \int_{v=z_j-x}^{z_{j+1}-x} f_{XV}(x, v)dvdx. \qquad (14.18)$$

Dividing the equation in (14.18) by $Pr\{Y \in I_j\}$ gives the conditional probability given Y in I_j. Now averaging over I_j, $j = 0, 1, \ldots, 2^{R_s} - 1$ yields the overall probability of success

$$P_S = \sum_{j=0}^{2^{R_s}-1} \int_{x=z_j-d/2}^{z_j+d/2} \int_{v=z_j-x}^{z_{j+1}-x} f_{XV}(x, v)dvdx. \qquad (14.19)$$

Example 14.3 Extension of example for calculation of P_S for independent Laplacian noise.

Using the same cosets and decision thresholds as the previous example, we consider the case of V independent of X in $Y = X + V$ and Laplacian with probability density function

$$f_V(v) = \frac{1}{\sqrt{2}\sigma_v} \exp -\sqrt{2}|v|/\sigma_v.$$

We assume, however, that X has the same Gaussian density

$$f_X(x) = \frac{1}{\sqrt{2\pi\sigma_x^2}} e^{-x^2/2\sigma_x^2}.$$

Statistical independence of X and V means that

$$f_{XV}(x, v) = f_X(x)f_V(v).$$

The decision thresholds are

$$z_j = (-4 + j)\sigma_x, \quad j = 0, 1, 2, \ldots, 7$$

and the inter-element distance in the cosets is $d = 4\sigma_x$. Substituting the probability densities and parameters into Equation (14.19) yields the following formula

$$P_S = \sum_{j=0}^{7} \int_{-6+j}^{-2+j} \frac{1}{\sqrt{2\pi}} e^{-u^2/2} \left(\int_{\sqrt{2}(-4+j-u)\beta}^{\sqrt{2}(-3+j-u)\beta} \frac{1}{2} e^{-|v|} dv \right) du, \qquad (14.20)$$

with $\beta = \sigma_x/\sigma_v$.

We now tabulate the results of calculating P_S for several values of β and present them in Table 14.3.

As expected, the greater the signal-to-noise ratio β, the closer the probability of success is to certainty. Notice that P_S is smaller here than in the Gaussian additive noise case in Table 14.2 for the same β. For example, 99% success rate is reached here for $\beta = 2.1$, corresponding to the correlation coefficient $r = 0.903$. The same 99% rate is reached for $\beta = 1.8$, $r = 0.874$ in the Gaussian additive case. The Laplacian noise

Table 14.3 Probability of successful reconstruction for Laplacian additive model, $R_S = 3$ and $R_C = 1$.

β	1	2	2.1	3	4	5
P_S	0.91787	0.0.98923	0.0.99106	0.99814	0.99958	0.99985

probability density does not fall off for larger magnitudes as rapidly as Gaussian, so the probability of success does not rise as rapidly with increasing signal-to-noise ratio.

14.3 Concluding remarks

The basic principles behind distributed source coding were explained in this chapter. One can conclude that joint decoding can be almost as efficient as joint encoding for correlated sources. In both the lossless (Slepian–Wolf) and lossy (Wyner–Ziv) cases, the penalty is a small, non-zero probability of failure. For the lossy case, there is an additional penalty, because the rate-distortion limit can be achieved only for joint Gaussian statistical dependence between the sources. These conclusions also generalize to more than two input sources, but we have chosen not to present this generalization in order to expose the basic principles. The reader interested in the theory may consult Chapter 14 of the textbook of Cover and Thomas [2]. The presentation here presumed little to no knowledge of channel coding theory, which would describe the means to partition the input sample space at the encoding side and decode the received coset label or syndrome to determine members of the cosets. Distributed source coding has been proposed in a variety of scenarios for sensor networks and mobile, handheld devices for audio, video, and data transmission. It was not the purpose here to catalog the many systems that have been proposed, so our apologies go out to the many researchers who have made contributions to this research area. At the time of writing, we know of no book or article that has presented the principles of this subject and techniques of design that attempt to reach a target of small probability of reconstruction failure. Most works just present results in terms of signal-to-noise ratio in Wyner–Ziv coding of video or image data reconstruction. The designs of the code parameters seem to be just ad hoc: pick a channel code for some source code and view the results. We hope that this chapter will help people to enter the field with understanding of the principles underlying proper system design.

Problems

14.1 Two correlated binary source words $\mathbf{x}$ and $\mathbf{y}$ are transmitted independently and are to be decoded jointly at the destination under the statistical model that $\mathbf{y} = \mathbf{x} \oplus \mathbf{v}$, with $\mathbf{v}$ an i.i.d. binary noise sequence, independent of $\mathbf{x}$ with probability $p = 0.05$ of a "1" causing a transition. The decoder uses a BCH (15, 7, 5) double error-correcting code to partition the signal space.

(a) Give the number of partitions and the number of bits needed to index the elements in a partition.

(b) Calculate the probability of incorrect reconstruction of **y**.

14.2 A continuous-amplitude signal Y is scalar quantized by a uniform step size, midpoint reconstruction quantizer. At the decoder, a signal X is available as side information to aid the decoding of $\hat{Y}$, the quantized version of Y. Assume X and Y have equal variances, i.e., $\sigma_x^2 = \sigma_y^2$. The rate of the quantizer is $R_s = 4$; and $R_c = 2$ bits index the partitions.

(a) Draw a figure analogous to Figure 14.6 to describe the quantization bins and cosets. Use the normalized step size of 1. What is the distance between coset elements?

(b) Assume that X and Y are joint Gaussian with correlation coefficient $r = 0.90$. Also assume that the quantizer step size is $\Delta = \sigma_x/2$.

Calculate the probability that the quantizer indices are correctly reconstructed in the decoder, when only the coset number of $\hat{Y}$ is sent to the decoder. How many bits per sample are saved by willingness to take the risk of incorrect reconstruction?

14.3 Consider the previous problem with a different correlation model. Suppose that the independent, additive noise model is $Y = X + V$, where V is a zero mean, Laplacian random variable with variance $\sigma_v^2 = \sigma_x^2/10$.

Calculate the probability that the quantizer indices are correctly reconstructed in the decoder, when only the coset number of $\hat{Y}$ is sent to the decoder. How many bits per sample are saved by willingness to take the risk of incorrect reconstruction?

Notes

1. The codewords of a systematic code comprise the k message symbols followed by the $n - k$ parity symbols.
2. The Golay code is one example of an uncommon perfect code, where all the elements of every correctable region are within the same distance from its codeword.
3. The reconstruction point can also be chosen to be the midpoint of 1.5, with the consequence of only a small increase in average distortion.

References

1. D. Slepian and J. K. Wolf, "Noiseless coding of correlated information sources," *IEEE Trans. Inf. Theory*, vol. IT-19, no. 4, pp. 471–480, July 1973.
2. T. M. Cover and J. A. Thomas, *Elements of Information Theory*. New York, NY: John Wiley & Sons, 1991, 2006.
3. A. Wyner, "Recent results in the Shannon theory," *IEEE Trans. Inf. Theory*, vol. IT-20, no. 1, pp. 2–10, Jan. 1974.
4. A. Wyner and J. Ziv, "The rate–distortion function for source coding with side information at the decoder," *IEEE Trans. Inf. Theory*, vol. IT-22, no. 1, pp. 1–10, Jan. 1976.

Further reading

B. Girod, A. M. Aaron, S. Rane, and D. Rebollo-Monedero, "Distributed video coding," *Proc. IEEE*, vol. 91, no. 1, pp. 71–83, Jan. 2005.

S. S. Pradhan, J. Kusuma, and K. Ramchandran, "Distributed compression in a dense microsensor network," *IEEE Signal Process. Mag.*, pp. 51–60, Mar. 2002.

S. S. Pradhan and K. Ramchandran, "Distributed source coding using syndromes (discus): design and construction," *IEEE Trans. Inf. Theory*, vol. 49, no. 3, pp. 626–643, Mar. 2003.

Z. Xiong, A. D. Liveris, and S. Cheng, "Distributed source coding for sensor networks," *IEEE Signal Process. Mag.*, pp. 80–94, Sept. 2004.

Index

Printed in the United States
by Baker & Taylor Publisher Services